Continuous and Discrete Signal and System Analysis

Holt, Rinehart and Winston Series
in Electrical Engineering, Electronics, and Systems

Shu-Park Chan, Introductory Topological Analysis of Electrical Networks
Chi-Tsong Chen, Introduction to Linear System Theory
George R. Cooper and Clare D. McGillem, Methods of Signal and System Analysis
George R. Cooper and Clare D. McGillem, Probabilistic Methods of Signal and System Analysis
Woodrow W. Everett, Jr., editor, Topics in Intersystem Electromagnetic Compatibility
D. J. Hamilton, F. A. Lindholm, A. H. Marshak, Principles and Applications of Semiconductor Device Modeling
Benjamin J. Leon and Paul A. Wintz, Basic Linear Networks for Electrical and Electronics Engineers
Clare D. McGillem and George R. Cooper, Continuous and Discrete Signal and System Analysis
Richard Saeks, Generalized Networks
Amnon Yariv, Introduction to Optical Electronics

Continuous and Discrete Signal and System Analysis

CLARE D. MCGILLEM/GEORGE R. COOPER

School of Electrical Engineering
Purdue University

HOLT, RINEHART AND WINSTON, INC.

New York   Chicago   San Francisco   Atlanta   Dallas
Montreal   Toronto   London   Sydney

McGillem, Clare D.
    Continuous and discrete signal and system analysis.
    (HRW series in electrical engineering, electronics, and systems)
    Includes bibliographies.
    1. Signal theory (Telecommunication) 2. System
analysis. 3. Electric engineering–Mathematics.
I. Cooper, George R., joint author. II. Title.
TK5102.5.M22   621.38'0433   74–5059
ISBN 0–03–084293–X
4   5   6   7   8   038   9   8   7   6   5   4   3   2   1

Preface

The purpose of this book is to present the most widely used techniques of signal and system analysis in a manner appropriate for instruction at the junior or senior level in electrical engineering. Although these techniques are mathematical in nature, the discussion tends to be heuristic rather than rigorous and includes many examples to illustrate the important features. The typical undergraduate course in calculus provides adequate mathematical background, and the only new mathematics introduced is that necessary to permit a logical development of the various new concepts. It is assumed, of course, that the student is already familiar with the elements of circuit theory and can write loop and node equations for both passive and active circuits. Although the emphasis in this book is on input-output relations for systems in general, simple circuits provide excellent examples of systems and, because of the student's background, tend to make the study of systems less abstract.

There has been no attempt to make this book exhaustive in the sense of providing the last word on any particular method of system analysis. Instead, the effort has been devoted to discussing as many different methods as practicable within the confines of a single volume. At the same time, the author's have been careful to avoid a superficial treatment that gives the illusion of providing the student with a great deal of knowledge while omitting the detail that would make that knowledge useful to him. Considerable attention has been devoted to covering every topic in enough detail to make that topic usable in the solution of engineering problems, while avoiding so much detail that the student cannot sort out the important features. Furthermore, constant reference is made to the interrelationships among the various topics so that the student will at least be aware that a unity does exist, even though he may not be able to appreciate it fully on first exposure. Finally, the student is repeatedly encouraged to look at other literature to extend his knowledge and increase his depth of understanding. It is believed that this text provides an adequate background for most of the graduate-level literature in this area.

The various derivations and examples presented in the text include enough of the detailed steps so that the entire procedure can be followed by the student. An important adjunct to mastering the text material are the exercises included at the ends of most sections. These exercises are relatively straightforward applications of the material immediately preceding them. Their purpose is to reinforce the learning process by permitting an immediate application of this material before proceeding to a new topic. It is intended that all of the exercises be completed as an integral part of the text material. The answers to most exercises are given, although their order may be scrambled relative to the various parts of the exercise in order to provide an additional challenge.

The first three chapters are introductory in nature. They provide a broad overview of the problems of signal and system analysis and introduce the terminology that will be used in subsequent discussions. Most of the ideas introduced in these chapters are covered again in later chapters and applied to practical problems.

Chapter 4, "Convolution," begins the development of the basic material of the text. In contrast to most previous texts, the concept of convolution is introduced *before* transform methods are studied. There are several reasons why the authors feel that this sequence is preferable. First, the convolution integral is of fundamental importance in the study of linear systems, and an ability to graphically visualize convolution aids greatly in the interpretation of certain transform operations. Second, convolution provides an excellent opportunity for representing signals in terms of elementary functions and provides an excellent way to familiarize the student with the manipulation of singularity functions. Third, convolution is extended readily to cover discrete systems and time-varying systems and, thus, provides a simple and direct introduction to this area of analysis.

The treatment of Fourier transforms in Chapter 5 is more extensive than is usually found in undergraduate texts. This is because of their great importance in signal theory and the fact that there is a readily understood physical interpretation that can be associated with them. The first sections of this chapter relate to Fourier series and are included primarily for completeness. This material normally is covered in mathematics courses or in earlier engineering courses and may be omitted without loss of continuity in the development. The material on the discrete Fourier transform in Chapter 5 is particularly important because of its relationship to the fast Fourier transform algorithm, a computation technique that virtually has revolutionized certain aspects of signal and system analysis.

Chapter 6 provides a broad coverage of the Laplace transform and its use in system analysis. Both the one-sided and two-sided transforms are included, and careful consideration is given to the relationship between the Laplace transform and the Fourier transform. Near the end of Chapter 6 the inversion of the Laplace transform by means of contour integration in the complex plane is considered. A discussion of this technique of integration is given in Appendix B. A clear picture of the significance of pole locations in rational transforms is presented in terms of the inversion integral.

Analysis of discrete signals and systems is considered in Chapter 7 by means of the z-transform. The concept of the discrete transfer function is introduced, and a substantial introduction to digital filtering is given. The inclusion of this latter topic in the detail given is a departure from the basic philosophy of the text, which is to consider methods of analysis rather than methods of design. However, this material is included because of its timeliness and relative unavailability in other textbooks.

There is more material contained in the text than can be covered in a single semester so a selection of topics must be made. In all chapters the basic material is covered in the early sections, and more specialized topics are covered in later sections. Generally, sections near the end of a chapter can be omitted without loss of continuity in the development. A typical three-hour course might include Chapters 1–3; the first seven sections of Chapter 4; Sections 5–7 through 5–15 and 5–18 of Chapter 5; and Sections 6–1 through 6–11, part of 6–13 and all of 6–14 and 6–15 of Chapter 6. Other selections may be made readily to meet particular requirements.

The authors make grateful acknowledgement of the considerable help received from their colleagues and students during preparation of the manuscript. In particular, special mention should be made of the suggestions of Professor George Saridis regarding the material of Chapters 6 and 7, and the considerable help provided by Lewis Thurman in proofreading and production of the manuscript.

CLARE D. MCGILLEM
GEORGE R. COOPER

Contents

Continuous and Discrete Signal and System Analysis

Signals and Systems

1-1 Introduction

This is a book about the use of mathematical methods to study systems and signals. If one thinks of a system as a collection of objects interacting with each other to accomplish some purpose, then it is evident that there are systems on every side. Both natural and man-made systems exist in amazing variety. Frequently, natural systems are characterized by enormous complexity. They range from "simple" atoms and molecules to vast ecological systems involving myriads of living and non-living subsystems, all of which are interacting with each other to a greater or lesser extent. From one point of view, the entire universe can be thought of as a single gigantic system. Generalizations on such a large scale, however, are of no use in the solution of engineering problems. Rather, it is necessary to narrow the scope of the problem, even at the expense of completeness and accuracy. Only in this way is it possible to obtain meaningful and useful answers to the behavior of a system under various conditions.

Systems can be studied in many ways. The detailed structure of the component parts and subsystems can be examined and this information used to build up a comprehensive description of system operation, including the interactions of all the various internal components. This might properly be termed a *microscopic system analysis*, meaning that the fine structure of the system is taken into account in the study. Many types of biological studies of subsystems of living things are of this nature. In general this type of analysis is extremely difficult in that it involves almost unlimited numbers of variables and enormous complexity in any mathematical description of the system.

The most useful and most frequently employed type of system analysis is what may be termed *macroscopic system analysis*. In this type of analysis a system is characterized in terms of subsystems and components and their interactions with each other, and no account is taken of the details of their internal operations. It is this type of analysis that has proven highly effective in the solution of engineering problems, and it is this type of analysis that is the subject of this text.

Macroscopic system analysis requires that a system be broken down into a number of individual components. The behavior of the various components is then described in sufficient detail and in such a manner that the overall system operation can be predicted from appropriate calculations of the component behavior. The crux of this type of analysis is the description of the component behavior. For system analysis this is done in terms of a mathematical model, that is, a set of mathematical relationships that sufficiently characterizes the component so that its interaction with other components can be computed. For many components of engineering interest the mathematical model is an expression of the response of the component (or subsystem) to some type of forcing function. The forcing function is a variable in the mathematical model, as is the system response. All these variables are categorized as *signals* in this text, although in an actual system they might be called displacement, voltage, BTU per hour, or they might even be such quantities as the price of a stock or the number of items maintained in an inventory.

Much light is shed on the subject of mathematical models as a direct result of attempting to use them in system analysis. In fact it is highly desirable for a system analyst to have a good understanding of the basis and validity of the models he is using in order to avoid serious errors in their application. However, the subject of developing good models is itself very involved and frequently requires study of a component on a microscopic scale to obtain useful results. This subject is most properly deferred until after the subject of system analysis has been covered, so that a clearer understanding of the model requirements is available.

1-2 Systems and System Analysis

An engineer's analysis of a system is usually made to determine the response of a system to some input signal or excitation. Such studies are used to establish performance specifications, to aid in selection of components, to uncover system deficiencies, to explain unusual or unexpected behavior, or to meet any of a variety of other needs for quantitative data about system operations. In order to clarify some of the concepts relating to systems and system analysis it is helpful to consider some specific examples.

The ignition system in an automobile is representative of a small system. It might consist of an energy source made up of the battery and alternator, the ignition switch, the distributor and points, the spark coil, the capacitor, the interference supressors, the spark plugs, the interconnecting wires, and the general environment within which the system operates. By representing the various components by suitable mathematical models, an engineer could study analytically the behavior of the system under various ambient operating conditions and determine the effects of changing parameters of the components. For example, the power drain from the energy source could be calculated as a function of speed, or the effect of changing inductance on the spark energy could be determined. In many instances a final system design requires a combined analytical and experimental approach. The analytical studies establish the proper direction and ranges of variables for the experimental program and are invaluable in the interpretation of results. For example, in the case of the ignition system, the erosion of the breaker points for different materials and operating conditions could probably best be determined experimentally. Once a satisfactory operating range for the current and voltage has been determined, this can be used as a design specification to be preserved even though other parameters in the system are changed. Such a combined analytical and experimental approach is essential to the successful accomplishment of any engineering problem in which accurate and complete models do not exist.

When the system being considered becomes large, it is necessary to break it up into a number of subsystems, each of which can be analyzed separately. Appropriate models of these subsystems are then used to study the overall system. An example of this type of system is the control and communication system of a small earth-orbiting satellite, such as illustrated by the block diagram in Fig. 1-1. Most of the blocks in this diagram represent functional subsystems and are studied themselves by the methods of system analysis. Their design must be coordinated with overall system requirements, and these are based on an overall system study in which only the external performance characteristics of the subsystems are considered. In carrying out designs of this complexity some type of iterative procedure is always used. A preliminary design is made based on previous experience or engineering judgment. The implications of specifications resulting from this design are then determined by analysis, and modifications made to improve performance or feasibility. The modified design is then reanalyzed, and further changes are made as required. This refining operation continues throughout the design phase and may extend into the fabrication phase if specifications cannot be met.

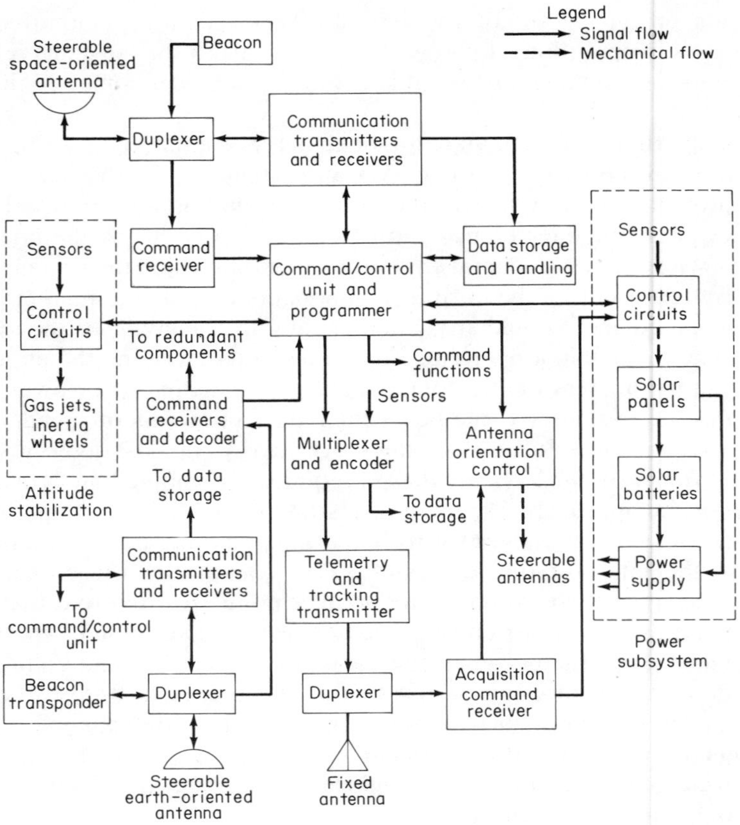

Fig. 1-1 Spacecraft control and communication system.

Many systems of practical engineering interest cannot be analyzed as a whole. A number of factors lead to this situation, two of the most common being the enormous complexity of the system and lack of satisfactory mathematical models for the components or subsystems. The power system for a large city is a good example of a system that is too large and complex for a comprehensive analysis. Such a system has several inputs from generating plants and coupling with other power systems and has a huge quantity and variety of outputs consisting of users. The nature of the outputs varies with such things as time of day, week, or year; weather; whims of consumers; and installation of new or different equipment at the output terminals. Because it is impossible to determine exactly what the system loading will be at any time, it is usually necessary to rely on past experience and to use probabilistic and statistical methods of analysis to arrive at suitable criteria for selecting system parameters. Other systems such as urban transportation systems involving social, economic, political, and technological problems are even less susceptible to presently known techniques of system analysis. Research is in progress on methods for handling such systems, and in the years to come it seems likely that new tools will be made available to the system analyst for handling such problems.

The system analysis methods discussed in the following chapters have proved remarkably effective in a wide range of system applications. Their direct application, or some kind of modification or extension of them, provides the basis of most engineering system analysis today. Most of the detailed examples of systems considered in later chapters are drawn from the field of electrical engineering. The reason for this is that such examples are simple to visualize and are easily extended to cover a wide variety of problems. It is through studies of electrical engineering problems that some of the most successful techniques of system analysis have evolved. However, even though the examples relate primarily to electrical signals and systems, the techniques are very general and can be applied to any system that can be described by comparable mathematical models.

1-3 Signals

In order to study and analyze a system properly, it is necessary to study the means by which energy is propagated through the system. In most systems this can be done by specifying how varying quantities within the system change as a function of time. Such a varying quantity is referred to as a *signal* in this text, even though it may actually be a force, voltage, power, volume per unit time, or any of an almost unlimited number of other variables. These signals measure the excitations and responses of systems and are indispensable in describing the interactions among various components and subsystems making up the complete system.

In complicated systems there are frequently many inputs and outputs. The number of inputs and outputs does not have to be the same. For example, in an aircraft control system the motion of a single lever can change several different aerodynamic surfaces in a manner designed to produce an optimum result.

Besides their usefulness in analyzing system performance, signals also have importance in their own right as a means of carrying information from one point to another. Oftentimes a system is designed for the purpose of transmitting or processing signals. In such cases the signals themselves are of primary concern, and the analyst's problem is to determine how a particular component or subsystem affects a signal propagating through the system. The problem of selecting signals for various engineering applications is of major importance and has led to the development of an extensive body of knowledge referred to as *signal theory*. Designing radar or sonar signals or choosing the proper modulation for a communication system are typical applications of signal theory. It is because of the intrinsic importance of signal theory, in addition to its role in system studies, that it is given a prominent place in this text.

1-4   System Analysis and System Design

The principal problem in system analysis is finding the response of a particular system to a specified input or range of inputs. Such results represent an important part of system studies by meeting the following needs:

1. When the system does not physically exist (as in the case of feasibility studies, for example), only mathematical analysis is possible.
2. Experimental evaluation of systems is often more difficult and expensive than analytical studies.
3. In some cases it is necessary to study systems under conditions that are too dangerous for actual experimentation. Examples of this are the operation of a nuclear power plant at fission rates that are too high for safety or the response of an aircraft flight control system under conditions of severe turbulence.

In addition to the direct use of the results of system analysis indicated above, there is the equally or perhaps more important application to system design. The system design problem is that of determining the necessary system characteristics to yield a desired response to a specified input. Frequently system designs are accomplished by means of parametric studies in which system performance is computed for a variety of cases in which parameters of the system are changed over appreciable ranges. From these analytical results decisions are made as to the parameter values required to give the desired performance. This method can generally be used when specific synthesis techniques are not available. It is in the area of design that a system engineer finds some of his most challenging and rewarding work. In order to be successful he must be creative, resourceful, and knowledgeable about the great variety of methods available for his use. It is this latter area that is the main content of the following chapters.

References

The following books present elementary discussions of system engineering and a number of interesting examples:

1. BEAKLEY, G. C., and H. W. LEACH, *Engineering.* New York: The Macmillan Company, 1968.

2. GIBSON, J. E., *Introduction to Engineering Design.* New York: Holt, Rinehart and Winston, Inc., 1968.

3. WILSON, W. E., *Concepts of Engineering System Design.* New York: McGraw-Hill Book Company, Inc., 1968.

Problems

1-1. List the principal components in the ignition system of an automobile and state the input, output, and purpose of each.

1-2. In Fig. 1-1 describe the nature and purpose of all the mechanical signals shown. What are the mechanical quantities involved in each instance?

1-3. Analysis is an indispensable part of system design. As an example of how analysis is used in design, consider the following example of a temperature controller. A temperature-sensitive resistor is immersed in a fluid that is to be heated to a preselected temperature and is then maintained at that temperature by switching the heating element off and on in accordance with the temperature as determined by the temperature-sensitive resistor. A circuit for accomplishing this operation is shown in the diagram. The circuit is essentially a Wheatstone

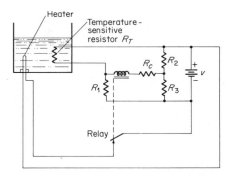

bridge with the unbalance current of the bridge passing through the relay that actuates the heating element. For purposes of analysis, assume that the temperature-sensitive resistor has a nominal value of $100\ \Omega$ at 300°K and has a positive temperature coefficient of $0.1\ \Omega/°K$. Let R_1 and R_3 be $100\ \Omega$ each and assume that the relay coil resistance is $20\ \Omega$. The drop-out current of the relay is 1 mA and the pull-in current is adjustable from 1.2 to 2.0 mA. The voltage v can be taken as 24 V.

  a. Write a general expression for the current in the relay coil as a function of R_T and R_2. This represents a *preliminary analysis* of the problem. (*Hint*: Use nodal analysis.)

b. Choose a value of R_2 such that the heater will turn off when the temperature reaches 350°K.

c. To what value of pull-in current must the relay be set so that the heater will turn on when the temperature drops below 340°K?

d. If R_2 is made adjustable, draw a calibration curve of resistance of R_2 versus temperature at which the heater turns off. Limit consideration to a temperature range of 300 to 400°K.

e. Draw a curve for the difference in temperature between "heater on" and "heater off" for the temperature settings of part **d.**

f. Explain qualitatively what the effects on overall system performance would be of altering the resistance of the relay coil, altering the voltage v, or reducing by a factor of 2 the resistances R_T, R_c, R_1, R_2, and R_3.

2

Representation of Systems

2-1 Introduction

This chapter introduces the concept of the mathematical model of a system, notes some of its limitations, and explains the terminology used to describe systems. The examples considered are primarily electrical systems but the concepts, terminology, and methods apply equally well to other types of systems.

2-2 System Representation

The most widely used method of representing a system is by means of specified relationships among the system variables. Such representations can take the form of graphs, tabular values, differential equations, difference equations, or various combinations of these. The methods considered in the following chapters relate almost exclusively to various analytical procedures that can be used when the system representation is in terms of ordinary linear differential or difference equations with constant coefficients. Such representations encompass an enormous variety of systems and often can be used as approximate representations for the analysis of systems outside these categories.

An important question that frequently faces the system engineer, and in fact almost any engineer, is: What is the appropriate model for a particular system and how good a representation does it provide? This is actually a very difficult question to answer in many instances and often requires decisions based on judgment or experience of the engineer. This problem is greatly simplified when standardized components are used as, for example, in electrical circuit design. In such cases the mathematical model is often supplied by the component manufacturer by specification of the model parameters.

As an example of component specification, consider a capacitor. A minimum sort of specification would merely give the value of the capacitance along with a tolerance figure and the maximum voltage at which the capacitor can be safely operated. Such a component is represented in Fig. 2-1(a). When more precise calculations of performance are required, or when the capacitor is to be used at higher frequencies, it is necessary to use a more accurate model, such as that given in Fig. 2-1(b). In this model the leakage and the dielectric loss have been taken into account by adding the resistor R_1 to the model. If the capacitor is to be used at still higher frequencies, the effects of lead inductance and lead loss may become important, and the model would have to be altered to take these effects into account. One representation that takes into account the inductance and resistance associated with the leads is shown in Fig. 2-1(c), where an inductance L and a resistance R_2 have been added to the previous model.

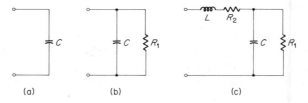

(a)   (b)   (c)

Fig. 2-1 Three models of a capacitor.

In other applications it may be necessary to employ an even more complicated model if such things as variations of capacitance or losses with temperature or losses due to radiation at high frequency are to be taken into account.

In every case the mathematical model of a component or system is an approximation. It may be a good approximation or a poor approximation, and in a particular system the degree of approximation may have a greater or lesser effect on the accuracy or validity of the analysis, depending on the manner in which the component interacts with the rest of the system. The engineer must decide on the degree of accuracy required and select the appropriate mathematical model to meet that requirement. To do this the engineer must know the degree of confidence that can be placed in the results of his analysis using a particular model, and he must be able to appreciate the difficulties that might be encountered from trying to obtain a more accurate model.

The development of mathematical models for specific components is a very difficult but generally very fascinating endeavor. In many instances obtaining the correct model for a component is more difficult than carrying out analysis of the system containing the component. Before such modeling can be done successfully it is necessary for the modeler to be familiar with the types of analytical techniques that are used in studying system performance. Accordingly the study of model development will not be considered further now. Rather, the mathematical forms of the most frequently used models will be assumed and the techniques of system analysis considered in detail. The discussion will be based primarily on electrical systems in which commonly accepted models of components are used. The actual selection of the most appropriate model in a particular instance will not be considered, although this is a major responsibility of an engineer and is one of the essential ingredients that distinguishes engineering analysis from applied mathematics.

Exercise 2-2.1

Analogous to Fig. 2-1, sketch four models for an inductor. These should be models that are useful at very low frequencies, medium frequencies, high frequencies, and very high frequencies. The effects of coil resistance and capacitance should be included where it is appropriate to do so. For simplicity, assume that this is an air-core inductor.

2-3 Forms of Mathematical Models

Initial discussion of mathematical models is limited to systems with a single input and a single output. The concepts employed in this simple case are then extended to include more general systems having many inputs and many outputs and to systems operating with a discrete rather than a continuous time variable.

A convenient way to decribe systems in general terms is by means of a block diagram, as shown in Fig. 2-2. The input signal $x(t)$ and the output signal $y(t)$ are shown as arrows and the system itself as a block, or box. When several components are interconnected, there will be several blocks and several lines interconnecting the blocks. Such a block diagram

Fig. 2-2 System block diagram.

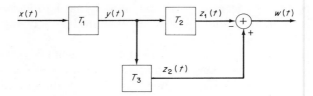

Fig. 2-3 System with several blocks.

is shown in Fig. 2-3. The various subsystems or components are designated with individual blocks identified by the symbols T_1, T_2, and T_3. It is often convenient to think of each of the blocks as transforming the signal at its input into the signal at its output. This can be represented mathematically by the operational notation

$$y(t) = T_1\{x(t)\}$$

Similar representations can be given for $z(t)$ and $w(t)$. The nature of the operation $T\{\cdot\}$ depends on the nature of the component or subsystem involved and might be a simple scaling or a complicated nonlinear mapping of one function into another. It usually turns out that the transformation $T\{\cdot\}$ for a block depends on what is connected to the input and output terminals of the block. In the mathematical model of a system, this effect of loading takes the form of simultaneous equations that must be satisfied. The input-output transformations for several ideal electrical components are given in Table 2-1. By means of the component models given in Table 2-1 and Kirchhoff's laws, it is possible to write a set of equations describing the behavior of an electrical network or system. As a simple example, consider the network shown in Fig. 2-4. The input signal comes from an ideal voltage source $v_1(t)$, and the output signal is the voltage $v_3(t)$ measured across the capacitor C. This network has four nodes, and its performance can therefore be completely described by the three equations obtained by conventional nodal analysis using $v_1(t)$, $v_2(t)$, and $v_3(t)$ as the variables. Since $v_1(t)$ is assumed known (the input signal), there are actually only two unknowns, and the corresponding equations are

$$\frac{1}{L}\int_{-\infty}^{t}[v_2(\xi) - v_1(\xi)]d\xi + \frac{1}{R_1}v_2(t) + \frac{1}{R_2}[v_2(t) - v_3(t)] = 0 \qquad \text{(2-1)}$$

$$\frac{1}{R_2}[v_3(t) - v_2(t)] + C\frac{dv_3(t)}{dt} = 0 \qquad \text{(2-2)}$$

Eliminating $v_2(t)$ from these two equations gives the following single equation relating $v_3(t)$ and $v_1(t)$.

$$\left(\frac{R_2LC}{R_1} + LC\right)\frac{d^2v_3}{dt^2} + \left(R_2C + \frac{L}{R_1}\right)\frac{dv_3}{dt} + v_3 = v_1 \qquad \text{(2-3)}$$

Table 2-1   Mathematical Models of Electrical Components

Component	Input	Output and transformation
Resistor	$v(t)$	$i(t) = \frac{1}{R} v(t)$
	$i(t)$	$v(t) = R\, i(t)$
Capacitor	$v(t)$	$i(t) = C\, \frac{dv(t)}{dt}$
	$i(t)$	$v(t) = \frac{1}{C} \int_{-\infty}^{t} i(\xi)\, d\xi$
Inductor	$v(t)$	$i(t) = \frac{1}{L} \int_{-\infty}^{t} v(\xi)\, d\xi$
	$i(t)$	$v(t) = L\, \frac{di(t)}{dt}$
Amplifier	$x(t)$	$y(t) = A\, x(t)$
Integrator	$x(t)$	$y(t) = \int_{-\infty}^{t} x(\xi)\, d\xi$
Delay	$x(t)$	$y(t) = x(t - T)$
Summer	$x_1(t), x_2(t)$	$y(t) = x_1(t) + x_2(t)$

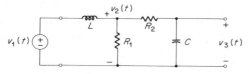

Fig. 2-4 A simple electrical system.

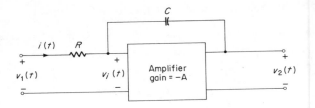

Fig. 2-5 Operational amplifier used as an integrator.

where v_3 is to be taken as $v_3(t)$ and v_1 as $v_1(t)$. The equation (2-3) represents the mathematical model of the system of Fig. 2-4.

As a second example, consider the system shown in Fig. 2-5, which corresponds to an operational amplifier connected in such a manner as to behave as an integrator. The amplifier is assumed to have infinite input impedance, zero output impedance, and a voltage gain of $-A$. It is easily seen from the diagram that

$$v_1(t) = v_2(t) + \frac{1}{C} \int_{-\infty}^{t} i(\xi)d\xi + Ri(t)$$

$$v_i(t) = v_1(t) - Ri(t)$$

$$v_2(t) = -Av_i(t)$$

After elimination of $i(t)$, the following single equation relating fhe input signal and output is obtained:

$$(A + 1)RC\frac{dv_2}{dt} + v_2 = -Av_1 \tag{2-4}$$

This is the mathematical model of the system. Operation of the system as an integrator is seen by considering the case of very large gain A, such that $A/(A + 1) \simeq 1$ and $A \gg 1$. Under these conditions (2-4) can be approximated as

$$v_1 = -RC\frac{dv_2}{dt}$$

from which it is clear that the output is proportional to the integral of the input.

Both of the mathematical models just considered are ordinary differential equations with constant coefficients. The analysis of the performance of these systems requires that the differential equations of the mathematical models be solved for the range of system parameters of interest and the different input signals likely to be encountered. It is the study of various methods of carrying out such solutions that makes up much of the material in the following chapters.

Exercise 2-3.1

Write a differential equation that relates the output $y(t)$ of the above system to the input $x(t)$.

ANS. $dy(t)/dt + y(t - T) = x(t)$

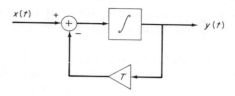

2-4   System Classification

Before proceeding to more detailed consideration of methods for solving the equations corresponding to the mathematical models of systems, it is essential to define more precisely some of the terms used in describing and specifying systems. To this end it is useful to consider a very general mathematical model that encompasses a wide range of systems. Such a model for a system having an input $x(t)$ and output $y(t)$ is

$$a_n(t)\frac{d^n y}{dt^n} + a_{n-1}(t)\frac{d^{n-1} y}{dt^{n-1}} + \cdots + a_1(t)\frac{dy}{dt} + a_0(t)y$$

$$= b_m(t)\frac{d^m x}{dt^m} + b_{m-1}(t)\frac{d^{m-1} x}{dt^{m-1}} + \cdots + b_1(t)\frac{dx}{dt} + b_0(t)x$$

(2-5)

Although it is beyond the scope of this discussion to prove it, the above equation is applicable to almost all linear systems. Its applicability can be verified easily in any specific case.

The mathematical relationship between the input and output of a system as given in (2-5) is not unique unless the time interval of the response is stated also. For example, suppose that the system input $x(t)$ is zero for all time prior to some time t_0. It is possible to find solutions to (2-5) that are not zero for $t < t_0$. Such solutions *cannot* correspond to the response of a physical system because no physical system can anticipate an excitation and begin to respond before this excitation is applied. Thus, when (2-5) is used as the mathematical model for any physical system, the further condition is imposed that if the excitation $x(t) = 0$ for all $t < t_0$, then the system response to this $x(t)$ also becomes $y(t) = 0$ for all $t < t_0$. With this assumption, the mathematical model ordinarily becomes unique in the sense that any specific input signal results in a corresponding unique output signal.

The mathematical model specified by (2-5) expresses the behavior of the system in terms of a single nth-order differential equation. An alternative method of expressing the same information is in terms of a set of first-order differential equations. This is known as the *normal form* of the system equations. This method has an advantage of notational simplicity (since matrix notation can be employed to express the entire set of equations in a very compact form), and it is also the form that is most easily extended to systems with many inputs and outputs. However, before considering the normal form, some of the nomenclature used in describing systems will be discussed in terms of (2-5).

Order of the system. The differential equation (2-5) is of the nth order, since this is the highest-order derivative of the response to appear. The corresponding system is said to be of nth order also.

Causal system, noncausal system. A *causal* (or *physical*, or *nonanticipatory*) system is one whose present response does not depend upon future values of the input. As noted previously, this condition is assumed to apply to the system equation. A noncausal system is one for which this condition is not assumed. Noncausal systems do not exist in the real world but can be approximated by the use of time delay, and they frequently occur in system analysis problems.

Linear system, nonlinear system. The system equation (2-5) represents a *linear* system, since all derivatives of the excitation and response are raised to the first power only, and since there are no products of derivatives. One of the most important consequences of linearity is that *superposition* applies. In fact, this may be used as a definition of linearity. Specifically, if

$$y_1(t) = \text{system response to } x_1(t)$$

$$y_2(t) = \text{system response to } x_2(t)$$

and if

$$ay_1(t) + by_2(t) = \text{system response to } ax_1(t) + bx_2(t)$$

for all a, b, $x_1(t)$, and $x_2(t)$, then the system is linear. If this is not true, then the system is not linear.

In the case of *nonlinear* systems, it is not possible to write a general differential equation of finite order that can be used as the mathematical model for all systems, because there are so many different ways in which nonlinearities can arise, and they cannot all be described mathematically in the same form. It is also important to remember that superposition does not apply in nonlinear systems.

A linear system usually results if none of the components in the system changes its characteristics as a function of the magnitude of the excitation applied to it. In the case of an electrical system, this means that resistors, inductors, and capacitors do not change their values as the voltages across them or the currents through them change. Other electrical devices such as amplifiers, motors, or transducers would likewise not change their properties with voltage or current. In all cases, however, the concept of linearity is an approximation, since any system component will change its characteristics if the forces applied to it are large enough. Hence, when we speak of a linear system, what we really mean is that, with normal input magnitudes, the system does not change significantly; therefore, linearity may be assumed, and the methods of linear system analysis may be employed. Just what changes are significant and what inputs will produce them are matters of engineering judgment.

There are many devices, such as rectifiers, junction diodes, gaseous conduction tubes, and saturating magnetic devices, for which the assumption of linearity is not valid for the normal range of excitation magnitude. Usually, if even one such device is included in the system, then the system must be treated as nonlinear.

Fixed system, time-varying system. Equation (2-5), as written, represents a *time-varying* system, since the coefficients $a_i(t)$ and $b_i(t)$ are indicated as being functions of time. The analysis of time-varying devices is difficult, since differential equations with nonconstant coefficients cannot be solved except in special cases.[1] The systems of greatest concern for the present discussion are characterized by a differential equation having constant coefficients. Such a system is known as *fixed, time-invariant,* or *stationary.* An alternative way of defining a fixed system is to state that if excitation $x(t)$ leads to response $y(t)$, then excitation $x(t - \tau)$ leads to response $y(t - \tau)$ for any $x(t)$ and any τ. That is, the shape of the system response depends only upon the shape of the system excitation and not upon the time of application.

Fixed systems usually result when the physical components in the system, and the configuration in which they are connected, do not change with time. Most systems that are not exposed to natural environments can be considered fixed unless they have been deliberately designed to be time-varying. A time-varying system results when any of its components, or their manner of connection, do change with time. In many cases this change is a result of environmental conditions. For example, an aircraft flight control system has greatly different parameter values at sea level than it has at 40,000 feet because of differences in the air density, and a medium-frequency radio communication system changes from day to night due to changes in the reflection characteristics of the ionosphere.

There is an intermediate situation that occurs when the system changes as a consequence of switches opening or closing in negligible times. Strictly speaking, such a system is time-varying, but in most cases the system analysis can be carried out by considering the system to be

---

[1] That is, closed-form analytic solutions are usually not possible. Computer solutions for specific inputs can almost always be obtained.

fixed and by replacing the switches by appropriate voltage or current sources.

Lumped-parameter system, distributed-parameter system.   Equation (2-5) represents a *lumped-parameter* system by virtue of being an ordinary differential equation. The implication of this condition is that the physical size of the system is of no concern, since excitations propagate through the system instantaneously. This assumption is usually valid if the largest physical dimension of the system is small compared to the wavelength of the highest significant frequency considered. A *distributed-parameter* system is represented by a partial differential equation and generally has dimensions that are not small compared to the shortest wavelength of interest. Transmission lines, waveguides, antennas, and microwave tubes are typical examples of distributed-parameter electrical systems.

In very large and complex systems, both attributes may exist at the same time. This is true, for example, in power systems, telephone systems, and radio communication systems, in which the terminal equipment is essentially lumped-parameter but the connection between terminals is distributed-parameter. Fortunately, the analysis of the various parts of such systems can usually be handled separately.

Continuous-time system, discrete-time system.   Equation (2-5) represents a *continuous-time* system by virtue of being a differential equation rather than a difference equation. That is, the inputs and outputs are defined for all values of time rather than just for discrete values of time. Since time itself is inherently continuous, all physical systems are actually continuous-time. However, there are situations in which one is interested solely in what happens at certain discrete instants of time. In many of these cases, the system itself is composed of a digital computer, which is performing certain specified computations and producing its answers at discrete time instants. If no changes (in input or output) take place between time instants, then system analysis is simplified by considering the system to be *discrete-time* and having a mathematical model that is a difference equation. Discrete-time systems can be either linear or nonlinear and either fixed or time-varying. In fact, such systems provide one of the most convenient ways to construct a system with a prescribed nonlinearity or time variation.

Instantaneous system, dynamic system.   An *instantaneous system* is one in which the response at time t_1 depends only upon the excitation at time t_1 and not upon any future or past values of the excitation. This may also be called a *zero-memory* system. A typical example is a resistance network or a nonlinear device without energy storage. If the response does depend upon past values of the excitation, then the system is said to be *dynamic* and to have *memory*. Any system that contains at least two different types of elements, one of which can store energy, is dynamic. If the present output can be completely specified by the input during a finite period of time, then the system has finite memory. The physical construction of continuous-time, finite-memory systems usually requires the use of time delay of some sort, whereas discrete-time systems are almost always finite-memory because of limited data-storage capacity.

Except where specifically otherwise stated, all systems discussed in later chapters will be assumed to be causal, linear, fixed, lumped-parameter, continuous-time, and dynamic. As such, their basic mathematical model is an ordinary, linear, differential equation with constant coefficients.

Exercise 2-4.1

A system is characterized by the differential equation

$$\frac{d^2y(t)}{dt^2} + 5\frac{dy(t)}{dt} + y(t) = 3x(t) + t\frac{dx(t)}{dt}$$

Classify this system according to the classifications discussed in this section.

2-5   The Normal Form of System Equations

It is noted in the preceding section that an alternative method of formulating the mathematical model for an nth-order system is by means of a set of n simultaneous first-order differential equations, known as the normal form of the system equations. This method is discussed in much greater detail in Chapter 8, but it is useful to take one more step in that direction now, in order to use the examples just discussed.

First of all, if there are to be n simultaneous equations, there must also be n unknown time functions to be solved for. These unknown time functions are called *state variables*, and a specification of their value at some time instant is the state of the system at that instant. The manner in which they vary, as a function of time, in reponse to some excitation, provides a description of the system response. The output of the system can be represented as some combination of these state variables and the system input.

In an electrical system the state variables may represent voltages and currents that occur at various places in the system, or they may be a set of abstract quantities that have no physical counterparts in the actual system. It should also be noted that there are many different ways in which the state variables can be selected for a given system, but that these cannot be selected completely arbitrarily if one wishes to use the smallest number of state variables.

For the general nth-order linear system, let the state variables be designated as $q_1(t), q_2(t), \ldots, q_n(t)$. If the system has only one input $x(t)$, then a set of simultaneous, first-order, linear differential equations can be written in the form

$$\frac{dq_1(t)}{dt} = a_{11}(t)q_1(t) + a_{12}(t)q_2(t) + \cdots + a_{1n}(t)q_n(t) + b_1(t)x(t)$$

$$\frac{dq_2(t)}{dt} = a_{21}(t)q_1(t) + a_{22}(t)q_2(t) + \cdots + a_{2n}(t)q_n(t) + b_2(t)x(t)$$

$$\vdots$$

(2-6)

$$\frac{dq_n(t)}{dt} = a_{n1}(t)q_1(t) + a_{n2}(t)q_2(t) + \cdots + a_{nn}(t)q_n(t) + b_n(t)x(t)$$

If there is also a single output $y(t)$, then it often can be represented as

$$y(t) = c_1(t)q_1(t) + c_2(t)q_2(t) + \cdots + c_n(t)q_n(t) \qquad \text{(2-7)}$$

although this may not be adequate for some choices of state variables. Equations (2-6) and (2-7), taken together, form a mathematical model for the general nth-order, linear, time-varying system and are equivalent to (2-5). If the system is fixed, then the coefficients $a_{ij}(t)$, $b_i(t)$, and $c_i(t)$ for $i = 1, 2, \ldots, n$, are constants instead of being functions of time.

Since the above formulation of the mathematical model appears to be much more complicated and much more abstract than that of (2-5), the student may question why it is useful. This method is useful for several reasons, and these are all related to the desire to make the mathematical model as general as possible. Some of these reasons are:

1. By defining appropriate matrices, (2-6) and (2-7) can be written in matrix form and then applied to systems of all orders with equal ease. Thus, increasing the complexity of the system would not result in a corresponding increase in notational complexity.

2. The same general form of equations can also be applied to many types of nonlinear systems. The major differences would be that the right-hand side of (2-6) would be a nonlinear rather than a linear function of $x(t)$ and the $q_i(t)$.

3. In the case of time-varying systems, the solution of a set of first-order equations with nonconstant coefficients may be more straightforward than the solution of the corresponding nth-order equation with nonconstant coefficients. However, in either form, only certain special cases can be solved.

4. If all of the state variables are solved for, then one has a better idea of what is going on inside the system than would be available from a solution for the output function only. This may be important in determining if any element within the system has such large voltages or currents associated with it that the assumption of linear behavior is of doubtful validity.

5. If either analog or digital computer methods are used to determine the system response, it is usually more convenient to work with the normal form of the equations than with the single nth-order equation.

6. The same formulation of equations applies to multiple-input, multiple-output systems when the normal form is used.

It is not intended to dwell at length on the state-variable representation of systems in this chapter, since it is covered in detail later. The foregoing discussion is primarily for motivation, rather than information, and no use is made of this material until Chapter 8. However, it is desirable to demonstrate at this time that different mathematical models can be derived for the same physical system, and the examples discussed in Section 2-3 are adequate for this purpose.

It is simplest to consider the operational amplifier of Fig. 2-5 first. From (2-4) it is clear that this is a first-order system, and only one state variable is necessary. One possibility is to select $v_2(t)$ as the state variable, and this is the most natural choice. Thus, (2-4) can be written in the form of (2-6) simply by transposing one term and dividing through by the coefficient of the derivative term. Hence,

$$\frac{dv_2(t)}{dt} = -\frac{1}{(A+1)RC}v_2(t) - \frac{A}{(A+1)RC}v_1(t) \qquad \text{(2-8)}$$

The relationship to (2-6) is made more explicit by noting that

$$n = 1$$

$$q_1(t) = v_2(t)$$

$$x(t) = v_1(t)$$

$$a_{11}(t) = -\frac{1}{(A+1)RC} \quad \text{(a constant)}$$

$$b_1(t) = -\frac{A}{(A+1)RC} \quad \text{(a constant)}$$

The corresponding output equation, analogous to (2-7), is

$$y(t) = v_2(t) \qquad \text{(2-9)}$$

where $c_1(t) = 1$, a constant. Hence, (2-8) and (2-9) represent another mathematical model for the system of Fig. 2-5.

A somewhat more involved situation results when the simple electrical system of Fig. 2-4 is considered. In this case the system is second order, and two state variables are required. Although various choices for the state variables are possible, it usually proves most practical to use the voltages across capacitors and the currents through inductors as the state

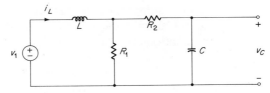

Fig. 2-6 Simple electrical system with state variables indicated.

variables in electrical circuits. Accordingly, let the state variables be

$$q_1(t) = i_L(t) \tag{2-10}$$

$$q_2(t) = v_C(t) \tag{2-11}$$

For convenience the circuit is redrawn and relabeled in Fig. 2-6. The state equations can be written by noting that the voltage across the inductor is $L(di_L/dt)$ and that the current through R_2 is $C(dv_C/dt)$. Using these values and summing the currents leaving the node of L. R_1, and R_2 gives

$$-i_L + \left(v_1 - L\frac{di_L}{dt}\right)\frac{1}{R_1} + \left(v_1 - L\frac{di_L}{dt} - v_C\right)\frac{1}{R_2} = 0 \tag{2-12}$$

The other required equation is obtained by summing the voltages around the loop containing R_1, R_2, and C, giving

$$-R_1\left(C\frac{dv_C}{dt} - i_L\right) - R_2 C\frac{dv_C}{dt} - v_C = 0 \tag{2-13}$$

Solving (2-12) and (2-13) for di_L/dt and dv_C/dt gives the desired set of normal form equations

$$\frac{di_L}{dt} = -\frac{R_1 R_2}{L(R_1 + R_2)}i_L - \frac{R_1}{L(R_1 + R_2)}v_C + \frac{1}{L}v_1 \tag{2-14}$$

$$\frac{dv_C}{dt} = \frac{R_1}{R_1 + R_2}i_L - \frac{1}{(R_1 + R_2)C}v_C \tag{2-15}$$

In the present instance the output variable is one of the state variables, v_C, and, therefore, no special output equation is required. In order to complete the identification of (2-14) and (2-15) with (2-6), the following equivalences should be noted:

$$n = 2$$

$$x(t) = v_1(t)$$

$$a_{11}(t) = -\frac{R_1 R_2}{L(R_1 + R_2)}$$

$$a_{12}(t) = -\frac{R_1}{L(R_1 + R_2)}$$

$$b_1(t) = \frac{1}{L}$$

$$a_{21}(t) = \frac{R_1}{R_1 + R_2}$$

$$a_{22}(t) = -\frac{1}{(R_1 + R_2)C}$$

$$b_2(t) = 0$$

Hence, (2-14) and (2-15) represent another (equivalent) mathematical model for the system of Fig. 2-4.

In the two examples just discussed, the state variables were apparently selected on an arbitrary basis with no consideration given as to whether or not they lead to equations of the proper form. In actuality, the selection is not arbitrary, and there are rules for selecting state variables in an appropriate manner. Some of these rules are discussed in Chapter 8. In order to see the problem more clearly, the student should attempt to write normal equations for the system of Fig. 2-4 when the state variables are chosen to be the voltages across the two resistors.

Exercise 2-5.1

A system is characterized by the normal equations

$$\frac{dq_1(t)}{dt} = q_1(t) + q_2(t) + x(t)$$

$$\frac{dq_2(t)}{dt} = q_1(t) - q_2(t)$$

and the output equation

$$y(t) = q_1(t) + q_2(t)$$

Write a single second-order differential equation relating the output, $y(t)$, to the input, $x(t)$.

ANS. $d^2y(t)/dt^2 - 2y(t) = dx(t)/dt + 2x(t)$

2-6 Discrete-Time Systems and Difference Equations

The study of discrete-time systems has become especially important in recent years because of the advent of the digital computer and of various related techniques using samples of continuous-time signals rather than the signal itself. One very direct application of discrete-time analysis techniques is in the approximate solution of differential equations by computer simulation. In such cases derivatives are approximated by finite differences of the form

$$\frac{dy}{dt} \doteq \frac{y(t+\Delta) - y(t)}{\Delta}$$

(2-16)

where Δ is a suitably chosen small time interval. A first-order differential equation of the form $a_1(dy/dt) + a_2y(t) = bx(t)$ can then be written as

$$a_1\frac{y(t+\Delta) - y(t)}{\Delta} + a_2y(t) = bx(t)$$

(2-17)

Starting with an appropriate set of initial conditions, the solution for subsequent times can be computed by repetitive applications of (2-17). If Δ is selected to be sufficiently small, good results can be obtained in this manner.

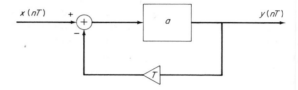

Fig. 2-7 Simple discrete system.

   Difference equations also arise directly from the mathematical model of the system. Figure 2-7 shows a simple digital filter in which each input sample, $x(nT)$, is added to a delayed sample of the output and the sum is multiplied by the factor a. The mathematical relationship between the output and input is readily found to be

$$y(nT) = a[x(nT) - y(nT - T)]$$

(2-18)

Taking the clock period, T, to be unity for convenience, this can be expressed as a first-order linear difference equation.

$$y(n) + ay(n-1) = ax(n)$$

(2-19)

The solution of difference equations parallels that of differential equations and involves both a complementary and a particular solution corresponding to the complementary function and particular integral encountered in ordinary differential equations. The input, $x(n)$, determines the particular solution, whereas solution of the homogeneous equation [that is, $x(n) \equiv 0$], along with the initial conditions, determines the complementary solution. Methods appropriate to the solution of difference equations are discussed in Chapters 5, 6, and 7. For the present it is sufficient to examine briefly the solution of the difference equation (2-19) and note the similarity to solutions of comparable differential equations.

   A straightforward solution to (2-19) can be obtained by starting from a known state of the system (that is, the initial conditions) and then iteratively computing succeeding values of the variables from (2-19). In order to do this, the inputs as well as the initial conditions must be known. Let the input be

$$x(n) = 0 \qquad n \le 0$$
$$= 1 \qquad n > 0$$

and let the initial condition be that $y(0) = 0$. $y(n)$ can be computed directly from (2-19) to give

$$y(0) = 0$$

$$y(1) = a$$

$$y(2) = a(1 - a)$$

$$y(3) = a(1 - a + a^2)$$

$$\vdots$$

$$y(n) = a[1 - a + a^2 - \cdots (-1)^{n-1} a^{n-1}]$$

This is seen to be a geometric progression and can be summed to give

$$y(n) = \frac{a}{1 + a}[1 - (-a)^n]$$

The operation of the system is seen to be quite different for different values of a. For example, when a is less than -1, the output continues to increase in magnitude and for very large n is given by

$$y(n) \doteq \frac{-|a|}{|a| - 1}|a|^n \qquad n \text{ large}$$

For $-1 < a < 1$ the system output converges to

$$y(n) \doteq \frac{a}{1 + a} \qquad n \text{ large}$$

For $a > 1$ the output becomes oscillatory and reverses sign at the output

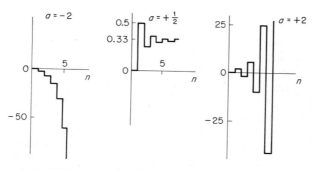

Fig. 2-8 Response of a discrete system to a step input.

from sample to sample. $y(n)$ for this case is given by

$$y(n) \doteq \frac{a}{1+a}(-1)^{n+1}a^n \qquad n \text{ large}$$

The behavior of $y(n)$ for three values of a is shown in Fig. 2-8. The responses shown in Fig. 2-8 indicate clearly that this system behaves in an unstable manner for large positive and negative values of the gain constant a.

Exercise 2-6.1

   For the discrete system of Fig. 2-7, evaluate and sketch the output $y(n)$ for $a = -1$, $a = 0$, and $a = 1$.

ANS. $-n$; 0; 0 for n even, 1 for n odd.

2-7   Initial Conditions

   The specification of a mathematical model for a system is the first step in carrying out any problem of system analysis, but it is not sufficient. It is also necessary to specify the conditions of energy storage within the system at the time the input signal is applied. There are various ways in which these initial conditions can be established, but as far as system analysis is concerned, it makes no difference. The effect of any set of initial conditions is exactly the same, regardless of how they came into being. A viewpoint that is consistent with the concept that only causal systems can exist is that the initial conditions were established by previous inputs, possibly so far in the past that equilibrium conditions have been established.

   There are a variety of ways of incorporating the initial conditions into the calculation of system response, and it is difficult to single out any one procedure and state that it will always be the easiest one to use. The following procedures are in common use, and each has certain advantages to recommend it:

1. Probably the most basic procedure is to employ the concepts of superposition and the foregoing concept of how initial conditions are established. Because of superposition, the system response to the past inputs that established the initial conditions and the system response to the desired present input can be added directly for all time. Therefore, the system can be analyzed by assuming that all initial conditions are zero and then adding the response due to the initial conditions to the result. This method is particularly convenient when the system configuration is known.

2. When the mathematical model for the system is in the form of an nth-order differential equation, as in (2-5), an appropriate method of specifying initial conditions is to give the value of the output

signal and $n-1$ of its derivatives at the time the desired input signal is applied. This procedure must be used with care when the mathematical model involves derivatives of the input signal, but when properly applied, it is probably the easiest one to use in this case.

3. When the mathematical model is given in terms of the normal equations, the appropriate procedure is to specify the values of the state variables at the time the desired signal is applied. As will be seen in Chapter 8, this form of initial conditions arises quite naturally in the solution of the normal equations.

Although concrete examples might be desirable to illustrate these procedures, they are not needed to achieve the objectives of the present chapter, since solutions for the mathematical models are not being sought. Further discussion is deferred, therefore, until specific systems are considered in the following chapters.

References

Many books discuss methods for obtaining mathematical models for systems. The following six represent some of those suitable for students at a junior or senior level.

1. BROWN, R. G., and J. W. NILSSON, *Introduction to Linear System Analysis.* New York: John Wiley & Sons, Inc., 1962.
   Chapter 2 of this book is concerned with translating the physical problem into mathematical language. This discussion includes mechanical systems as well as electrical systems.
2. CHENG, D. K., *Analysis of Linear Systems.* Reading, Mass.: Addison–Wesley Publishing Company, Inc., 1959.
   This book has several chapters on the mathematical models for linear systems, including an extensive discussion of mechanical and electromechanical systems.
3. CHUA, L. O., *Introduction to Nonlinear Network Theory.* New York: McGraw-Hill Book Company, Inc., 1969.
   This book, written for undergraduate electrical engineers, provides an extensive and very readable treatment of modeling and an analysis of nonlinear networks. It is an excellent reference for anyone wishing to pursue these subjects further.
4. KUO, FRANKLIN F., *Network Analysis and Synthesis*, 2d ed. New York: John Wiley & Sons, Inc., 1966.
   This book contains a chapter on the solution of linear differential equations and another chapter on the solution of network equations. There is an excellent discussion of initial conditions.
5. LEON, B. J., and P. A. WINTZ, *Basic Linear Networks for Electrical and Electronic Engineers.* New York: Holt, Rinehart and Winston, Inc., 1970.
   This book presents a modern treatment of linear network analysis using non-transform methods. Considerable discussion of writing and solving network equations is included, and the level is appropriate for sophomores or juniors in electrical engineering.

6. LYNCH, W. A., and J. G. TRUXAL, *Signals and Systems in Electrical Engineering.*
New York: McGraw-Hill Book Company, Inc., 1963.
This book contains an extensive discussion of systems of many types, with
interesting illustrative examples. The examples are often complex systems
treated in an elementary fashion, rather than just simple electrical circuits.

Problems

2-1. A two-stage, $R-C$, lowpass filter has the following configuration:

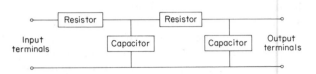

a. Sketch a circuit diagram that is suitable as the basis for a mathematical model for this filter at relatively low frequencies.
b. Sketch a circuit diagram that might form the basis for a mathematical model for this filter at intermediate frequencies, that is, at frequencies where the lead inductance and resistance are significant and where losses in the dielectric of the capacitor are not negligible.
c. What precautions can be taken to make the high-frequency model look more like the low-frequency model?

2-2. For each of the following differential equations determine the order of the system and classify it with respect to linearity, time dependence, and memory. The system input is $x(t)$ and the output is $y(t)$.

a. $\dfrac{dy}{dt} + y(t) = x(t) + 5$

b. $\dfrac{d^2y}{dt^2} + t\dfrac{dy}{dt} + 5y(t) = \dfrac{dx(t)}{dt} + x(t)$

c. $y(t) = 10x^2(t) + 10$

d. $\dfrac{d^2y}{dt^2} - \dfrac{dy}{dt}y(t) = 10x(t)$

2-3. In carrying out an engineering analysis, it is frequently necessary to approximate the true behavior of a system by the theoretical behavior of some appropriate idealized model that can be easily modeled mathematically. Comment on the appropriateness of assuming that the following systems are linear and time-invariant:
a. A loudspeaker in a hi-fi system.
b. The propagation path of a radio signal.
c. A television receiver.
d. The electrical power system of a small city, including generators, transmission equipment, and the load.

e. The thermostatic control system for a home heating system.

f. The guidance system of the launch booster for a spacecraft.

2-4. a. Write a single differential equation that relates the input and output for the system shown.

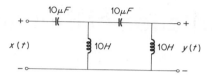

b. What is the order of this system?

2-5. a. Write a single differential equation that relates the input, $i_1(t)$, and output, $i_2(t)$, of the system shown.

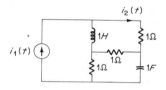

b. What is the order of this system?

2-6. Write the differential equations that represent the system of Problem 2-4 in normal form.

2-7. Write the differential equations that represent the system of Problem 2-5 in normal form.

2-8. The input, $x(t)$, and output, $y(t)$, of a system are known at discrete time instants, $t = nT$, where $-\infty < n < \infty$. The output and input are related by the following equation:

$$y(nT) = x[nT] + \frac{1}{2}x[(n-1)T] + \frac{1}{3}x[(n-2)T]$$

$$+ \frac{1}{4}x[(n-3)T] \qquad -\infty < n < \infty$$

a. Classify the system represented by this mathematical model with respect to as many of the descriptors given in Section 2-4 as you can.

b. If $x(nT) = 0$ for $n < 0$, and $x(nT) = 1$ for $n \geq 0$, plot $y(nT)$ for $0 \leq n < \infty$.

CHAPTER

3

Representation of Signals

3-1 Introduction

Now that some methods of representing a linear system by means of a mathematical model have been considered, it is necessary to consider methods of representing the input and output signals. Thus, we shall seek mathematical models that can be used to describe these variables in a quantitative manner.

For much of the discussion that follows, the major reason for representing signals by mathematical models is to be able to carry out system analysis under specified conditions that are of practical importance. There are, however, other reasons for being interested in signal representation that may be equally important in the long run. For example, in most systems studies, the important aspect of a signal is the information that it carries and not its energy or average power. This is because most signals are used to represent some other physical quantity such as force, distance, pressure, temperature, angle, and time. Not all signal forms are equally efficient for such a representation, and not all mathematical models for the signal are equally useful in making the information content evident. Thus, while any valid mathematical model for the signal could be used to carry out the system analysis, some models might be more useful than others in describing the information-bearing aspects of the signal.

Another important reason for studying different mathematical models for signals arises in connection with *signal design*. In the design of systems to accomplish specific tasks, one frequently has the option of specifying the forms of the signals to be used. It is desirable to be able to select those forms that are best, as judged by an appropriate performance criterion. In such cases it is usually necessary to impose some limitations on the signal in terms of duration, bandwidth, average power, peak power, or some other constraint dictated by the physical situation. The ability to represent signals by a variety of mathematical models is a vital aid in carrying out the actual problem of signal design under these conditions.

A final reason for studying signal representation is the physical insight that such a study provides. This insight is essential for both the system analyst and the signal designer because it provides a viewpoint that greatly simplifies thought processes and thereby increases the generality with which problems can be attacked. There are a number of relatively simple concepts, such as orthogonality, signal dimensionality, and time-bandwidth product that lead to useful approximations of signal capabilities but can also be made mathematically precise. Thus, the engineer has at his disposal some rules of thumb to use for preliminary judgments, which can also form the basis of an exact analysis. The importance of being able to predict what is feasible and what is not can hardly be overemphasized, since without this a great deal of painstaking exact analysis would be wasted on exploring unprofitable approaches. Indeed, the development of this type of insight may well be the most important reason for studying signal representation.

3-2 General Methods of Signal Representation

A signal is a function of time, and a mathematical representation of this function is necessary to be able to carry out any type of system analysis. The most general representation is by means of an abstract symbol, such as $x(t)$, and many general analytical results can be obtained without resorting to more specific representations. For example, it is shown in Chapter 2 that the mathematical model for a linear system can be obtained by representing the input and output signals by abstract symbols. This is possible because the system characteristics do not depend in any way upon the explicit form of the signals.

It is clear, however, that such an abstract representation is not quantitative; that is, it does not specify the numerical value of the signal at any instant of time. In order to be able to specify such values, it is necessary to represent the signal in terms of one or more explicit functions of time whose numerical values are exactly defined for all instants of time.

It is important to recognize that any quantitative representation of a real signal is necessarily an approximation. The actual variation of signal amplitude with time is so complex that it can never be represented exactly by any finite number of explicit time functions. When the signal waveform is intended to have a simple form, such as an exponential or sinusoid, the approximation may be very good, but in more general cases the approximation usually leaves a significant error. The problem of deciding how good an approximation must be to obtain meaningful results in any specific application of system analysis calls for the exercise of engineering judgment.

3-3   Classification of Signals

Before some of the more useful methods for representing signals explicitly are discussed, it is desirable to consider some of the various classes of signals that will be of importance. To a considerable extent, the most useful method of signal representation for any given situation depends upon the type of signal being considered and one's interest in system analysis or in some other aspect of the signal. There are many different ways in which signals can be separated into categories, and only a few of the more important ones are discussed here.

Periodic signal, nonperiodic signal.   A *periodic signal* is one that repeats the sequence of values exactly after a fixed length of time known as the *period.* More precisely, a signal $x(t)$ is periodic if there is a number T such that

$$x(t) = x(t + T) \tag{3-1}$$

for all t. The smallest positive number T that satisfies (3-1) is the period, and it defines the duration of one complete *cycle.* The *fundamental frequency* of a periodic signal is

$$f = \frac{1}{T} \tag{3-2}$$

A *nonperiodic*, or *aperiodic*, *signal* is one for which no value of T satisfies (3-1). This is an extremely important class of signals that must include all *actual* signals (since they must start and stop at finite times). It will be shown later, however, that even nonperiodic signals can be represented in terms of periodic ones; thus, from the standpoint of the mathematical representation of signals, the periodic class undoubtedly has the greatest theoretical importance.

Typical examples of periodic signals are the sinusoidal signals used in power systems and for some types of system testing, and periodically repeated, nonsinusoidal signals such as the rectangular pulse sequences used in radar and the sawtooth waveform used as a time base in oscilloscopes. In almost all such cases it is possible to write an explicit mathematical expression for the periodic signal involved.

Typical nonperiodic signals include speech waveforms, transients due to switching, and random signals arising from unpredictable disturbances of all kinds. In some cases it is possible to write explicit mathematical expressions for nonperiodic signals, and in other cases it is not.

Another class of signals represents a borderline case between periodic and nonperiodic signals. These *almost-periodic* signals are actually the sum of two or more periodic signals having incommensurate periods. The resultant signal is nonperiodic, since no *T* will satisfy (3-1), but the signal has many of the properties of periodic signals and can be represented by a finite (or countably infinite) number of periodic signals. The almost-periodic signal arises in the analysis of many types of communication systems.

Random signal, nonrandom signal. Another method of classifying signals is based on whether or not there is any randomness associated with them. This is a rather subtle point that cannot be discussed in detail here. For the time being, it is sufficient to state that a *random signal* is one about which there is some degree of uncertainty *before* it actually occurs. This terminology is most often applied to signals that are random functions of time in the sense that their magnitude varies in an erratic and unpredictable manner. For example, the output of a radio receiver responding to atmospheric disturbances, man-made interference, and internal noise sources will have such a random behavior. If future values of the signal cannot be predicted, even after observation of past values, then it is not possible to write an explicit mathematical expression for the signal.

It is possible, however, to have random signals that have a great deal of regularity and whose future values can be predicted, once the past values are observed. For example, a sinusoid, whose phase is unknown before it is observed, can be considered random if the particular phase that actually occurs is purely a matter of chance. Once it has been observed for a sufficient period of time, the phase is known, and all future values can be predicted. Random signals of this type can be represented by an explicit mathematical expression, and the randomness is associated with the parameters of the representation.

A *nonrandom signal* is one about which there is no uncertainty before it occurs, and in almost all cases an explicit mathematical expression can be written for it. All of the signals considered in this book are nonrandom to avoid the subtleties that would otherwise confuse the basic concepts.

Energy signals, power signals. For electrical systems, the signal is usually a voltage or a current. The energy dissipated by a voltage in a resistance during a given time interval t_1 to t_2 is simply

$$E = \int_{t_1}^{t_2} \frac{v^2(t)}{R} \, dt \qquad \text{watt seconds} \qquad (3\text{-}3)$$

For a current, it would be

$$E = \int_{t_1}^{t_2} Ri^2(t)\,dt \qquad \text{watt seconds} \tag{3-4}$$

In each case the energy is proportional to the integral of the *square* of the signal.

It is often convenient to talk about the energy "on a one-ohm basis." For this case, the R in (3-3) and (3-4) becomes unity, and the equations assume the same form. Because of this convenience, it has become customary to speak of the "energy" associated with any signal $x(t)$, regardless of units, as

$$E = \int_{t_1}^{t_2} x^2(t)\,dt \tag{3-5}$$

despite the fact that this does not appear to be dimensionally correct. This causes no confusion if it is remembered that equations of this type implicitly contain a "1" with the appropriate dimensions. This convenient practice will be followed here and in subsequent chapters.

An *energy signal* is defined here to be one for which (3-5) is finite, even when the time interval becomes infinite. Specifically, if $x(t)$ satisfies the condition

$$\int_{-\infty}^{\infty} x^2(t)\,dt < \infty \tag{3-6}$$

it is said to have finite energy and will be called an *energy signal*. Specific examples of energy signals are decaying exponentials and exponentially damped sinusoids (in the semi-infinite interval $t > 0$), rectangular pulses, or any signal that is nonzero in a finite time interval only and finite within that interval.

However, there are many interesting signals that do not satisfy (3-6). Examples include all periodic signals and many aperiodic signals as well. In these cases it is often more appropriate to consider the *average power* of the signal.

The average power associated with (3-5), for example, is simply

$$P = \frac{1}{t_2 - t_1} \int_{t_1}^{t_2} x^2(t)\,dt \tag{3-7}$$

where, as in (3-5), this computation of power is understood to be on a one-ohm basis. If this remains greater than zero when the time interval becomes infinite, then the signal has finite average power and will be called a *power signal*. More specifically, a power signal satisfies the condition

$$0 < \lim_{T \to \infty} \frac{1}{2T} \int_{-T}^{T} x^2(t)\,dt < \infty \tag{3-8}$$

Upon comparing (3-6) and (3-8), it is clear that an energy signal has zero average power and a power signal has infinite energy. Thus, a signal may be classified as one or the other but not both. There are, of course, some signals that may not be classified as either, since both the energy and average power may be infinite. The signal

$$x(t) = \epsilon^{-\alpha t} \qquad -\infty < t < \infty$$

is an example of this.

Exercise 3-3.1

a. Find the period of the signal represented by

$$x(t) = \epsilon^{\cos^2 10\pi t}$$

b. Is the signal in (a) an energy signal, a power signal, or neither?
c. Classify the following signal as an energy signal or power signal and calculate its energy or power.

$$x(t) = \epsilon^{-10|t|} \qquad -\infty < t < \infty$$

ANS. 0.1, 0.1

3-4   Representation in Terms of Elementary Signals

It is noted in Section 3-2 that in order to have a quantitative description of a signal, it is necessary to represent the signal in terms of explicit time functions whose numerical values are exactly defined. There are, of course, a great many ways in which this representation might be realized, even for the same physical signal. The choice of any particular method is usually motivated by mathematical convenience, ease of visualization, or specific application.

Mathematical convenience usually dictates that a signal $x(t)$ be represented as a *linear* combination of a set of elementary time functions. These elementary functions, usually called *basis functions*, are selected to have certain convenient properties that are discussed later. The signal representation can be formulated in a very general way by use of an abstract notation for the basis functions. This general formulation can be employed to indicate desirable characteristics for the basis functions. It will then be possible to consider some specific types of basis functions within the same mathematical framework and, thus, achieve a degree of versatility that would not be possible if the specific types were considered separately and independently.

Accordingly, let the set of basis functions be designated as $\phi_0(t)$, $\phi_1(t)$, $\phi_2(t), \ldots, \phi_N(t)$, where N may be infinity in some cases, and write the linear combination of these as

$$x(t) = \sum_{n=0}^{N} a_n \phi_n(t) \qquad (3\text{-}9)$$

The use of only positive integers for indexing the basis functions is an arbitrary choice. In some cases, examples of which are shown later, both positive and negative integers are employed. The problem of selecting the best set of basis functions $\phi_n(t)$ for a given application and of determining the corresponding coefficients a_n has not been solved in general. Instead, a great many specific results are available for functions that have specific properties, and these serve as an aid to making an appropriate selection in any given case.

One property that is desired for a set of basis functions is known as *finality of coefficients*. This property allows one to determine any given coefficient without needing to know any other coefficient. Stated another way, more terms can be added to the representation (to obtain greater accuracy, for example) without making any changes in the earlier coefficients. To achieve finality of coefficients, it is necessary that the basis functions be *orthogonal* over the time interval for which the representation is to be valid.

The condition of orthogonality for real basis functions requires that

$$\int_{t_1}^{t_2} \phi_n(t)\phi_k(t)dt = 0 \qquad k \neq n$$

$$= \lambda_k \qquad k = n \tag{3-10}$$

for all k and n.[1] If $\lambda_k = 1$, for all k, the basis functions are *orthonormal*. The limits of integration in (3-10) can define either a finite interval or an infinite (or semi-infinite) interval, depending upon the nature of the problem.

In order to demonstrate how the coefficients can be determined, multiply both sides of (3-9) by $\phi_j(t)$, for any j, and integrate over the specified interval. This gives

$$\int_{t_1}^{t_2} \phi_j(t)x(t)\, dt = \int_{t_1}^{t_2} \phi_j(t)\left[\sum_{n=0}^{N} a_n\phi_n(t)\right]dt$$

$$= \sum_{n=0}^{N} a_n \int_{t_1}^{t_2} \phi_j(t)\phi_n(t)dt \tag{3-11}$$

From the orthogonality condition of (3-10), this equation may be written as

$$\int_{t_1}^{t_2} \phi_j(t)x(t)\, dt = a_j\lambda_j \tag{3-12}$$

---

[1] If the basis functions are *complex* functions of time, and $\phi_k^*(t)$ is the complex conjugate of $\phi_k(t)$, then the condition for orthogonality is

$$\int_{t_1}^{t_2} \phi_n(t)\phi_k^*(t)\, dt = \begin{cases} 0 & k \neq n \\ \lambda_k & k = n \end{cases}$$

where the λ_k are real. The general discussion here will assume that the basis functions are real. An example of complex basis functions is given in the next section.

since all the terms on the right side of (3-11) will be zero except for $n = j$. Thus, the coefficient a_j may be expressed quite generally as

$$a_j = \frac{1}{\lambda_j} \int_{t_1}^{t_2} \phi_j(t)x(t)dt \tag{3-13}$$

when the basis functions are orthogonal and real.[2]

It is of interest to note that the energy or average power of a signal can be expressed in terms of the coefficients of the orthogonal representation. To see this, consider the energy in a finite time interval as given by

$$E = \int_{t_1}^{t_2} x^2(t)\,dt = \int_{t_1}^{t_2} x(t) \sum_{n=0}^{N} a_n\phi_n(t)dt$$

$$= \sum_{n=0}^{N} a_n \int_{t_1}^{t_2} x(t)\phi_n(t)dt \tag{3-14}$$

$$= \sum_{n=0}^{N} a_n^2 \lambda_n$$

in which (3-12) is used to evaluate the integral.[3] Since λ_n is the energy in the nth basis function, each term of the summation in (3-14) is simply the energy associated with the nth component of the representation. Thus, (3-14) says that the energy of a signal is the *sum* of the energies of its individual *orthogonal* components, regardless of what form they may have. This is one form of a quite general theorem that is usually referred to as *Parseval's theorem*.

The discussion so far has been so general as to give no clue to desirable types of functions to use as basis functions. Even the requirement of orthogonality is not very restrictive, since a great many different classes of functions can be made orthogonal over some interval by suitable definitions. To a large extent, the choice depends upon the eventual use that is to be made of the signal representation. When the application is system analysis, as it will be here, the sinusoidal functions are extremely useful because they remain sinusoidal after the various mathematical operations that are needed in such an analysis are performed. Specifically, the sum or difference of two sinusoids of the same frequency is still a sinusoid, and the derivative or integral of a sinusoid is still a sinusoid.

---

[2]For complex basis functions, this becomes

$$a_k = \frac{1}{\lambda_k} \int_{t_1}^{t_2} \phi_k^*(t)x(t)dt$$

and a_k may be complex. Use of (3-13) as given yields a_k^*.

[3]When the coefficients are complex, (3-14) becomes

$$E = \sum_{n=-N}^{N} |a_n|^2 \lambda_n$$

These properties, combined with the superposition properties of linear systems, imply that representing a signal as a sum of sinusoids may be a very convenient technique. This method is used for periodic signals in the next section. Chapter 5 describes how the same thing can be accomplished for aperiodic signals.

Before leaving this general discussion of signal representation to consider some specific cases, it is instructive to discuss another property of orthogonal functions that is of great usefulness in many situations. This property has to do with the accuracy of the representation when not all of the terms needed for an exact representation can be used. In almost all cases the value of N in (3-9) required for an exact representation will be infinity, but the values of a_n tend to become smaller as n becomes large. Since it is not possible to use an infinite number of terms in most practical numerical examples, the series must be terminated after some finite number of terms, and the resulting expression is an *approximation* to $x(t)$. Thus, the approximation $\hat{x}(t)$ may be expressed as[4]

$$\hat{x}(t) = \sum_{n=0}^{M} \hat{a}_n \phi_n(t) \tag{3-15}$$

where the value of M is determined by practical computational considerations, and the $\hat{a}_n$ are yet to be determined.

The practical question that now arises is how to select the coefficients $\hat{a}_n$ so that for a given, finite M the approximation $\hat{x}(t)$ is as close as possible to the true value $x(t)$. The use of (3-13) is still possible, of course, but it was derived to ensure finality of coefficients, and the matter of accuracy of the representation was not considered. Is it possible to find another set of coefficients, which may be more difficult to obtain but yield a better approximation? The answer to this question, fortunately, is "no."

In order to investigate this statement, it is necessary to have some measure of the closeness of the approximation. One measure that is frequently used is the integral of the *square* of the differences between $x(t)$ and $\hat{x}(t)$. This quantity, called the *integral squared error*, is given by

$$I = \int_{t_1}^{t_2} [x(t) - \hat{x}(t)]^2 dt \tag{3-16}$$

and is zero only when $\hat{x}(t) = x(t)$. Otherwise, it is always positive, and the smaller it is, the better the approximation.

In connection with this integral squared error, it should be mentioned that it is usually applicable *only* when $x(t)$ is an energy signal or when $x(t)$ is periodic. If $x(t)$ is an energy signal, then the limits of integration

---

[4] When the basis functions are complex, this approximation may be expressed as

$$\hat{x}(t) = \sum_{n=-M}^{M} \hat{a}_n \phi_n(t)$$

may extend from $-\infty$ to $+\infty$. When $x(t)$ is periodic, then $t_2 - t_1$ is usually chosen to be equal to the period T.

The coefficients $\hat{a}_n$, whose optimum values are being sought, may be introduced into (3-16) by using (3-15). Thus, for the case of real basis functions, the integral squared error becomes

$$I = \int_{t_1}^{t_2} \left[x(t) - \sum_{n=0}^{M} \hat{a}_n \phi_n(t) \right]^2 dt \tag{3-17}$$

It is now desired to find the set of coefficients $\hat{a}_n$ that will make the integral have its *minimum* value.

Various methods might be employed to minimize I. A straightforward procedure would be to differentiate I with respect to each of the coefficients and set these derivatives equal to zero. It could then be shown that the resulting values of $\hat{a}_n$ do, in fact, lead to a minimum rather than a maximum. An alternative procedure, which is followed here, is to rearrange the terms in (3-17) in such a way that the result becomes obvious.

If the squared term in the integrand of (3-17) is expanded, then I may be expressed as

$$I = \int_{t_1}^{t_2} \left[x^2(t) - 2x(t) \sum_{n=0}^{M} \hat{a}_n \phi_n(t) + \sum_{n=0}^{M} \hat{a}_n \phi_n(t) \sum_{k=0}^{M} \hat{a}_k \phi_k(t) \right] dt$$

Upon interchanging the sequence of integration and summation and writing as separate integrals, this equation becomes

$$I = \int_{t_1}^{t_2} x^2(t)\, dt - 2 \sum_{n=0}^{M} \hat{a}_n \int_{t_1}^{t_2} x(t)\phi_n(t) dt$$
$$+ \sum_{n=0}^{M} \sum_{k=0}^{M} \hat{a}_n \hat{a}_k \int_{t_1}^{t_2} \phi_n(t)\phi_k(t) dt \tag{3-18}$$

It may be noted that the first integral does not depend upon $\hat{a}_n$ at all and, hence, for purposes of minimizing I, can be replaced by a constant, say K. The second integral is, from (3-12), simply the previous coefficient, a_n, multiplied by λ_n. From the orthogonality condition (3-10), the last integral is zero when $n \neq k$ and λ_n when $n = k$. Then (3-18) may be written as

$$I = K - 2 \sum_{n=0}^{M} \hat{a}_n a_n \lambda_n + \sum_{n=0}^{M} \hat{a}_n^2 \lambda_n \tag{3-19}$$

(Note that the double summation in the third term is no longer needed, since only the terms for $n = k$ remain.)

Equation (3-19) can now be rewritten by adding and subtracting terms involving $\lambda_n a_n^2$. Thus,

$$I = K - \sum_{n=0}^{M} \lambda_n a_n^2 + \sum_{n=0}^{M} \lambda_n (a_n - \hat{a}_n)^2 \tag{3-20}$$

It is now clear that the quantity being sought, $\hat{a}_n$, appears *only* in the term $(a_n - \hat{a}_n)^2$ and that this term can never be negative. Hence, I will have its minimum value when the term is as small as it can be (which is zero), and this condition occurs when

$$\hat{a}_n = a_n \qquad n = 0, 1, 2, \ldots, M$$

From this result it may be concluded that the previously defined coefficients are also the best ones from the standpoint of minimizing the approximation error when only a finite number of terms is used.[5] Thus, the use of orthogonal basis functions is not only convenient from the standpoint of finality of coefficients, but these same coefficients also minimize the integral squared error of the representation. Because of the ease of computing the coefficients, orthogonal basis functions are almost always employed whenever a series representation of the type given by (3-9) is used.

It is also of interest to note that the foregoing results can be used to determine the goodness of the approximation for any value of M or the value of M needed to obtain any approximation of a given quality. The constant K is, from Parseval's theorem, simply the energy E of the actual signal in the time interval from t_1 to t_2. Similarly, I can be interpreted as the energy associated with the error in the approximation in the same time interval. The ratio of this energy to the signal can be obtained by dividing both sides of (3-20) by E. Since the last term of (3-20) is zero, this ratio becomes

$$\eta_M = \frac{\text{error energy}}{\text{signal energy}} = 1 - \frac{1}{E} \sum_{n=0}^{M} \lambda_n a_n^2 \qquad \text{(3-21)}$$

Note that η_M must lie between 0 and 1 so that its actual value gives a realistic impression of the goodness of the approximation.[6]

If the basis functions are such that

$$\lim_{M \to \infty} \eta_M = 0 \qquad \text{all } x(t)$$

then the set of basis functions is said to be *complete*. It is possible to test any set of basis functions to determine if it is complete, but these tests are beyond the scope of the present discussion. It is assumed that all basis functions discussed in the following sections are complete.

The significance of having a complete set of basis functions is that the error in the approximation can be made arbitrarily small by making M sufficiently large. There may not always be a straightforward method of

---

[5]Although only real basis functions were considered in reaching this conclusion, the same result can be obtained for complex basis functions.
[6]For complex basis functions, (3-21) becomes

$$\eta_M = 1 - \frac{1}{E} \sum_{n=-M}^{M} \lambda_n |a_n|^2$$

determining the size of M required to achieve a given accuracy, but it can be determined by trial-and-error if necessary. In some cases a direct solution is possible, as illustrated by the following exercise.

Exercise 3-4.1
In a given signal representation problem it is found that $\lambda_n = 1$ and $a_n^2 = 2^{-n}$, $n = 0, 1, 2, \ldots$.
a. Find the signal energy E. [*Hint*: Use the binomial theorem on $(1-x)^{-1}$.]
b. Find value of η_M for $M = 2, 5, 10$.
c. Find the value of M for which $\eta_M \leq 1/100$.

ANS. 1/8, 2, 1/2048, 6, 1/64.

(*Note:* In this and other exercises, answers may not be given in the same order as the questions, in order to provide an additional challenge.)

3-5 An Example: Walsh Functions

A set of basis functions that is particularly easy to handle consists of mutually orthogonal rectangular waveforms that exist in a finite time interval. These basis functions are usually referred to as *Walsh functions* and have found considerable practical use in performing digital operations on signals. A set of Walsh functions that is both orthonormal and restricted to the time interval from 0 to 1 is shown in Fig. 3-1. Other time intervals can be used by suitable translation and time-scaling. Since each

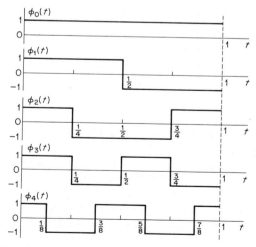

Fig. 3-1 Walsh functions indexed according to the number of zero crossings.

function has only two possible values (+1 and −1), they are generated readily with digital circuitry, and the formation of products is extremely simple.

It is clear from the figure that

$$\phi_0(t) = 1 \qquad 0 \le t \le 1$$
$$= 0 \qquad \text{elsewhere}$$

and that this waveform satisfies the condition for unit energy (that is, orthonormality). Other waveforms in the set can be found one at a time by using the conditions for orthonormality. This procedure, which is usually referred to as a *Gram-Schmidt procedure*, can always be used to find successively higher orders of basis functions—although it is usually not as simple as in this particular case. However, this simple situation is a good one to illustrate the use of a powerful and important technique.

In order to obtain $\phi_1(t)$, it is only necessary to find a function that satisfies the two equations resulting from orthonormality; that is,

$$\int_0^1 \phi_0(t)\phi_1(t)dt = 0$$

$$\int_0^1 \phi_1^2(t)dt = 1$$

It is evident that this will be true if

$$\phi_1(t) = 1 \qquad 0 \le t < 1/2$$
$$= -1 \qquad \tfrac{1}{2} < t \le 1$$

The next basis function, $\phi_2(t)$, must satisfy three equations:

$$\int_0^1 \phi_0(t)\phi_2(t)dt = 0$$

$$\int_0^1 \phi_1(t)\phi_2(t)dt = 0$$

$$\int_0^1 \phi_2^2(t)dt = 1$$

It is clear that the waveform for $\phi_2(t)$ shown in Fig. 3-1 satisfies these three equations. This procedure can be continued in the same fashion by noting that for any $\phi_k(t)$ there will be $(k + 1)$ equations that must be satisfied. The student should verify that $\phi_3(t)$ and $\phi_4(t)$ satisfy all conditions to be orthornormal basis functions.

Having obtained a set of basis functions, it is now appropriate to use them to represent some time function. Suppose that the desired time function is

$$x(t) = 6t \qquad 0 \le t < 1$$
$$= 0 \qquad \text{elsewhere}$$

Since $\lambda_j = 1$ for all j, the first coefficient, a_0, is given by (3-13) as

$$a_0 = \int_0^1 (1)6t\,dt$$

$$= 3$$

Similarly, the second coefficient, a_1, becomes

$$a_1 = \int_0^{1/2} (1)6t\,dt + \int_{1/2}^1 (-1)6t\,dt$$

$$= -\frac{3}{2}$$

In a like manner it can be shown that

$$a_2 = 0 \qquad a_3 = -\tfrac{3}{4} \qquad a_4 = 0$$
$$a_5 = 0 \qquad a_6 = 0 \qquad a_7 = -\tfrac{3}{8}$$

The approximate mathematical expression for $x(t)$ may now be written as

$$\hat{x}(t) = 3\phi_0(t) - \tfrac{3}{2}\phi_0(t) - \tfrac{3}{4}\phi_3(t) - \tfrac{3}{8}\phi_7(t) - \cdots$$

The approximate representation for these four terms is shown in Fig. 3-2 along with the true time function.

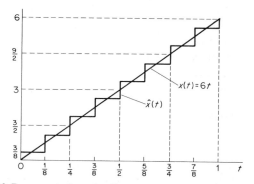

Fig. 3-2 Representation of a linear function by Walsh functions.

It is also of interest to use (3-21) to compute the fractional squared error in the approximation. The energy, E, of the true signal is

$$E = \int_0^1 (6t)^2\,dt = 12$$

Since $\lambda_n = 1$ for all n, and $M = 7$ in this case, (3-21) becomes

$$\eta_7 = 1 - \frac{1}{2}\left[(3)^2 + \left(-\frac{3}{2}\right)^2 + \left(-\frac{3}{4}\right)^2 + \left(-\frac{3}{8}\right)^2\right] = \frac{1}{256}$$

It is clear from this result that although the approximation looks fairly crude when presented graphically, the fractional squared error is really quite small. A little thought will also reveal that one more term in the approximation will cut the maximum error in half and reduce the fractional squared error by a factor of 4. Considerations of this sort may be of great value in enabling the engineer to estimate the number of terms required to obtain a desired accuracy in any signal approximation problem.

Exercise 3-5.1
It is desired to represent the function

$$x(t) = \frac{\pi}{2} \sin \pi t \qquad 0 \le t \le 1$$

by Walsh functions. For this representation find:
a. The coefficients a_0, a_1, and a_2.
b. The fractional error, η_2.

ANS. -0.414, 0, 0.0505, 1.0

3-6 Fourier Series Representation

It is noted in Section 3-4 that the use of sinusoidal basis functions is convenient because of the invariance of these functions with respect to summation, differentiation, and integration. Hence, this type of basis function is very commonly used in the representation of periodic signals, and the resulting representation is called a *Fourier series.*

It is certainly possible to use sines and cosines as basis functions, but this calls for the use of two summations, one for sine terms and the other for cosine terms. A more convenient method is to use complex exponentials and to let the index of summation be negative as well as positive. The resulting series can be converted easily into sines and cosines, if desired, by the familiar relationship

$$\epsilon^{\pm jn\omega_0 t} = \cos n\omega_0 t \pm j \sin n\omega_0 t \qquad (3\text{-}22)$$

Consider the time function shown in Fig. 3-3. The Fourier series representation that is obtained will be valid in the time interval from t_1 to t_2 for almost any time function. If, however, $x(t)$ is periodic with a period

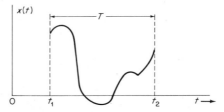

Fig. 3-3 Time function to be represented by a Fourier series.

$T = t_2 - t_1$, then the representation is valid for all times and not just those within the interval.

The term "almost any time function" implies that $x(t)$ must satisfy certain mathematical conditions if the resulting series is to converge to the true value (which must be suitably defined at discontinuities). These are the *Dirichlet conditions*, which require that within the interval t_1 to t_2 $x(t)$ be single-valued, have only a finite number of maxima and minima in a finite time, have only a finite number of finite discontinuities, and satisfy the inequality

$$\int_{t_1}^{t_2} |x(t)| \, dt < \infty$$

The actual time function, $x(t)$, corresponding to *any physical* signal will satisfy these conditions, although some common mathematical representations do not.

For the exponential Fourier series the basis functions may be defined as

$$\phi_n(t) = \epsilon^{jn\omega_0 t} \qquad n = 0, \pm 1, \pm 2, \ldots, \pm \infty$$

where

$$\omega_0 = \frac{2\pi}{T}$$

It is easy to show that these functions are orthogonal so that[7]

$$\int_{t_1}^{t_1+T} \epsilon^{jn\omega_0 t} \epsilon^{-jk\omega_0 t} dt = 0 \qquad n \neq k$$
$$= T \qquad n = k \tag{3-23}$$

If coefficients for the Fourier series are designated as α_n, then from (3-13) they may be expressed as

$$\alpha_n = \frac{1}{T} \int_{t_1}^{t_1+T} x(t) \epsilon^{-jn\omega_0 t} dt \tag{3-24}$$

and they are usually complex. In any case, however, $\alpha_{-n} = \alpha_n^*$. Then $x(t)$ is given by

$$x(t) = \sum_{n=-\infty}^{\infty} \alpha_n \epsilon^{jn\omega_0 t} \tag{3-25}$$

Evaluation of the accuracy of an approximate representation using a finite number of terms can be obtained from the modification of (3-21) noted in footnote 6. It is clear from (3-23) that $\lambda_n = T$ for all values of n. Also, since $\alpha_{-n} = \alpha_n^*$, it follows that $|\alpha_n|^2 = |\alpha_{-n}|^2$. Thus, if n ranges from

---

[7]Note that the basis functions are complex and that the complex conjugate of exp $(jk\omega_0 t)$ is exp $(-jk\omega_0 t)$. See footnote 2.

$-M$ to M, the fractional error becomes

$$\eta_M = 1 - \frac{1}{E} \sum_{n=-M}^{M} T|\alpha_n|^2$$

$$= 1 - \frac{T}{E}\left[\alpha_0^2 + 2\sum_{n=1}^{M} |\alpha_n|^2\right] \tag{3-26}$$

As a specific example of the exponential Fourier series, consider the periodic sequence of rectangular pulses shown in Fig. 3-4. The time func-

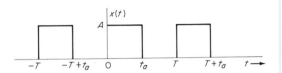

Fig. 3-4 A periodic sequence of rectangular pulses.

tion $x(t)$ may be defined during the time interval from 0 to T (one period) as

$$x(t) = A \qquad 0 < t < t_a$$
$$= 0 \qquad t_a < t < T$$

Hence, the coefficients become

$$\alpha_n = \frac{1}{T}\int_0^{t_a} A\epsilon^{-jn\omega_0 t}\, dt$$

$$= \frac{A}{T}\left(\frac{1 - \epsilon^{-jn\omega_0 t_a}}{jn\omega_0}\right)$$

$$= \frac{A}{T}\exp\left(\frac{-jn\omega_0 t_a}{2}\right)\left[\frac{\exp{(jn\omega_0 t_a/2)} - \exp{(-jn\omega_0 t_a/2)}}{jn\omega_0}\right]$$

$$= \frac{At_a}{T}\left[\frac{\sin{(n\omega_0 t_a/2)}}{n\omega_0 t_a/2}\right]\exp\left(\frac{-jn\omega_0 t_a}{2}\right)$$

This may be written in slightly different form by replacing ω_0 by its equivalent, $2\pi/T$. Thus,

$$\alpha_n = \frac{At_a}{T}\left[\frac{\sin{(n\pi t_a/T)}}{n\pi t_a/T}\right]\exp\left[-j\frac{2\pi n}{T}\left(\frac{t_a}{2}\right)\right]$$

The complete Fourier series expression for $x(t)$ now becomes

$$x(t) = \sum_{n=-\infty}^{\infty} \frac{At_a}{T}\left[\frac{\sin{(n\pi t_a/T)}}{n\pi t_a/T}\right]\exp\left[j\frac{2\pi n}{T}\left(t - \frac{t_a}{2}\right)\right] \tag{3-27}$$

This representation of $x(t)$ is in terms of sinusoids having frequencies that are multiples of the fundamental frequency $1/T$. The coefficients α_n give the magnitude and phase of these sinusoids and, hence, are said to constitute a *frequency-domain* description of the signal. The explicit time function $x(t)$ is said to be a *time-domain* description of the signal. The properties and uses of frequency-domain descriptions are extended and considered in more detail in Chapter 5, where the *Fourier transform* is introduced to describe signals over all time.

Exercise 3-6.1

It is desired to represent the function

$$x(t) = 6t \quad 0 \le t \le 1$$
$$= 0 \quad \text{elsewhere}$$

by an exponential Fourier series.
a. Find the coefficients α_n for $n = 0, \pm1, \pm2, \pm3$.
b. Evaluate the fractional error η_M for $M = 3$.
c. Compare the accuracy of this representation with the representation of this function using Walsh functions as discussed in Section 3-5.

ANS. $0.0430, 3, +j3/n\pi$

Exercise 3-6.2

It is desired to represent the function

$$x(t) = (\pi/2) \sin \pi t \quad 0 \le t \le 1$$
$$= 0 \quad \text{elsewhere}$$

by an exponential Fourier series.
a. Find the coefficients α_n for $n = 0, \pm1$.
b. Evaluate the fractional error, η_M, for $M = 1$.
c. Compare with the result of Exercise 3-5.1.

ANS. $-\frac{1}{3}, -\frac{1}{3}, 0.00931, 1$

3-7 Other Representations

Although the Fourier series is by far the most commonly used orthogonal representation, there are many other possibilities, and some of these may be more convenient in a particular situation. An important factor determining convenience is the number of terms required to achieve a desired accuracy for the representation, and there may be striking differences in this respect for different sets of basis functions. For example, the function $x(t) = 6t, 0 \le t \le 1$, was represented by seven Walsh functions in Section 3-5, and the resulting approximation shown in Fig. 3-2 is not impressively close. This same function was represented by a seven-term Fourier series in Exercise 3-6.1, and the fractional error was even greater. Yet it is clear that basis functions composed of polynomials in t would require only *one* term (the linear one) to provide an *exact* representation of this function. Hence, it is important to be aware of the existence of other basis functions and the circumstances in which they may be useful.

Before listing the properties of other basis functions, it is worth repeating these properties for the two sets of basis functions that have already been considered. The Walsh functions described in Section 3-5 are important primarily because of their piecewise constant nature and the fact that they take on only two values. Generation of such waveforms is accomplished readily by digital circuitry, and multiplication of any signal by a Walsh function is done by a polarity-reversing switch actuated at the proper instants. Thus, if one wishes to find the coefficients for representing a signal by experimental (rather than by analytical) methods, and to do this at the same time the signal is occurring, the use of Walsh functions provides a very convenient technique for accomplishing this. In this case the ease of implementation is more important than achieving maximum accuracy with a minimum number of terms.

As noted previously, the outstanding importance of Fourier series representations lies in the fact that sinusoids (or their equivalent complex exponentials) remain sinusoids under all the operations of linear circuit and system analysis. Thus, the response of the system to any one basis function is also a single basis function of the same type and can be completely described by an appropriate change in the coefficient of that one basis function. This useful property is not exhibited by any other class of basis functions. In all other cases the response of a linear system to any *one* basis function will require a *complete set* of basis functions to decribe it. Hence, the ease with which system analysis can be carried out using the Fourier series is the significant factor and in most cases is more important than the accuracy achievable with a finite number of terms.

Legendre functions. Legendre functions constitute a complete orthonormal set of basis functions in the time interval from -1 to $+1$, although other intervals can be used by suitable translation and time-scaling. They can be generated by the Gram-Schmidt procedure, and the first few basis functions are

$$\phi_0(t) = \frac{1}{\sqrt{2}} \qquad\qquad -1 \le t \le 1$$

$$\phi_1(t) = \sqrt{\frac{3}{2}}t \qquad\qquad -1 \le t \le 1$$

$$\phi_2(t) = \sqrt{\frac{5}{2}}\left(\frac{3}{2}t^2 - \frac{1}{2}\right) \qquad -1 \le t \le 1$$

$$\phi_3(t) = \sqrt{\frac{7}{2}}\left(\frac{5}{2}t^3 - \frac{3}{2}t\right) \qquad -1 \le t \le 1$$

The general basis function is given by

$$\phi_n(t) = \sqrt{\frac{2n+1}{2}} P_n(t) \qquad -1 \le t \le 1 \tag{3-28}$$

where

$$P_n(t) = \frac{1}{2^n n!} \frac{d^n}{dt^n}(t^2 - 1)^n \qquad n = 0, 1, 2, \ldots$$

are the *Legendre polynomials*. These basis functions may be convenient for representing signals over a finite time interval when the signals have a predominate linear or quadratic term.

Laguerre functions.   When the time interval of the representation is from 0 to ∞, the Laguerre functions form a complete orthonormal set. These functions are obtained from

$$\phi_n(t) = \frac{1}{n!} e^{-t/2} L_n(t) \qquad 0 \le t < \infty \tag{3-29}$$

where

$$L_n(t) = e^t \frac{d^n}{dt^n}(t^n e^{-t}) \qquad n = 0, 1, 2, 3, \ldots \tag{3-30}$$

are the *Laguerre polynomials*. The first few basis functions of this set are

$$\phi_0(t) = e^{-t/2} \qquad\qquad 0 \le t \le \infty$$
$$\phi_1(t) = e^{-t/2}(1 - t) \qquad\qquad 0 \le t < \infty$$
$$\phi_2(t) = 1/2\, e^{-t/2}(2 - t)^2 \qquad\qquad 0 \le t < \infty$$
$$\phi_3(t) = \tfrac{1}{6} e^{-t/2}(6 - 18t + 9t^2 - t^3) \qquad 0 \le t < \infty$$

The Laguerre functions have a special significance in some applications because they can be generated in relatively simple linear systems of finite order. One method of generating them uses the *transversal filter*, a concept that is discussed in more detail in Chapter 7.

Hermite functions.   Hermite functions form a complete orthonormal set over the range from $-\infty$ to ∞. They can be obtained from

$$\phi_n(t) = (2^n n! \sqrt{\pi})^{-1/2} e^{-t^2/2} H_n(t) \qquad -\infty < t < \infty \tag{3-31}$$

where

$$H_n(t) = (-1)^n e^{t^2} \frac{d^n}{dt^n}(e^{-t^2}) \qquad n = 0, 1, 2, 3, \ldots \tag{3-32}$$

are the *Hermite polynomials*. The first few basis functions are

$$\phi_0(t) = (\sqrt{\pi})^{-1/2} e^{-t^2/2}$$
$$\phi_1(t) = \sqrt{2}(\sqrt{\pi})^{-1/2} t e^{-t^2/2}$$
$$\phi_2(t) = (2\sqrt{\pi})^{-1/2}(2t^2 - 1) e^{-t^2/2}$$
$$\phi_3(t) = (3\sqrt{\pi})^{-1/2}(2t^3 - 3t) e^{-t^2/2}$$

The Hermite functions are useful because they are valid for all time, a property that is not true of most other representations. They also have some useful mathematical properties (for example, continuous derivatives of all orders) that make them desirable for more theoretical investigations into signal theory.

Cardinal functions.   Another basis function that is valid for all time is derived from the *sinc function* defined by

$$\text{sinc } t = \frac{\sin \pi t}{\pi t} \qquad -\infty < t < \infty \tag{3-33}$$

The resulting basis functions have the form

$$\phi_n(t) = \frac{\sin \pi(2Wt - n)}{\pi(2Wt - n)} \qquad n = 0, \pm 1, \pm 2, \ldots \tag{3-34}$$

in which different values of n correspond to a time translation by an amount $n/2W$ from the original sinc function. It is shown in Chapter 5 that any signal that has no frequencies higher than W can be exactly represented as an infinite sum of these basis functions and that the coefficients are simply the values of the signal at time instants equal to $n/2W$. This representation is referred to as a *cardinal function* and leads to the extremely important *sampling theorem* that makes it possible to represent a continuous time function by a discrete set of numbers. It should be noted, however, that although the basis functions defined by (3-34) are orthogonal, they do not form a complete set because signals having bandwidths greater than W cannot be exactly represented with even an infinite set of components. This point is considered in more detail in Chapter 5.

Exercise 3-7.1
It is desired to represent the function

$$x(t) = 6t \qquad -1 \le t \le 1$$
$$= 0 \qquad \text{elsewhere}$$

by Legendre polynomials. Find *all* the coefficients of the representation.

ANS. $a_n = 0, n \ne 1; a_1 = 6\sqrt{2/3}$.

3-8   Time-Bandwidth Product and Signal Dimensionality

The discussion so far has been aimed at representing signals, and the accuracy of this representation has been an important consideration. There are many situations in system analysis, however, in which this degree of specification is more complete than is necessary or desirable. This is particularly true in the early stages of system planning or design in which a complete specification of the signals is impossible or premature, but a *characterization* of the signals by a few significant parameters is extremely useful. Some parameters that are helpful in this type of preliminary study are the signal energy, average power, bandwidth, or time duration. The first two parameters are considered in the earlier sections of this chapter; the last two parameters are discussed briefly in this section. A complete discussion of these aspects of signal theory involves concepts that are much more advanced than any that can be considered here.

In defining the *time duration* of a signal, it is necessary to consider the signal's value for all time from $-\infty$ to ∞, rather than over just a prescribed interval in which we choose to represent the signal. A particularly simple situation arises in the case of *time-limited* signals that are defined to be zero outside of a prescribed interval. That is, if

$$x(t) \equiv 0 \qquad t_1 > t > t_1 + T$$

time $x(t)$ is said to be time-limited to T seconds and the time duration of this signal is clearly T. The waveform shown in Fig. 3-5(a) is an example of a time-limited signal.

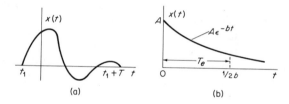

(a)   (b)

Fig. 3-5 Examples of signal time duration. (a) A time-limited signal. (b) Time duration of a general signal.

When the signal has no definite starting or stopping times, the problem of defining a time duration is not so clear-cut, and many definitions of time duration have been suggested. One of the more common definitions is to use the normalized second-central moment of the square of the time function as a measure of the time duration. This is defined as

$$T_e = \left[\frac{\int_{-\infty}^{\infty} (t - t_0)^2 x^2(t)\,dt}{\int_{-\infty}^{\infty} x^2(t)\,dt} \right]^{1/2} \tag{3-35}$$

where

$$t_0 = \frac{\int_{-\infty}^{\infty} t x^2(t)\,dt}{\int_{-\infty}^{\infty} x^2(t)\,dt} \tag{3-36}$$

is the normalized first moment. The quantity T_e is a measure of the spread of $x(t)$ about its center of gravity as given by t_0.

As an example of the calculation of time duration, consider the function

$$x(t) = Ae^{-bt} \qquad t \geq 0$$
$$= 0 \qquad t < 0$$

as shown in Fig. 3-5(b). The normalizing factor (which is just the signal energy) is

$$E = \int_{-\infty}^{\infty} x^2(t)dt = A^2 \int_0^{\infty} e^{-2bt}dt = \frac{A^2}{2b}$$

and the normalized first moment is

$$t_0 = \frac{2b}{A^2} \int_0^{\infty} A^2 t e^{-2bt}dt = \frac{1}{2b}$$

The time duration then becomes

$$T_e = \left[\frac{2b}{A^2} \int_0^{\infty} \left(t - \frac{1}{2b}\right)^2 e^{-2bt}dt\right]^{1/2} = \frac{1}{2b}$$

The *bandwidth* of a signal is another quantity that can be defined in many different ways. The bandwidth is intended to indicate a range of frequencies within which most of the signal energy lies. For example, if a Fourier series representation of a signal is made, and it is found that all coefficients, α_n, for n-values larger than $2\pi W/\omega_0$ are small, then it would be reasonable to say that the bandwidth of this signal is W hertz (Hz). If these coefficients are all exactly zero, then the signal would be *band-limited* to W Hz.

The bandwidth of a general signal is more easily defined in terms of its *Fourier transform*, which is considered in Chapter 5, than it is in terms of its Fourier series representation. Nevertheless, it is possible to define a bandwidth in terms of the *derivative* of the time function as

$$W_e = \frac{1}{2\pi} \left[\frac{\int_{-\infty}^{\infty} \left(\frac{dx(t)}{dt}\right)^2 dt}{\int_{-\infty}^{\infty} x^2(t)dt}\right]^{1/2} \qquad \text{Hz} \qquad (3\text{-}37)$$

It is shown later that this is equivalent to the normalized second moment of the Fourier transform representation of the signal. This definition is not entirely satisfactory because it leads to values of infinity whenever the signal has a discontinuity. For example, the exponential function shown in Fig. 3-5(b) is just such a case. However, the definition is adequate for the present discussion.

The concept of *time-bandwidth product* is a useful one for describing the complexity of a signal. Using the above definitions of time duration and bandwidth, this product is simply $T_e W_e$. It can be shown by relatively simple arguments that no signal can have a time-bandwidth product smaller than $1/4\pi$, although any larger value is possible. Small values of time-bandwidth product are associated with very simple signals, whereas large values imply that the signal has much more structure. Since the amount of information that can be conveyed by a signal depends upon the complexity of its structure, it follows that large time-bandwidth products also imply a large information content. Thus, the time-bandwidth product of a signal is a very significant parameter for judging the usefulness of a given signal in conveying information. Conversely, if a system study indicates the need for a great deal of information from every signal, the corresponding requirement on time-bandwidth product gives a measure of the complexity of the signals that must be used.

A concept that is closely related to time-bandwidth product is that of *signal dimensionality*. It was noted previously that although an infinite number of basis functions are required for an exact representation of an arbitrary signal, reasonable approximations can be obtained with a finite number. The smallest number of basis functions needed to achieve this approximation is said to be the dimensionality of the signal. This is clearly an ill-defined quantity in most cases because it depends upon the accuracy of approximation desired and whether or not the basis functions chosen are the ones that will lead to the minimum number. Nevertheless, the concept is a useful one because it makes it possible to define a *signal space* whose coordinates are the coefficients of the various basis functions representing a signal. The dimensionality of this space is the same as that of the signals. Thus, every signal that can be represented by the specified set of basis functions can also be represented as a *point* in signal space. The geometrical concepts of a multidimensional space can then be employed for the purpose of relating one signal to another.

In order to relate signal dimensionality to time-bandwidth product, it is convenient to look at the Fourier series representation. If a signal has a time duration of T and if the resulting Fourier coefficients can be neglected for $|n| > 2\pi W/\omega_0$, then the number of significant coefficients is just $2WT + 1$, since $\omega_0 = 2\pi/T$ and both positive and negative values of n (and $n = 0$) are used. Thus, the signal dimensionality would be approximately

$$D \simeq 2WT + 1 \tag{3-38}$$

Although this result was obtained by considering a specific set of basis functions, the general conclusion is valid and is a common way of relating dimensionality and time-bandwidth product.

Although this elementary and incomplete discussion of signal dimensionality and time-bandwidth product has been based on ideas that may seem imprecise, the concepts introduced are extremely important in advanced studies of signal theory. Furthermore, they can be made more precise by employing methods that are beyond the scope of this discussion.

Exercise 3-8.1

Find the time-bandwidth product of the signal

$$x(t) = e^{-at^2} \qquad -\infty < t < \infty$$

Comment on the significance of this result.

ANS. $1/4\pi$

3-9 Singularity Functions

There is a class of elementary signals whose members have very simple mathematical forms but are either discontinuous or have discontinuous derivatives. Because such signals do not have finite derivatives of all orders, they are usually referred to as *singularity functions*. Two of the most common singularity functions are the unit ramp function and the unit step function, as shown in Fig. 3-6.

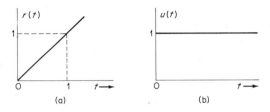

(a)   (b)

Fig. 3-6 Examples of singularity functions. (a) Unit ramp function. (b) Unit step function.

Although signals such as these are mathematical idealizations and cannot really occur in any physical system, they serve several useful purposes for system analysis. In the first place, they serve as good approximations to the signals that actually do arise in systems when switching operations take place. Secondly, their simple mathematical forms make it possible to carry out system analysis much more easily than could be done with more complicated signals. Furthermore, many complicated signals can be represented as sums of these elementary ones. Finally, they can be approximated easily in the laboratory, so one can determine experimentally whether or not a given system behaves the way the mathematical analysis says it should.

The *unit ramp function*, designated as $r(t)$, is defined to start at $t = 0$ and have unit slope thereafter. Hence, it may be represented mathematically as

$$r(t) = t \qquad t \geq 0$$
$$ = 0 \qquad t < o$$

(3-39)

If a slope other than unity is desired, it is necessary only to multiply by a constant. Thus for $b > 0$, the function $br(t)$ is a ramp having a slope of b.

An alternative way of changing the slope is to change the time scale of the argument. Since $r(t)$ has unit slope, its value must be unity whenever the argument is unity. Thus, both $br(t)$ and $r(bt)$ represent ramps with slopes of b $(b > 0)$.

The *unit step function*, designated as $u(t)$, is defined to be zero before zero time and unity thereafter. Thus, it may be represented mathematically as

$$u(t) = 1 \qquad t \geq 0$$
$$= 0 \qquad t < o$$

(3-40)

It may be noted that the unit ramp function is just the integral of the unit step function; that is,

$$r(t) = \int_{-\infty}^{t} u(\lambda)d\lambda$$

(3-41)

It is also true that at all times except $t = 0$, where a unique derivative does not exist,

$$u(t) = \frac{dr(t)}{dt} \qquad t \neq 0$$

(3-42)

A step change of value other than unity can be obtained by multiplying by a constant. Thus, $cu(t)$ is a step change of magnitude c. There is no time scale change that will accomplish the same result.

All of the singularity functions just discussed were assumed to start at $t = 0$. It is often necessary to consider other starting times, and this can be done by translating the argument of the function in time. Thus, $u(t - a)$ is zero whenever $(t - a)$ is negative, unity when it is positive, and represents a step starting at $t = a$. Some examples of translated functions are illustrated in Fig. 3-7.

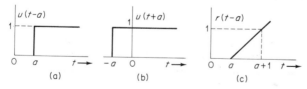

Fig. 3-7 Some translated singularity functions. (a) Step function translated to right. (b) Step function translated to left. (c) Translated ramp function.

By using combinations of ramps and step functions, it is possible to represent many other types of functions. For example, a rectangular pulse of width a can be considered as the difference between a step function at the origin and one at $t = a$. This is illustrated in Fig. 3-8. Hence, the mathematical representation of a unit rectangular pulse, $p_a(t)$, could be written as

$$p_a(t) = u(t) - u(t - a)$$

(3-43)

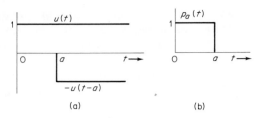

(a)   (b)

Fig. 3-8 Two steps combining to form a rectangular pulse. (a) Step-function components. (b) Resulting pulse function.

It should also be clear that

$$cp_a(t) = c[u(t) - u(t - a)] \tag{3-44}$$

is a rectangular pulse with magnitude c and duration a.

As another example, consider the finite ramp function shown in Fig. 3-9(a). In Fig. 3-9(b) are shown the two ramps whose difference will yield the finite ramp.

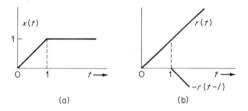

(a)   (b)

Fig. 3-9 (a) Finite ramp. (b) Its components.

Thus, a mathematical representation for this finite ramp is

$$x(t) = r(t) - r(t - 1) \tag{3-45}$$

As a final example of a more complicated waveform, consider the time function shown in Fig. 3-10. It is left as an exercise for the student to show that this time function can be represented by

$$x(t) = -r(t + 1) + 2r(t) - r(t - 2) - u(t - 3) \tag{3-46}$$

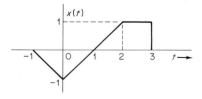

Fig. 3-10 A composite waveform.

Exercise 3-9.1

Sketch the following time functions:
a. $u(\sin t)$
b. $r(\sin t)$
c. $u(t)\, u(a-t)$
d. $u(t)\, r(a-t)$
e. $r(t)\, u(a-t)$
f. $r(t)\, r(a-t)$

Exercise 3-9.2

Express the following time function in terms of singularity functions:

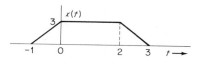

3-10 The Impulse Function

There is another singularity function, known as the *impulse*, or *delta, function*, which is of such great importance in system analysis that it is considered separately. This is not a well-behaved function in the sense that an explicit mathematical description can be written for it. Nevertheless, it has some well-defined properties that provide the best way of describing it. Before we discuss these properties, however, it may be helpful to discuss the impulse from an intuitive standpoint.

The intuitive interpretation of an impulse is that it is an idealization of a very narrow pulse having a finite total area. For convenience, the area is usually taken to be unity. A nonmathematical approach to this interpretation, which emphasizes its relationship to the step function, can be developed by considering the finite ramp function and its derivative.[8] The particular forms of these functions that will be used for this discussion are shown in Fig. 3-11.

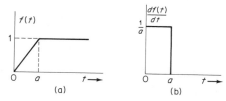

Fig. 3-11 Functions used to obtain the impulse function. (a) The finite ramp. (b) Derivative of the finite ramp.

---

[8]As used in this section, the term "derivative" is assumed to mean the actual derivative, except at those points at which a unique derivative does not exist. Hence, it corresponds to the geometrical "slope."

It is evident that the finite ramp function, $f(t)$, approximates a step function when a is small. In fact, one may write

$$u(t) = \lim_{a \to 0} f(t) \tag{3-47}$$

The derivative of the finite ramp is seen to be a rectangular pulse of duration a and magnitude $1/a$. As a becomes small, this pulse becomes narrower and taller, but its area remains constant at unity. Hence, the derivative of a finite ramp approaches an impulse as a approaches zero. If the impulse is designated as $\delta(t)$, then it may be expressed heuristically as

$$\delta(t) = \lim_{a \to 0} \frac{df(t)}{dt} \tag{3-48}$$

From these two relationships it is possible to write a "definition" of the impulse as

$$\delta(t) = \frac{du(t)}{dt} \tag{3-49}$$

even though the derivative of the step function does not exist in the strict mathematical sense.

The term "delta function," or "δ function," to designate the impulse function arises from the notation employed and from historical usage. It is also sometimes called a *Dirac delta function,* but this terminology is more properly reserved for a special form of the impulse, which is discussed later. The graphical representation of $\delta(t)$ is illustrated in Fig. 3-12(a); a magnitude-scaled and time-shifted version is shown in Fig. 3-12(b). It should be emphasized that the factor multiplying an impulse is really designating the *area* of the impulse and is not just scaling its magnitude. Thus, $A\delta(t - a)$ is an impulse with an area of A located at $t = a$.

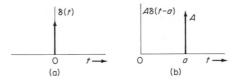

Fig. 3-12 Graphical representation of the impulse function. (a) The unit impulse. (b) Time-shifted impulse with area A.

The usefulness of the impulse, or delta, function in system analysis arises from the fact that the response of a linear system to a unit impulse at its input can be used to obtain the response of the system to *any* input signal. Thus, the *impulse response* of the system can be considered as another mathematical model for the system, since it can be used to relate input and output signals.

It can also be shown that the impulse is a good approximation to physical pulses of any shape, provided they are narrow compared to the time it takes the system to produce a significant response. The advantage of using the impulse rather than the actual pulse in system analysis is that the analysis is much simpler and that it is only necessary to specify the *area* of the physical pulse rather than its complete time function.

Although the foregoing approach provides a convenient visualization of the delta function and some of its properties, it presents serious mathematical difficulties that cannot be resolved within the framework of the mathematical background assumed for this discussion. Hence, it seems desirable to re-examine the definition of the delta function on the basis of an axiomatic description of its properties. The following list of properties, while not complete, does provide for all of the applications of delta functions that will arise in subsequent chapters. Hence, the delta function will be defined to be a function that satisfies the following conditions:

$$(1) \quad \delta(t - t_0) \qquad\qquad = 0 \qquad t \neq t_0 \qquad\qquad\qquad \text{(3-50)}$$

$$(2) \quad \int_{t_1}^{t_2} \delta(t - t_0)dt \quad = 1 \qquad t_1 < t_0 < t_2 \qquad\qquad \text{(3-51)}$$

$$(3) \quad \int_{-\infty}^{\infty} f(t)\delta(t - t_0)dt = f(t_0) \qquad f(t) \text{ continuous at } t_0 \qquad \text{(3-52)}$$

Some specific comments on these properties are in order. In the first place, conditions (1) and (2) are in complete agreement with the intuitive approach, in that they define an arbitrarily narrow pulse with unit area. Condition (3) is usually referred to as the "sifting property" of the delta function[9] and is seen to be a logical consequence of the first two conditions; that is, the integrand is zero everywhere except at $t = t_0$, where it becomes $f(t_0)\delta(t - t_0)$. Since $f(t_0)$ is a constant and the area of $\delta(t - t_0)$ is unity, the conclusion of (3) follows. The sifting property is undoubtedly the most important one from the standpoint of using the delta function in system analysis.

It is worth noting one property *not* included in this list: the value of $\delta(t - t_0)$ at $t = t_0$. The discussion on the delta function as the limit of a rectangular pulse with vanishing width would seem to imply that $\delta(0) = \infty$, but this same conclusion cannot be made from any of the defining conditions listed. In fact, it can be shown that $\delta(0)$ may have any value from $-\infty$ to $+\infty$, including zero. Fortunately, this lack of having a unique value is of little consequence in applications of the delta function, since it is seldom necessary to specify its value. In those cases in which it is necessary, the value of $\delta(0)$ is usually *defined* to be infinite.

Another property that is not included in the foregoing list has to do with how the area of the delta function is distributed with respect to the

---

[9]Actually, condition (3) is the only one necessary. Conditions (1) and (2) are included primarily for their heuristic value.

time t_0 at which it occurs. From Fig. 3-11(b) it would appear that, for this case, all of the area occurs *after* $t = 0$. This is not always true, however, and the delta function can be defined so that any fraction of its area can occur to the right of t_0. This can be expressed mathematically by the integral

$$\int_{t_0}^{t_2} \delta(t - t_0)dt = K \qquad t_2 > t_0$$

$$0 \le K \le 1$$

(3-53)

in which K is the fraction of the area to the right of t_0. For most applications in system analysis, the value of K is taken to be unity. In discussions involving random time functions, it is often necessary to define the delta function so that $K = \frac{1}{2}$; that is, the delta function would be symmetrical about t_0. It may be noted here that the term "Dirac delta function" is applied properly *only* to the case for $K = \frac{1}{2}$.

Before we leave the subject of delta functions, it is desirable to list some other properties that may be useful in some cases. These properties will be stated without proof.

1. Time scaling.

$$\delta(bt - t_0) = \frac{1}{|b|}\delta\left(t - \frac{t_0}{b}\right)$$

(3-54)

2. Multiplication by a time function.

$$f(t)\delta(t - t_0) = f(t_0)\delta(t - t_0) \qquad f(t) \text{ continuous at } t_0$$

(3-55)

3. Integral of product of delta functions.

$$\int_{t_1}^{t_2} \delta(\lambda - t)\delta(\lambda - t_0)d\lambda = \delta(t - t_0) \qquad t_1 < t_0 < t_2$$

(3-56)

4. General arguments.

$$\delta[f(t)] = 0 \qquad\qquad f(t) \ne 0$$

(3-57)

$$\int_{-\infty}^{\infty} \delta[f(t)]dt = \sum_k \frac{1}{|f'(t_k)|} \qquad f'(t_k) \ne 0$$

(3-58)

where $f'(t) = df(t)/dt$ and t_k is defined by $f(t_k) = 0$ [that is, all the zeros of $f(t)$].

3-11 Limiting Operations and Derivatives of Impulse Functions

In the previous section the delta function was visualized as being the limit approached as the width of a rectangular pulse of unit area approached zero. However, the delta function can also be obtained from limiting operations on many other types of function, and it is often convenient in system or signal analysis to recognize these when they occur.

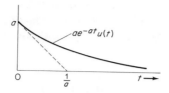

Fig. 3-13 An exponential pulse that approaches a δ function as a approaches infinity.

As a simple illustration, consider the exponential pulse shown in Fig. 3-13. It is clear that as the parameter, a, becomes large, the pulse decays more rapidly and, hence, becomes narrow. Furthermore, the area of this pulse is always unity, since

$$\int_0^\infty ae^{-at}dt = \left[\frac{ae^{-at}}{-a}\right]_0^\infty = 1$$

Thus, another definition of the delta function might be

$$\delta(t) = \lim_{a\to\infty} [ae^{-at}u(t)] \tag{3-59}$$

A number of representations of the delta function are presented in Table 3-1 along with the value of K that specifies the fraction of the area to the right of the singularity point, as discussed in connection with (3-53).

Table 3-1   Limiting Operations Leading to δ Functions

Representation		K
$\delta(t) = \lim\limits_{a\to 0} \left[\frac{1}{a}u(t)-\frac{1}{a}u(t-a)\right]$	$a \geq 0$	1
$\delta(t) = \lim\limits_{a\to 0} \left[\frac{1}{a}u(t+a)-\frac{1}{a}u(t)\right]$	$a \geq 0$	0
$\delta(t) = \lim\limits_{a\to\infty} ae^{-at}u(t)$	$a > 0$	1
$\delta(t) = \lim\limits_{a\to\infty} ae^{at}u(-t)$	$a > 0$	0
$\delta(t) = \lim\limits_{a\to\infty} \frac{a}{2}e^{-a|t|}$	$a > 0$	$\frac{1}{2}$
$\delta(t) = \lim\limits_{a\to 0} \frac{1}{\sqrt{2\pi}a}e^{-t^2/2a^2}$	$a \geq 0$	$\frac{1}{2}$
$\delta(t) = \lim\limits_{a\to\infty} \frac{\sin mat}{\pi t}$	$a > 0$ $m > 0$	$\frac{1}{2}$ $\frac{1}{2}$
$\delta(t) = \lim\limits_{a\to\infty} \frac{1}{2a}\left[\frac{\sin 2\pi at}{\pi t}\right]^2$	$a > 0$	$\frac{1}{2}$
$\delta(t) = \lim\limits_{a\to\infty} \frac{1}{2\pi}\int_{-a}^{a} e^{jtu}du$	$a > 0$	$\frac{1}{2}$

It is also possible to consider derivatives of a delta function, even though it does not exist in the usual sense. The first derivative (frequently called a *doublet*) is designated as $\delta'(t)$ and may be defined axiomatically by

$$\delta'(t - t_0) = 0 \qquad t \neq t_0 \tag{3-60}$$

$$\int_{t_1}^{t_2} \delta'(t - t_0)dt = 0 \qquad t_1 < t_0 < t_2 \tag{3-61}$$

$$\int_{t_1}^{t_2} f(t)\delta'(t - t_0)dt = -f'(t_0) \qquad \text{for } f(t) \text{ and } f'(t) \text{ continuous at } t_0 \tag{3-62}$$

Note that in this case the sifting property of the doublet leads to the *negative* of the derivative of $f(t)$ at the point of singularity. The reason for this becomes clear if one integrates (3-62) by parts. Higher-order derivatives can be defined in a similar way. Their sifting properties are expressed by

$$\int_{t_1}^{t_2} f(t)\delta^{(k)}(t - t_0)dt = (-1)^k f^{(k)}(t_0) \qquad t_1 < t_0 < t_2 \tag{3-63}$$

where the superscript (k) implies the kth derivative and all derivatives of $f(t)$ up to and including the kth exist at $t = t_0$.

Exercise 3-11.1

Evaluate the following integrals.

a. $\displaystyle\int_0^\infty (t^2 + 3t - 1)\delta(t - 1)dt$

b. $\displaystyle\int_0^\infty (t^2 + 3t - 1)\delta'(t - 1)dt$

c. $\displaystyle\int_0^\infty (t^2 + 3t - 1)\delta''(t - 1)dt$

d. $\displaystyle\int_0^\infty (t^2 + 3t - 1)\delta^{(3)}(t - 1)dt$

ANS. $-5, 0, 2, 3$

References

Except for Fourier series, much of the material in Chapter 3 on signal representation is available only at an advanced level. However, some aspects of this material are treated at a less-advanced level in the following books:

1. DAVIS, R. F., *Fourier Series and Orthogonal Functions*. Boston, Mass.: Allyn and Bacon, Inc., 1963.
   This book contains an introductory treatment of orthogonal functions and discusses some applications of them.
2. LAITHI, B. P., *Signals, Systems, and Communication*. New York: John Wiley & Sons, Inc., 1965.
   Lathi treats orthogonal signal representation methods at a more advanced level than this text and uses some matrix notation in doing so. However, it should be readable by junior-level students interested in extending their knowledge.

3. MARSHALL, J. L., *Introduction to Signal Theory.* Scranton, Pa.: International Textbook Company, 1965.
   This book contains an expanded treatment of the topics in Chapter 3 at an elementary level.
4. FRANKS, L. E., *Signal Theory.* Englewood Cliffs, N.J.: Prentice-Hall, Inc., 1969.
   This is a more advanced book that treats multidimensional signal representation in great detail.

Problems

3-1. For each of the following engineering systems, classify the signals involved with respect to periodicity, randomness, power, and so on. Specify the time interval you are considering.
  a. power system **d.** radar system
  b. telephone system **e.** television system
  c. digital computer **f.** automobile ignition system

3-2. State if each of the following signals is periodic or nonperiodic, and if it is an energy signal or a power signal. Calculate its energy or average power.
  a. $x(t) = 5 \cos(10\pi t)$ $\quad t \geq 0$
  $\quad\quad = 0 \quad\quad\quad\quad\quad t < 0$
  b. $x(t) = 8e^{-4t} \quad t \geq 0$
  $\quad\quad = 0 \quad\quad\quad t < 0$
  c. $x(t) = 5 \sin 20\pi t + 10 \cos 30\pi t \quad -\infty < t < \infty$
  d. $x(t) = 20 e^{-10|t|} \cos 20\pi t \quad -\infty < t < \infty$
  e. $x(t) = \cos 5\pi t + 2 \cos 2\pi^2 t \quad -\infty < t < \infty$

3-3. Determine whether each of the following statements is true or false:
  a. The sum of two periodic signals is always almost-periodic.
  b. All nonperiodic signals are energy signals.
  c. All energy signals are nonperiodic.
  d. All random signals are nonperiodic.
  e. The product of two power signals is always a power signal.
  f. The sum of two power signals is always a power signal.
  g. The product of an energy signal and a power signal is always an energy signal.

3-4. A signal $x(t)$ is scaled in both magnitude and time to become $ax(bt)$.
  a. If $x(t)$ is an energy signal with energy, E, over all time $(-\infty < t < \infty)$, find the energy of $ax(bt)$.
  b. If $x(t)$ is a power signal with average power, P, find the average power of $ax(bt)$.

3-5. Derive the result shown in footnote 2, page 37, for obtaining the coefficients of an orthogonal expansion when the basis functions are complex.

3-6. An energy signal is represented in terms of a set of orthonormal basis functions. The coefficients are found to have the form

$$a_n = \left(\frac{2^n}{n!}\right) \qquad n = 0, 1, 2, \ldots$$

Find the energy of this signal. (*Hint*: Use the series expansion for e^x.)

3-7. For the signal and basis functions of Problem 3-6 find the number of terms in the representation that are required to reduce the fractional error, η_M, to less than 0.1.

3-8. A set of basis functions, $\phi_n(t)$, are orthonormal on the time interval from 0 to 1. These functions are to be time-scaled to form a new set of basis functions, $\psi_n(t)$, that are orthonormal on the time interval from 0 to T. Write an explicit expression that relates $\psi_n(t)$ to $\phi_n(t)$.

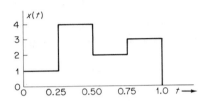

3-9. The time function shown above, $x(t)$, is to be represented in terms of Walsh functions indexed according to the number of zero crossings.

  a. Find the coefficients a_0, a_1, a_2, and a_3 for this representation.
  b. Find the fractional error, η_3, for this representation.

3-10. Consider a set of basis functions whose first two members are

$$\phi_1(t) = Ae^{-t} \qquad\qquad t \geq 0$$
$$\qquad\quad = 0 \qquad\qquad\quad t < 0$$

$$\phi_2(t) = Be^{-t} + Ce^{-2t} \qquad t \geq 0$$
$$\qquad\quad = 0 \qquad\qquad\qquad\quad t < 0$$

  a. Determine the values of A, B, and C such that $\phi_1(t)$ and $\phi_2(t)$ are orthonormal on the interval 0 to ∞. (*Hint*: Use the Gram-Schmidt procedure.)
  b. Explain why $\phi_0(t)$ is not included as a member of this set of basis functions.

3-11. A time function of the form

$$x(t) = 10e^{-3t/2} \qquad t \geq 0$$
$$\qquad\quad = 0 \qquad\qquad t < 0$$

is to be represented in terms of the basis functions derived in Problem 3-10.

a. Determine the coefficients a_1 and a_2 for this representation.

b. Determine the fractional error, η_2, for this representation.

3-12. Prove the orthogonality relation for complex exponentials as given in (3-23).

3-13. For the time function $x(t)$ shown in Problem 3-9,

   a. find the coefficients α_n for the exponential Fourier series representation for $n = 0, \pm 1, \pm 2, \pm 3$.

   b. evaluate the fractional error, η_3, for this representation.

3-14. The Fourier series representation for periodic rectangular pulses is given in (3-27). Write this representation in its simplest form for the following special cases:

   a. $t_a = \dfrac{T}{2}$ **b.** $t_a = T$

3-15. Modify (3-27) so that it is applicable to each of the following functions:

   a. $x(t) - A/2$ **c.** $x(10t)$
   b. $x(t + t_a/2)$ **d.** $x^2(t)$

3-16. The time function

$$x(t) = 4(1 - |t|) \qquad |t| \le 1$$
$$= 0 \qquad\qquad |t| > 1$$

is to be represented by Legendre polynomials. Find the coefficients a_0, a_1, and a_2.

3-17. The time function $x(t) = te^{-t/2}u(t)$ is to be represented by Laguerre functions. Find the coefficients a_0, a_1, and a_2.

3-18. Find the effective time duration, T_e, for the signal of Problem 3-16 and compare it with the actual time duration.

3-19. Find the effective time duration, T_e, for the signal of Problem 3-17.

3-20. Find the effective bandwidth, W_e, for the signal of Problem 3-16 and compute the time-bandwidth product.

3-21. Find the effective bandwidth, W_e, for the signal of Problem 3-17 and compute the time-bandwidth product.

3-22. Sketch each of the following time functions:

   a. $r(t + 2)$ **f.** $u(t + 2)$
   b. $r(2 - t)$ **g.** $u(1 - t)$
   c. $r(2t + 1)$ **h.** $u(2t + 1)$
   d. $-2r(1 - t)$ **i.** $-2u(1 - t)$
   e. $r(t) - 2r(t - 1)$ **j.** $u(t)u(1 - t)$

3-23. Write a mathematical representation (in terms of unit ramps and unit steps) for each of the following time functions:

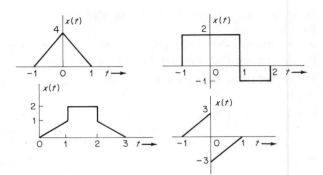

3-24. Sketch each of the following time functions:

a. $e^{-at}\delta(t-1)$ e. $(\cos t)\delta(t-\pi)$

b. $10\delta(10t-1)$ f. $(\sin t)\delta(t-\pi/2)$

c. $2u(t)+3\delta(t-2)$ g. $u(t)u(1-t)\delta(t-2)$

d. $\delta(2t)\delta(t-1)$ h. $u(1-t)\delta(t+2)$

3-25. Evaluate each of the following integrals:

a. $\displaystyle\int_{0}^{\infty} \delta(t-2)\cos[\omega(t-3)]dt$

b. $\displaystyle\int_{0}^{\infty} \delta(t+3)e^{-j\omega t}dt$

c. $\displaystyle\int_{0}^{\infty} \delta(t/2-1)(t^{2}+2)dt$

d. $\displaystyle\int_{0}^{\infty} tu(2-t)u(t)dt$

e. $\displaystyle\int_{0}^{\infty} e^{-2t}\delta(\lambda-t)\delta(\lambda-2)d\lambda$

3-26. It is often convenient in system analysis to replace short pulses having complicated shapes with impulses having an equal area. For each of the following waveforms, find the equivalent δ functions:

a. $\dfrac{d}{dt}\left(3\tan^{-1}\dfrac{t}{2\times 10^{-6}}\right)$

b. $6\exp(-2\times 10^{6}|t-3|)$

c. $2\left[\dfrac{\sin(\pi\times 10^{6}t)}{\pi\times 10^{6}t}\right]$

3.27. Write each of the following limits in terms of δ functions:

a. $\displaystyle\lim_{b\to 0}\dfrac{3}{b}\exp(-|t-2|/b)$

b. $\lim\limits_{a \to \infty} 10a^2 t e^{-at} u(t)$

c. $\lim\limits_{a \to \infty} 5 \dfrac{\sin ma(t-2)}{\pi(t-2)}$

3-28. Evaluate each of the following integrals:

a. $\displaystyle\int_0^\infty (t^4 + 6t^3 - 5t^2 - 2)\delta'(t-1)dt$

b. $\displaystyle\int_0^\infty \delta'(t-2) \cos [\omega(t-3)]dt$

c. $\displaystyle\int_0^\infty \delta'(t-1)e^{-t}u(t)dt$

d. $\displaystyle\int_0^\infty \delta''(t-2) \cos [\omega(t-3)]dt$

e. $\displaystyle\int_{-\infty}^\infty \delta'(t)\dfrac{\sin 10t}{10t}dt$

CHAPTER

4

Convolution

4-1 Introduction

One of the most frequently encountered problems in system analysis is determining the output response of a system to some specified input. Such information can be obtained experimentally with suitable measuring equipment or analytically through various mathematical techniques. The study of a number of these analysis techniques is the subject of much of the remainder of this book. The methods most widely used can be separated into three major categories: solution of the system differential equations by classical methods; use of elementary signals and the principle of superposition; and transform methods. Actually, all of these methods are intimately related to each other and merely represent different ways of looking at the same problem. In this chapter the superposition procedure is considered in some detail, with the use of a technique known as *convolution.*

The basic ideas behind convolution are very simple and may be stated briefly as follows: First, the input signal is represented (or "analyzed") as a continuum of impulses. Second, the response of the system to a single impulse is determined. Third, the response of the system to each of the elementary impulses representing the input is computed. Fourth, the total system response is obtained by superimposing the responses to all of the impulses used in the representation of the input. This latter superposition is accomplished by means of a convolution integral. For this procedure to work it is necessary to be able to analyze a signal into an appropriate set of impulses, to be able to characterize the response of a system by its impulse response, and to have the principle of superposition be valid. Fortunately these requirements are satisfied by linear systems and by virtually all signals likely to be encountered in practice.

In addition to relating the input and output of linear systems, the convolution integral also provides a powerful analytical tool for evaluating Fourier and Laplace transforms and for deriving a number of important signal and system properties. This integral also occurs in many other places in applied mathematics and probability, and a thorough understanding of its use and interpretation is invaluable to the system analyst.

4-2   Representation of Signals by a Continuum of Impulses

An arbitrary time function can be approximated in an interval from $-T$ to T by a closely spaced series of rectangular pulses. The height of each pulse is equal to the value of the function at the center of the pulse. This is illustrated in Fig. 4-1. As the width of the pulses, ΔT, decreases, it is evident

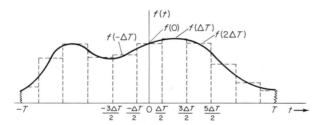

Fig. 4-1 Approximation of a signal by a series of pulses.

that the approximation improves. The whole approximation process can be examined analytically by writing out the mathematical expression for the series of pulses and then considering the behavior in the limit as $\Delta T \to 0$. The pulse whose center is located at $t = k\Delta T$ can be written as

$$f(k\Delta T)\left[u\left(t - k\Delta T + \frac{\Delta T}{2}\right) - u\left(t - k\Delta T - \frac{\Delta T}{2}\right)\right]$$

The approximation to $f(t)$ can be written as

$$f(t) \doteq \sum_{k=-N}^{N} f(k\Delta T)\left[u\left(t - k\Delta T + \frac{\Delta T}{2}\right) - u\left(t - k\Delta T - \frac{\Delta T}{2}\right)\right] \quad \text{(4-1)}$$

where the total number of pulses is $2N + 1 = 2T/\Delta T$. If (4-1) is multiplied by $\Delta T/\Delta T$, it can be written as

$$f(t) = \sum_{k=-N}^{N} f(k\Delta T)\ \frac{u\left(t - k\Delta T + \dfrac{\Delta T}{2}\right) - u\left(t - k\Delta T - \dfrac{\Delta T}{2}\right)}{\Delta T}\ \Delta T \qquad \text{(4-2)}$$

As ΔT is made smaller, it is seen that the bracketed factor in (4-2) approaches a δ function located at $t = k\Delta T$. In the limiting case as $\Delta T \to 0$, N becomes infinite. However, for a constant T, the product $(2N + 1)\Delta T$ remains constant and equal to $2T$. The product $k\Delta T$ takes on all possible values in the interval $-T < t < T$ and can be considered to be a continuous variable λ. Likewise, the increment ΔT approaches a differential $d\lambda$. In this limiting case the summation becomes an integral with respect to λ over the range $-T$ to T. Accordingly, $f(t)$ can be written as

$$f(t) = \int_{-T}^{T} f(\lambda)\delta(t - \lambda)\,d\lambda \qquad -T < t < T \qquad \text{(4-3)}$$

The complete time function can be obtained by letting $T \to \infty$, giving

$$f(t) = \int_{-\infty}^{\infty} f(\lambda)\delta(t - \lambda)\,d\lambda \qquad \text{(4-4)}$$

The function $f(t)$ is thus represented as the summation (integral) of a continuum of impulses having strengths at any time t given by $f(t)dt$. The relationship given in (4-4) is often used as the definition of the unit impulse. It was derived in the preceding manner to provide a physical basis for understanding the convolution integral derived in the next section.

4-3 System Impulse Response and the Convolution Integral

In the preceding section it is shown that a time function can be represented in terms of a continuum of impulses. By computing the response of a linear system to each member of this continuum of impulses and summing up these responses, the total system response is obtained. The validity of this approach rests on the superposition theorem, which states that the response of a linear system having a number of inputs can be computed by determining the response to each input considered separately and then summing the individual responses to obtain the total response. Superposition, as has been discussed previously, is applicable only to linear systems, and in the following sections only such systems are considered.

The impulse response, $h(t)$, of a time-invariant system is the output time function that results when the input signal is a unit impulse occurring at $t = 0$. It is assumed that the output is zero before application of the impulse and would have remained zero if the impulse had not been applied. As an example, let us compute the impulse response of the R–C circuit shown in Fig. 4-2.

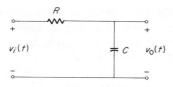

Fig. 4-2 *R-C* circuit.

The desired result is most easily found by considering the response to a unit area pulse and allowing the pulse width to go to zero, leading to a δ function. A unit area pulse of width ΔT has an amplitude of $1/\Delta T$. When such a pulse is applied, the output voltage exponentially increases toward a value of $1/\Delta T$. After ΔT seconds, the pulse terminates, and the voltage decays toward zero from the value $1/\Delta T(1 - \epsilon^{-\Delta T/RC})$. This is shown in Fig. 4-3.

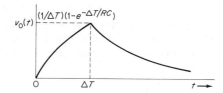

Fig. 4-3 Pulse response of circuit of Fig. 4-2.

In the limit as $\Delta T \to 0$, the initial rise occurs in zero time and reaches a value of

$$\lim_{\Delta T \to 0} \frac{1}{\Delta T}\left(1 - 1 + \frac{\Delta T}{RC} + \cdots\right) = \frac{1}{RC} \tag{4-5}$$

Therefore, the impulse response $h(t)$ is as shown in Fig. 4-4.

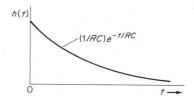

Fig. 4-4 Impulse response of circuit of Fig. 4-2.

This response can be explained physically as follows: Prior to the occurrence of the impulse the input terminals are short-circuited, since the impulse generator is assumed to be an ideal voltage source with zero internal resistance. At the instant the impulse occurs, a current flows in the circuit and charges the capacitor to a voltage of $1/RC$. Subsequent to the impulse, the input is again short-circuited, and the capacitor discharges through the resistor in an exponential fashion with a time constant of RC. The voltage across the capacitor during this sequence of events is the circuit impulse response.

Simpler and more powerful methods for determining a system's impulse response are considered later, particularly in connection with the Laplace transform. For the present, all that is necessary is to have a physical appreciation of what is meant by the impulse response of a system.

It should also be recalled that for a time-invariant system, a time shift of the input results in a corresponding time shift of the output. Therefore, the response of such a system to a δ function at $t = \lambda$ is just $h(t - \lambda)$.

Using the concept of impulse response $h(t)$ for a linear, time-invariant system, it is now possible to determine the system output for an arbitrary input. The input $x(t)$ can be resolved into a continuum of impulses, as in (4-4). Each of these impulses is of the form $x(\lambda)\delta(t - \lambda)d\lambda$. The response to each elementary impulse is $h(t - \lambda)$ multiplied by the strength of the impulse $x(\lambda)d\lambda$ and is properly positioned to coincide with the time of application of the impulse. Mathematically, this is expressed as $x(\lambda)h(t - \lambda)d\lambda$. The total response $y(t)$ is the summation of all the elementary responses and is given by

$$y(t) = \int_{-\infty}^{\infty} x(\lambda)h(t - \lambda)d\lambda \qquad \text{(4-6)}$$

The integral relationship expressed in (4-6) is called the *convolution* of $x(t)$ and $h(t)$ and relates the input and output of the system by means of the system impulse response. A simple change of variable shows that convolution is commutative for time-invariant systems, and therefore an equivalent expression is

$$y(t) = \int_{-\infty}^{\infty} h(\lambda)x(t - \lambda)d\lambda \qquad \text{(4-7)}$$

The convolution operation is generally denoted by an asterisk as follows:

$$f_1(t) * f_2(t) = \int_{-\infty}^{\infty} f_1(\lambda)f_2(t - \lambda)d\lambda \qquad \text{(4-8)}$$

The effective limits on the convolution integral will vary with the particular characteristics of the functions being convolved. As discussed in a later section, for physically realizable systems, $h(t) = 0$ for $t < 0$, and this requirement establishes the upper limit in (4-6) as t. Actually, it would not be incorrect to write the upper integration limit as ∞, since the function $h(t - \lambda)$ is zero when $\lambda > t$. Making the upper limit t merely places this property more clearly in evidence. Similarly, if the time function starts at a time t_0, the lower limit could be t_0. It is common practice to extend the lower limit to $-\infty$ to allow for input functions extending into the infinite past, such as constants and sinusoids. When this is done, the convolution integral becomes

$$y(t) = \int_{-\infty}^{t} x(\lambda)h(t - \lambda)d\lambda \qquad \text{(4-9)}$$

When the order of convolution is changed as in (4-7), the expression for physically realizable systems becomes

$$y(t) = \int_{0}^{\infty} h(\lambda)x(t - \lambda)d\lambda \qquad \text{(4-10)}$$

In (4-10) the lower limit is determined by the physical realizability constraint on $h(t)$, and the upper limit is set to include negative as well as positive time for the excitation $x(t)$.

As an example of the application of the convolution integral to a network problem, let us compute the response of the R-C circuit in Fig. 4-2 to a rectangular pulse of amplitude A and duration T. The expression for the pulse is $A[u(t) - u(t - T)]$, and the impulse response can be taken from Fig. 4-4. The output, $v_0(t)$, then becomes

$$v_0(t) = \int_{-\infty}^{\infty} A[u(\lambda) - u(\lambda - T)]\left(\frac{1}{RC}\right) \exp\left(-\frac{t - \lambda}{RC}\right)u(t - \lambda)d\lambda$$

$$= \frac{A}{RC}\left[\int_{-\infty}^{\infty} \exp\left(-\frac{t - \lambda}{RC}\right)u(\lambda)u(t - \lambda)d\lambda \right. \qquad \text{(4-11)}$$

$$\left. - \int_{-\infty}^{\infty} \exp\left(-\frac{t - \lambda}{RC}\right)u(\lambda - T)u(t - \lambda)d\lambda\right]$$

The limits on the integrals are determined by the regions of the λ-axis over which the step functions are nonzero and can be seen by sketching out the three functions involved, as shown in Fig. 4-5.

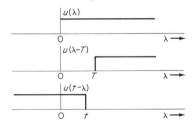

Fig. 4-5 Sketch for determining limits on a convolution integral.

It is evident that for t less than T the product $u(\lambda)u(t - \lambda)$ is unity over the range $0 < \lambda < t$ and zero elsewhere and that the product $u(\lambda - T)u(t - \lambda)$ is zero everywhere in the interval. Therefore, for $t < T$ the limits on the first integral become 0 and t, and the value of the second integral is zero. When $t > T$, the limits on the first integral are unchanged, but the limits on the second integral now become T and t. Accordingly, we can write (4-11) as

$$v_0(t) = \frac{A}{RC}\left[\int_0^t \exp\left(-\frac{t-\lambda}{RC}\right)d\lambda\right]u(t) - \frac{A}{RC}\left[\int_T^t \exp\left(-\frac{t-\lambda}{RC}\right)d\lambda\right]u(t-T)$$

Carrying out the integration gives

$$v_0(t) = A\left\{\left[1 - \exp\left(-\frac{t}{RC}\right)\right]u(t) - \left[1 - \exp\left(-\frac{t-T}{RC}\right)\right]u(t-T)\right\}$$

This can be written equivalently as

$$v_0(t) = A\left[1 - \exp\left(-\frac{t}{RC}\right)\right] \qquad 0 < t \le T$$

$$= A\left[1 - \exp\left(-\frac{T}{RC}\right)\right]\exp\left(-\frac{t-T}{RC}\right) \qquad t > T$$

This response is of the same form as that sketched in Fig. 4-3. Much simpler methods of evaluating the convolution integral are available and are considered next.

4-4 Evaluation and Interpretation of Convolution Integrals

Consider the convolution of the two time functions $f_1(t)$ and $f_2(t)$. Formally, the convolution operation is given by

$$f_3 = f_1 * f_2 = \int_{-\infty}^{\infty} f_1(\lambda)f_2(t-\lambda)d\lambda$$

The value of f_3 for any particular time t is seen to be the area under the product of $f_1(\lambda)$ and $f_2(t-\lambda)$. In order for the convolution technique to be used efficiently, it is necessary to be able to sketch rapidly (or visualize mentally) the functions $f_1(\lambda)$ and $f_2(t-\lambda)$. Visualizing function $f_1(\lambda)$ presents no difficulty, since it is identical with $f_1(t)$, except for a change in independent variable from t to λ. The function $f_2(t-\lambda)$ as a function of λ requires a little more thought, however. It can be visualized most readily as a combination of reflection and translation of the original function $f_2(\lambda)$. This process, called *folding*, is most easily described by means of an example. In Fig. 4-6(a), an arbitrary function $f_2(\lambda)$ is shown. The function $f_2(-\lambda)$, shown in Fig. 4-6(b), is merely a reflection of $f_2(\lambda)$ about the vertical axis. In order to sketch the reflected function, it is only necessary to start at the origin with the ordinate $f_2(0)$ and sketch on the right that portion of the function which was originally on the left and sketch on the left that portion of the function which was originally on the right. The function $f_2(t-\lambda)$ can be thought of as $f_2[-(\lambda - t)]$, in which case it is clear that the variable λ has been replaced by $\lambda - t$, which corresponds to a delay (or translation to the right) by an amount t when the function is

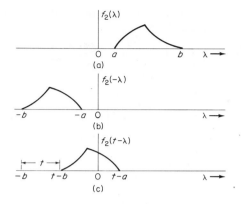

Fig. 4-6 Folding and sliding the function $f_2(\lambda)$.

plotted along the λ-axis. The function $f_2(t - \lambda)$ is shown in Fig. 4-6(c) for an arbitrary t. The amount of displacement t is measured from the position of $f_2(-\lambda)$, which corresponds to $f_2(t - \lambda)$ with $t = 0$. The convolution is then given as the area under the product of $f_1(\lambda)$ and $f_2(t - \lambda)$; it is generally a function of t, since the value of t determines the relative positions of $f_1(\lambda)$ and $f_2(t - \lambda)$. The convolution can be thought of as being obtained by folding or reflecting one function and then determining the area under the product as the folded function is slid along the horizontal axis to the right for positive time and to the left for negative time. Figures 4-7 and 4-8 show some examples of convolution.

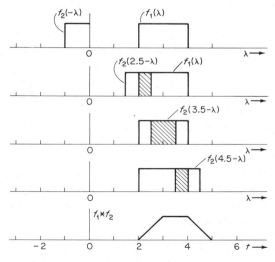

Fig. 4-7 Convolution of two rectangular pulses.

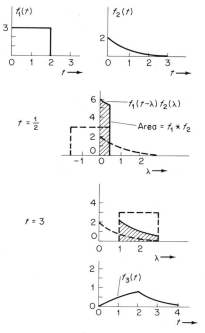

Fig. 4-8 Convolution of a rectangular pulse and an exponential pulse.

In Fig. 4-7 the two functions being convolved are of unit height, and so the product function is also of unit height and is shown with its area cross-hatched. In Fig. 4-8 the functions have unequal magnitudes, and the product function is quite distinct from either of the individual functions being convolved.

The action of the functions f_1 and f_2 on each other can be thought of as a smoothing or weighting operation, and one of the functions is often referred to as a *weighting function*. In the convolution integral expressing the output of a linear system, the weighting function is the system impulse response.

Convolution can also be carried out analytically. Such calculations are quite straightforward when the functions involved can be represented by simple analytical expressions. However, when the functions are discontinuous or when they require different analytic representations over different portions, difficulties may arise in keeping the limits straight on the integrals. As an illustration of direct calculation of a convolution, consider the example shown in Fig. 4-8. The two functions to be convolved can be expressed analytically as

$$f_1(t) = 3[u(t) - u(t-2)] \tag{4-12}$$

$$f_2(t) = 2\epsilon^{-2t}u(t) \tag{4-13}$$

The convolution is given by

$$f_3(t) = f_1(t) * f_2(t) = \int_{-\infty}^{\infty} 6[u(t-\lambda) - u(t-2-\lambda)]\epsilon^{-2\lambda}u(\lambda)d\lambda \quad \text{(4-14)}$$

This expression can be most easily evaluated by considering it over three separate time intervals. The limits are readily established from the sketches in Fig. 4-8.

$t < 0$: There is no overlap of the functions, and $f_3(t)$ is, therefore, zero.

$$f_3(t) = 0 \qquad t < 0$$

$0 \le t \le 2$: The folded function, $f_3(t - \lambda)$, partially overlaps the non-folded function, $f_2(\lambda)$, and the amount of overlap increases with t until $t = 2$. The integrand in this interval is just three times the nonfolded function, and the limits on the integral extend from the beginning of f_2 (that is, $\lambda = 0$) to the end of the folded function (that is, $\lambda = t$). Thus,

$$f_3(t) = \int_0^t 6\epsilon^{-2\lambda}d\lambda = 6\left(-\frac{1}{2}\epsilon^{-2t} + \frac{1}{2}\right)$$

$$= 3(1 - \epsilon^{-2t}) \qquad 0 < t < 2$$

$2 < t < \infty$: In this interval the folded function is completely overlapped by the nonfolded function. The integrand is three times the nonfolded function, and the limits on the integral are $t - 2, t$. Thus,

$$f_3(t) = \int_{t-2}^t 6\epsilon^{-2\lambda}d\lambda = 6\left(-\frac{1}{2}\epsilon^{-2t} + \frac{1}{2}\epsilon^{-2t+4}\right)$$

$$= 3\epsilon^{-2t}(\epsilon^4 - 1) = 161\epsilon^{-2t} \qquad 2 < t < \infty$$

The same result can be obtained by direct evaluation of (4-14) using the known properties of the three step functions to arrive at the correct limits. Making a simple sketch, however, usually leads to the solution more directly and often with less chance for error.

Exercise 4-4.1
Evaluate the convolution in (4-14) using the exponential pulse as the folded function and the rectangular pulse as the nonfolded function.

4-5 Numerical Convolution

Numerical evaluation of the convolution integral can be carried out in a straightforward manner. When either of the functions is of finite duration, there will be a finite number of terms in each summation, and the operation is readily handled. However, when the functions involved are of semi-infinite or infinite duration (for example, exponentials), the computations can become quite lengthy when the amount of overlap becomes large. Whenever the number of samples of the functions being convolved exceeds a relatively small number, it is advisable to carry out the calculation on a digital computer. In the next chapter an alternative method of evaluating convolution integrals is discussed that provides a substantial reduction in overall computation time compared to direct calculation when large numbers of computations are involved.

For hand calculations the following systematic procedure is convenient. Tabulate sample values of the two functions taken at equally spaced intervals on strips of paper that can be slid along relative to each other. The magnitude of each sample is taken as the mean ordinate of the function in a width equal to the sampling interval, which is centered at the sample point. When closely spaced points are employed, it is often sufficiently accurate to use the value of the function at the center of the sampling interval. Spacing the sample values close together (that is, using a short interval) improves the resolution and accuracy of the computa-

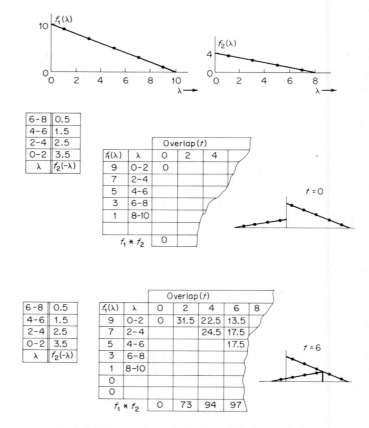

Fig. 4-9 Sliding strip method of numerical convolution.

tion. One set of samples should be tabulated in reverse order to corres-pond to the folded function $f_2(-\lambda)$. The strips are laid beside each other to give the correct registration of the samples corresponding to $f_1(\lambda)$ $f_2(t - \lambda)$ where the quantity, t, is the relative displacement between the origins of the two sets of samples. This procedure is illustrated in Fig. 4-9, where $f_1(t)$ and $f_2(t)$ are triangular waveforms of different amplitudes and durations. The relative positions of the strips for $t = 0$ and $t = 6$ are shown in the figure. After the samples of the product function $f_1(\lambda) f_2(t - \lambda)$ are computed, the integrated value is obtained by summing them and multiplying by the interval width.

Another scheme for carrying out numerical convolution is the so-called multiplication method. Values of the two functions being con-volved are obtained at equal time increments and are tabulated in rows, one above the other. The values in the upper row are then multiplied one at a time by the values in the lower row and are tabulated below the original rows. The multiplication starts at the left and proceeds to the right, and the beginning of each row of products is started immediately below the second-row element that is multiplying the first row. The con-volution is obtained by summing the column and multiplying by the inter-val width. An example is shown in Fig. 4-10.

One point in connection with numerical convolution that is sometimes confusing is establishing the proper relationship of the time variable to the

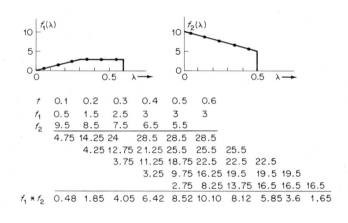

t	0.1	0.2	0.3	0.4	0.5	0.6				
f_1	0.5	1.5	2.5	3	3	3				
f_2	9.5	8.5	7.5	6.5	5.5					
	4.75	14.25	24	28.5	28.5	28.5				
		4.25	12.75	21.25	25.5	25.5	25.5			
			3.75	11.25	18.75	22.5	22.5	22.5		
				3.25	9.75	16.25	19.5	19.5	19.5	
					2.75	8.25	13.75	16.5	16.5	16.5
$f_1 * f_2$	0.48	1.85	4.05	6.42	8.52	10.10	8.12	5.85	3.6	1.65

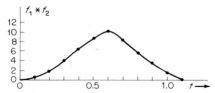

Fig. 4-10 Multiplication method of numerical convolution.

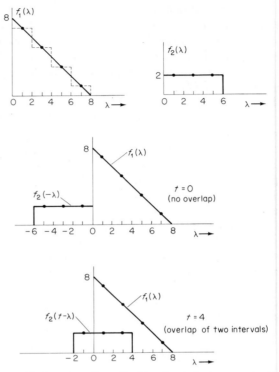

Fig. 4-11 Sampling with function discontinuities located at approximating step discontinuities.

sample values employed in the convolution calculations. The problem is illustrated in Figs. 4-11 and 4-12, in which two different origins for the sampling intervals are used. In both cases the abscissa is divided into equal intervals, and the average ordinate within each interval is taken as the sample value for that interval. This procedure amounts to replacing the original function by a staircase approximation as shown by the dashed lines in the figures. The amount of overlap (and therefore the value of t) of two functions being convolved is best determined from a sketch of the original functions. It is seen from the figures that when the original functions are sketched in the position of $t = 0$, there is a difference in the overlap present in the two cases. In Fig. 4-11, for $t = 0$, the product $f_1(\lambda)f_2(t - \lambda)$ is 0, since at any value of λ where $f_1(\lambda)$ is nonzero, the function $f_2(t - \lambda)$ is zero, and vice versa. The first nonzero product of $f_1(\lambda)f_2(t - \lambda)$ occurs when the displacement between the functions equals one sampling interval and, therefore, corresponds to a value of t equal to the duration of one sampling interval. For a computation involving k samples of each function the corresponding time will be k times the sampling interval. The important point here is that the amount of overlap

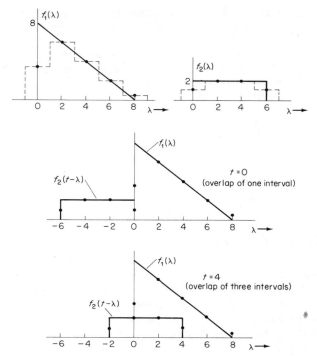

Fig. 4-12 Sampling with function discontinuities not coincident with approximating step discontinuities.

is equal to the time corresponding to the number of intervals used and not the time corresponding to the abscissa of the center of the farthest interval; these two times differ by an amount equal to one-half the sample interval.

For the sampling procedure shown in Fig. 4-12 the situation is somewhat different. In this case the staircase approximating function starts ahead of the actual function, as shown. Here there will be an overlap of one sample interval at $t = 0$, and for t equal to k times the sample interval there will be an overlap of $k + 1$ intervals.

The results of using either of these procedures on smooth functions are usually comparable. However, when discontinuities are present, the procedure that locates the discontinuities along one of the steps in the approximating function will give the better results. In the example shown in Figs. 4-11 and 4-12 it would be expected that the procedure shown in Fig. 4-11 would be the better of the two. That this is true may be seen from Table 4-1, in which the results of using each of the sampling procedures is shown along with the exact analytical result. By choosing very small sample intervals, such effects can be made negligible in most instances.

Table 4-1   Comparison of Sampling Procedures

$$f_1 * f_2$$

t	method a	method b	exact
0	0	7.5	0
2	28	27.0	28
4	48	47.0	48
6	60	51.5	60
10	36	36.5	36
12	16	17.0	16
14	4	5.0	4
15	0	0.5	0

4-6   Properties of Systems Impulse Responses

The impulse response of a linear system represents a very general description of the system's characteristics. In cases where the system parameters are varying with time, the particular instant at which an impulse is applied will affect the response of the system. In such systems the impulse response is written as a function of both the time of application of the impulse, ξ, and the independent variable, t. Symbolically such an impulse response is denoted as $g(t, \xi)$, and the system response to an excitation $x(t)$ becomes

$$y(t) = \int_{-\infty}^{\infty} g(t, \xi) x(\xi) d\xi \qquad \text{(4-15)}$$

As an example of a time-varying impulse response, consider the function

$$g(t, \xi) = \cos 2\pi \xi \epsilon^{-(t-\xi)} u(t - \xi)$$

Suppose that it is desired to compute the response of this system to an exponential input, $x(t) = \epsilon^{-2t} u(t)$. The resulting convolution becomes

$$y(t) = \int_{-\infty}^{\infty} \cos 2\pi \xi \epsilon^{-(t-\xi)} \epsilon^{-2\xi} u(t - \xi) u(\xi) d\xi$$

$$= \epsilon^{-t} \int_{0}^{t} \epsilon^{-\xi} \cos 2\pi \xi d\xi$$

$$= \epsilon^{-t} \left[\frac{\epsilon^{-\xi}}{1 + 4\pi^2} (-\cos 2\pi \xi + 2\pi \sin 2\pi \xi) \Big|_{0}^{t} \right]$$

$$= \frac{1}{1 + 4\pi^2} [\epsilon^{-2t} (2\pi \sin 2\pi t - \cos 2\pi t) + \epsilon^{-t}] \qquad t > 0$$

In this simple case it is possible to obtain an answer in a straightforward manner. However, in general the analysis of time-varying systems is quite difficult and will not be considered further at this time. Instead, the discussion will be limited to consideration of impulse responses of time-invariant systems. Such systems are characterized by the fact that if an excitation is applied at some particular time and a response $y(t)$ is observed, then an identical excitation applied t_0 seconds later will lead to the response $y(t - t_0)$, which is identical with the first response, except for the time shift, t_0. Another way of stating the same thing is to say that the amplitude of the response function depends only on the elapsed time after application of the excitation and not on the specific time of applying the excitation. Therefore, it may be concluded that the impulse response of a time-invariant linear system will depend only on the variable $t - \xi$, and not on t or ξ separately. Functionally, this becomes $g(t, \xi) = h(t - \xi)$. An example of such an impulse response is

$$g(t, \xi) = h(t - \xi) = \frac{1}{RC} \exp\left(-\frac{t - \xi}{RC}\right) u(t - \xi)$$

This is the response of the R–C circuit in Fig. 4-2 to an impulse applied at $t = \xi$. If the impulse were applied at $\xi = 0$, then the variable $t - \xi$ would become simply t, and the impulse response would be as given previously in Fig. 4-4.

For physically realizable systems—that is, systems that are not inherently impossible to construct—it is required that[1]

$$h(t) = 0 \qquad t < 0 \tag{4-16}$$

This corresponds to the intuitively obvious requirement that the response be zero prior to application of the excitation. This relationship is known as the *causality requirement*, and as noted in Chapter 2, systems having this property are called *causal*, or *nonanticipatory, systems*.

In addition to the causality requirement, it is further generally required that $h(t) \to 0$, as $t \to \infty$ for passive physical systems. This requirement reflects the practical (but not mathematical) impossibility of having lossless elements in a system. If lossless systems, or active systems, (that is, systems having internal energy sources) are being considered, then it is no longer necessary that $h(t) \to 0$ as $t \to \infty$, but to be practically useful, the system must be stable in the sense that the output must be bounded for a bounded input. A sufficient condition for this BIBO (bounded-input/bounded-output) stability is that the area under the magnitude of the system impulse response be finite. In equation form, this requirement is

$$\int_0^\infty |h(t)|\, dt \leq K_2 < \infty \tag{4-17}$$

where K_2 is constant.

---

[1] For the time-varying case, the requirement takes the form

$$g(t, \xi) = 0 \qquad t < \xi$$

This relationship may be proved in the following manner. Let the input, $x(t)$, be bounded by (that is, never exceed) a number K_1; in other words, let $|x(t)| < K_1$ for all t. The output, $y(t)$, may be expressed in terms of $x(t)$ as

$$y(t) = \int_0^\infty h(\lambda)x(t-\lambda)d\lambda$$

Using the property of integrals that the absolute value of an integral is less than or equal to the integral of the absolute value of the integrand, and making use of the boundedness of $x(t)$, we get

$$|y(t)| \le \int_0^\infty |h(\lambda)x(t-\lambda)|d\lambda$$

$$\le \int_0^\infty |h(\lambda)||x(t-\lambda)|d\lambda \qquad \textbf{(4-18)}$$

$$\le K_1 \int_0^\infty |h(\lambda)|d\lambda$$

From (4-18) it follows that if (4-17) is satisfied, then the output will be bounded by $K_1 K_2$, and the system will be BIBO stable. The question of system stability is discussed further in connection with transfer functions in Chapter 7.

The units of the impulse response vary with the input and output functions being considered. For example, if $h(t)$ is the current resulting from a voltage impulse, then the units of $h(t)$ are amperes/volt-second. The units are most easily found from the response equation. For the case just mentioned, we have

$$i_0(t) = \int_0^\infty h(\lambda)v_i(t-\lambda)d\lambda$$

It is seen that for the integral to have units of amperes, it is necessary that $h(t)$ have units of amperes/volt-second. For a case in which both the input and output are voltages, the units of $h(t)$ are volts/volt-second or seconds^{-1}.

4-7   Convolution Algebra

The general formulation of the convolution of the functions $f_1(t)$ and $f_2(t)$ is

$$f_1(t) * f_2(t) = \int_{-\infty}^\infty f_1(\lambda)f_2(t-\lambda)d\lambda \qquad \textbf{(4-19)}$$

where the asterisk separating the two functions on the left implies the integral on the right. This notation is widely used and allows ready handling of otherwise cumbersome expressions. Consider, for example, the convolution of one function with another function which is itself a convolution of two functions; thus

$$f_1(t) * [f_2(t) * f_3(t)]$$ (4-20)

The convolution contained in the brackets can be written as

$$f_2(t) * f_3(t) = \int_{-\infty}^{\infty} f_2(\lambda_1)f_3(t - \lambda_1)d\lambda_1$$

Combining this with the first function gives

$$f_1(t) * [f_2(t) * f_3(t)] = \int_{-\infty}^{\infty} f_1(\lambda_2)d\lambda_2 \left[\int_{-\infty}^{\infty} f_2(\lambda_1)f_3(t - \lambda_1 - \lambda_2)d\lambda_1 \right]$$

$$= \int_{-\infty}^{\infty} d\lambda_2 \int_{-\infty}^{\infty} f_1(\lambda_2)f_2(\lambda_1)f_3(t - \lambda_1 - \lambda_2)d\lambda_1$$

The advantage of the asterisk notation is readily apparent.

Consider again the defining relationship as given in (4-19). By making the change of variables $\lambda = t - \lambda_1$, we obtain

$$f_1 * f_2 = \int_{\infty}^{-\infty} f_1(t - \lambda_1)f_2(\lambda_1)(-d\lambda_1)$$

$$= \int_{-\infty}^{\infty} f_2(\lambda_1)f_1(t - \lambda_1)d\lambda_1$$ (4-21)

Therefore,

$$f_1 * f_2 = f_2 * f_1$$

It, therefore, follows that convolution is *cummutative* and that the order of the functions being convolved is immaterial.

As a direct result of the superposition property of integrals (that is, the integral of a sum of terms is equivalent to the sum of the integrals of the terms taken separately), it is shown readily that convolution is *distributive*. Accordingly,

$$f_1 * (f_2 + f_3) = f_1 * f_2 + f_1 * f_3$$ (4-22)

When functions are reasonably well-behaved, as they always are when they arise in physical problems, the order of integration can be changed, and it is shown readily that convolution is also *associative*; that is,

$$f_1 * (f_2 * f_3) = (f_1 * f_2) * f_3$$ (4-23)

In view of the associative property, it is unnecessary to use brackets to separate the functions being convolved, and (4-23) can be written in the equivalent forms

$$(f_1 * f_2) * f_3 = f_1 * (f_2 * f_3) = f_1 * f_2 * f_3$$ (4-24)

Also, because of the commutative property, the order of convolution is not important. In actually carrying out the integration involved, the limits on the integrals will be different for different orders of the convolved functions, but the final result will be the same for all orders.

Some interesting and useful additional properties of convolution integrals can be obtained by considering convolution with singularity functions, particularly the unit step, unit impulse, and unit doublet. From the defining relationships given in Chapter 3 it is seen readily that

$$z(t) * \delta(t) = \int_{-\infty}^{\infty} z(\lambda)\delta(t - \lambda)d\lambda = z(t) \tag{4-25}$$

$$z(t) * u(t) = \int_{-\infty}^{\infty} z(\lambda)u(t - \lambda)d\lambda = \int_{-\infty}^{t} z(\lambda)d\lambda \tag{4-26}$$

$$z(t) * \delta'(t) = \int_{-\infty}^{\infty} z(\lambda)\delta'(t - \lambda)d\lambda = z'(t) \tag{4-27}$$

Therefore,

$$u(t) * \delta'(t) = \int_{-\infty}^{\infty} u(\lambda)\delta'(t - \lambda)d\lambda = \delta(t) \tag{4-28}$$

Using these relationships and the commutative property of convolution, the following relationships may be obtained readily. Assume that

$$z(t) = x(t) * y(t)$$

Then

$$z(t) * \delta'(t) = x(t) * y(t) * \delta'(t)$$
$$z'(t) = x(t) * y'(t) \tag{4-29}$$
$$= x'(t) * y(t)$$

and

$$z(t) * \delta(t) = x(t) * y(t) * \delta'(t) * u(t) \tag{4-30}$$

$$z(t) = [x(t) * \delta'(t)] * y(t) * u(t) = x'(t) * \int_{-\infty}^{t} y(\lambda)d\lambda$$

$$= y'(t) * \int_{-\infty}^{t} x(\lambda)d\lambda \tag{4-31}$$

These relationships are particularly useful in the computation of Fourier transforms of the products of functions and are discussed in that connection in the next chapter.

As an example of the use of convolution algebra consider the following problem: A signal is recorded on film by varying the intensity of light passing through a narrow slit and falling on a moving film. The situation is shown schematically in Fig. 4-13. The input signal is $x(t)$, and the recorded

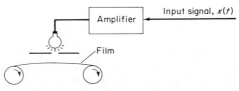

Fig. 4-13 Intensity recording on film.

intensity variations on the film are $y(t)$. Because of the finite width of the slit, each portion of the film will be exposed to the light variations occurring during the time required for a spot on the film to pass completely by the slit. If the slit width is a and the film velocity v, then the time required for a portion of the film to pass the slit will be $T = a/v$. Accordingly, the recorded signal, $y(t)$, will be the sum of the values of $x(t)$ occurring in the interval $t \pm T/2$, or

$$y(t) = \int_{t-T/2}^{t+T/2} x(\xi)d\xi \tag{4-32}$$

This can be rewritten as

$$y(t) = \int_{-\infty}^{t+T/2} x(\xi)d\xi - \int_{-\infty}^{t-T/2} x(\xi)d\xi$$

and making use of the expression for the convolution of a function with the unit step given in (4-26), this can be further simplified to

$$y(t) = x(t) * u\left(t + \frac{T}{2}\right) - x(t) * u\left(t - \frac{T}{2}\right)$$

$$= x(t) * \left[u\left(t + \frac{T}{2}\right) - u\left(t - \frac{T}{2}\right)\right] \tag{4-33}$$

Thus, the system output can be considered as being the convolution of the input signal with a unit amplitude pulse of duration T. This type of processing occurs frequently and is called a *running average*.

Suppose now that it is desired to recover $x(t)$ from the recorded $y(t)$. This may be done quite simply (in theory, at least) by the following artifice. Taking the derivative of both sides of (4-33) and using the identity given in (4-29), we obtain

$$y'(t) = x(t) * \frac{d}{dt}\left[u\left(t + \frac{T}{2}\right) - u\left(t - \frac{T}{2}\right)\right]$$

$$= x(t) * \left[\delta\left(t + \frac{T}{2}\right) - \delta\left(t - \frac{T}{2}\right)\right]$$

$$= x\left(t + \frac{T}{2}\right) - x\left(t - \frac{T}{2}\right)$$

Therefore, solving for $x(t)$, we have

$$x\left(t + \frac{T}{2}\right) = y'(t) + x\left(t - \frac{T}{2}\right)$$

$$x(t) = y'\left(t - \frac{T}{2}\right) + x(t - T) \tag{4-34}$$

The value of $x(t)$ is given by the derivative of $y(t)$ one-half a slit width earlier plus the value of $x(t)$ obtained one full slit width earlier.

4-8 Additional Properties of Convolution

There are a number of properties of the convolution operation that are quantitative and are frequently of use, either to check on the validity of a particular calculation or to gain some insight without actually carrying out the computation. For example, consider the relationship between the area of the convolution resultant and the areas of the two factors entering into the convolution. Mathematically we have

$$f_3(t) = \int_{-\infty}^{\infty} f_1(\lambda)f_2(t-\lambda)d\lambda \tag{4-35}$$

The area can be computed by integrating over the interval $-\infty < t < \infty$, giving

$$\int_{-\infty}^{\infty} f_3(t)dt = \int_{-\infty}^{\infty} \int_{-\infty}^{\infty} f_1(\lambda)f_2(t-\lambda)d\lambda dt$$

$$= \int_{-\infty}^{\infty} f_1(\lambda)\left[\int_{-\infty}^{\infty} f_2(t-\lambda)dt\right]d\lambda \tag{4-36}$$

$$= \int_{-\infty}^{\infty} f_1(\lambda)[\text{area under } f_2(t)]d\lambda$$

Area under $f_3(t)$ = area under $f_1(t) \times$ area under $f_2(t)$

Thus, the area of a convolution product is equal to the product of the areas of the factors. Another useful relationship can be found by considering the center of gravity of the convolution in terms of the centers of gravity of the factors. The center of gravity (or centroid) of a waveform is defined as

$$\eta = \frac{\int_{-\infty}^{\infty} tf(t)dt}{\int_{-\infty}^{\infty} f(t)dt} \tag{4-37}$$

This quantity can be simply expressed in terms of the moments of the waveform defined by

$$M_n(f) = \int_{-\infty}^{\infty} t^n f(t)dt \tag{4-38}$$

Thus, we have

$$\eta = \frac{M_1(f)}{M_0(f)} \tag{4-39}$$

If a waveform is symmetrical, the center of gravity will be located on the axis of symmetry. Computing now the first moment of the convolution, we have (to simplify the notation assume that all integrals have infinite limits)

$$M_1(f_3) = \int tf_3(t)dt = \int t\left[\int f_1(\lambda)f_2(t-\lambda)d\lambda\right]dt$$

$$= \int f_1(\lambda)\left[\int tf_2(t-\lambda)dt\right]d\lambda$$

$$= \int f_1(\lambda)\left[\int (\xi+\lambda)f_2(\xi)d\xi\right]d\lambda$$

$$= \int \lambda f_1(\lambda)d\lambda \int f_2(\xi)d\xi + \int \xi f_2(\xi)d\xi \int f_1(\lambda)d\lambda$$

$$= M_1(f_1)M_0(f_2) + M_1(f_2)M_0(f_1)$$

Noting from (4-36) that $M_0(f_3) = M_0(f_1) \cdot M_0(f_2)$, we obtain

$$\frac{M_1(f_3)}{M_0(f_3)} = \frac{M_1(f_1)}{M_0(f_1)} + \frac{M_1(f_2)}{M_0(f_2)} \tag{4-40}$$

or

$$\eta_3 = \eta_1 + \eta_2 \tag{4-41}$$

These calculations can be extended to show that the squares of the radii of gyration also add. The radius of gyration, s, is obtained from

$$\sigma^2 = \frac{M_2(f)}{M_0(f)} - \eta^2 \tag{4-42}$$

and can be thought of as a measure of the spread of a function around its centroid. The square of the radius of gyration corresponds to the variance in a probability distribution. The corresponding expression for the convolution operation is

$$\sigma_3^2 = \sigma_1^2 + \sigma_2^2 \tag{4-43}$$

It is this result that leads to the conclusion in probability theory that the variance of the sum of two independent random variables equals the sum of their variances, since the probability density function of the sum is the convolution of the individual probability density functions. Also the relationship in (4-43) lends more quantitative evidence to the intuitive concept that convolution is a smoothing or broadening operation—at least for pulse-type signals.

4-9 Discrete Convolution and Deconvolution

There is a close analogy between the mathematical operation of convolution as applied to a sequence of numbers and the multiplication of polynomials. This may be easily seen by considering a simple example. Let two sequences be given by $\{a\} = \{a_1, a_2, a_3\}$ and $\{b\} = \{b_1, b_2, b_3\}$. Their convolution is obtained by reversing one sequence and then sliding it along and multiplying the overlapping terms just as was done in the case of numerical convolution. Thus,

$$\{a\} * \{b\} = \{a_1, a_2, a_3\} * \{b_1, b_2, b_3\}$$
$$= \{a_1b_1, (a_1b_2 + a_2b_1), (a_1b_3 + a_2b_2 + a_3b_1), (a_2b_3 + a_3b_2), a_3b_3\}$$

Now compare this result to that obtained by multiplying the following polynomials.

$$(a_1 + a_2x + a_3x^2)(b_1 + b_2x + b_3x^2)$$
$$= a_1b_1 + (a_1b_2 + a_2b_1)x + (a_1b_3 + a_2b_2 + a_3b_1)x^2 + (a_2b_3 + a_3b_2)x^3 + a_3b_3x^4$$

The terms resulting from the convolution of two sequences are identical to the coefficients obtained from the multiplication of two polynomials whose individual coefficients are the same as the elements in the sequences.

The above relationship suggests that it might be possible to deconvolve a function by means of polynomial division. For example, consider the sequences {1, 1, 2, 1} and {2, 1, 2}. Their convolution product is readily found to be

$$\{1, 1, 2, 1\} * \{2, 1, 2\} = 2, 3, 7, 6, 5, 2$$

This can be thought of as corresponding to the polynomial product

$$(1 + x + 2x^2 + x^3)(2 + x + 2x^2) = 2 + 3x + 7x^2 + 6x^3 + 5x^4 + 2x^5$$

If one of the factors on the left is known, then the other can be recovered by polynomial division. Assume that the factor $(2 + x + 2x^2)$ is known; then the other factor can be obtained in the following manner:

$$
\begin{array}{r}
1 + x + 2x^2 + x^3 \\
2 + x + 2x^2 \overline{\big)\, 2 + 3x + 7x^2 + 6x^3 + 5x^4 + 2x^5} \\
2 + x + 2x^2 \\
\hline
2x + 5x^2 + 6x^3 \\
2x + x^2 + 2x^3 \\
\hline
4x^2 + 4x^3 + 5x^4 \\
4x^2 + 2x^3 + 4x^4 \\
\hline
2x^3 + x^4 + 2x^5 \\
2x^3 + x^4 + 2x^5 \\
\hline
\end{array}
$$

The quotient corresponds to the deconvolution and yields the correct sequence of {1, 1, 2, 1}. This method of deconvolution is quite sensitive to errors in the individual elements and, therefore, has not found much application in the solution of practical problems.

Exercise 4-9.1

Deconvolve the following two sequences: {1, 3, 6, 6, 6, 5, 3} and {1, 2, 3}.

ANS. Check by convolving.

References

    Most textbooks on system analysis derive the convolution integral in connection with transform methods of analysis. This may lead students to some difficulty in following the discussion if they are unfamiliar with the Laplace and Fourier transforms. The books listed below provide extensive discussions of convolution based on time-domain methods before tying in the ideas of frequency-domain analysis and are recommended for further study of this technique.

1. BRACEWELL, R. N., *The Fourier Integral and Its Applications.* New York: McGraw-Hill Book Company, Inc., 1965.
   Although written at a more advanced level than this text, Bracewell's book contains a wealth of useful material on convolution and related topics that can be readily understood by an average student who is willing to put forth a little extra effort.
2. MASON, S. J., and H. J. ZIMMERMAN, *Electronic Circuits, Signals, and Systems.* New York: John Wiley & Sons, Inc., 1960.
3. GUILLEMIN, E. A., *Theory of Linear Physical Systems.* New York: John Wiley & Sons, Inc., 1963.
   Although this is a graduate-level text, it contains many interesting physical interpretations of mathematical operations, including convolution, and is recommended as a source of new perspectives for some of the items discussed in Chapter 4.

Problems

4-1. Using the definition given in Section 2-4, show that convolution is a linear operation.

4-2. A system has an impulse response of the form $A\epsilon^{-\alpha t}u(t)$, where A and α are constants. Find the system output for an input of
    a. $B\epsilon^{-\beta t}u(t)$
    b. $C \sin \omega_0 t$

4-3. Sketch the function that results from convolving each pair of the following functions. Label the significant time and amplitude values.

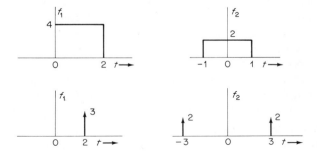

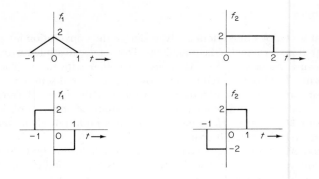

4-4. It is frequently suggested that the impulse response of a system can be approximated by measuring the response to an input pulse having a sufficiently short duration. In order to estimate how short the duration of such a testing pulse should be, consider the simple R-C circuit shown and the three inputs $v_1(t)$, $v_2(t)$, and $v_3(t)$.

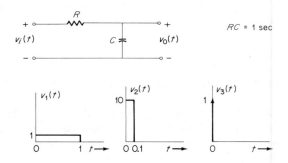

a. Compute and sketch the output, $v_0(t)$, for the three inputs shown.

b. What are the parameters of the input pulse that must be controlled to give a good approximation to the impulse response, and how should these parameters be selected?

c. If you were testing with a variable pulse duration, how could you tell when the testing pulse was short enough to give a good approximation to the impulse response?

d. If a pulse shape other than rectangular is employed for testing, how can you convert from the response to a narrow pulse of arbitrary shape to the system impulse response?

4-5. A system having an impulse response as shown is called a "finite time integrator." Using the convolution integral write an expression for the output of this system for an arbitrary input. By a change of variables show why this is an appropriate name for such a system.

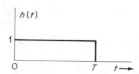

4-6. A system has the response shown when excited by a 1-ms duration, rectangular pulse having a peak amplitude of 10 V. Without carrying out any detailed computations, sketch the system response to each of the following inputs:

    a. $u(t) - u(t - 0.001)$
    b. $u(t)$
    c. $\sin(\pi t/5)$
    d. $\sin(20\,\pi t)$

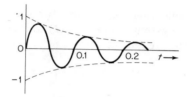

4-7. Using numerical methods, compute the convolution of the two functions shown. Sketch the result.

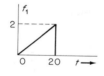 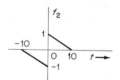

4-8. An experimentally determined impulse response and input signal for a system are as shown. Compute the system output at $t = 0.5$ sec.

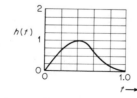 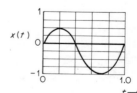

4-9. It is often convenient to use a step function as a test signal when studying or measuring system response. Such signals are easy to generate and lead to a response whose derivative is the system impulse response.

a. For the system shown, compute the unit step response and the impulse response.

b. For the system step response $w(t)$ shown, find the impulse response and sketch an electrical system having such a response for an input signal that is a voltage step.

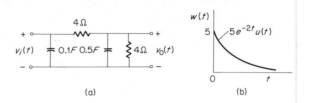

(a)   (b)

4-10. Derive the following expression for the "superposition integral" that gives the system response $y(t)$ to a causal input $x_c(t)$ in terms of the unit step response $w(t)$:

$$y(t) = x_e(0+)w(t) + \int_{0+}^{t} x_c'(\lambda)w(t-\lambda)d\lambda$$

4-11. Are the following systems stable? Why?

a. A system whose output is the derivative (d/dt) of the input.

b. A system whose output is the integral $\left(\int_{-\infty}^{t} \right)$ of the input.

4-12. Consider the following pairs of functions:

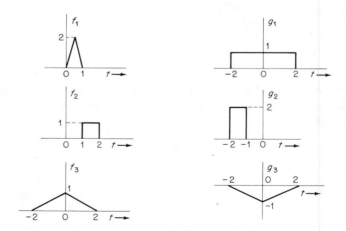

a. Find the duration and average value over the duration interval for the convolution of each pair.

b. Find the centroid of each convolution.

c. Find the radius of gyration of each convolution. Which pair of functions leads to a convolution resultant having the greatest spread around its centroid?

4-13. When signals have negative-going portions, the moments as defined in Section 4-8 are often difficult to interpret physically. However, the rules relating the moments of the convolved functions to the convolution resultant are still valid in such cases. Verify the relationships for the area, centroid, and radius of gyration for the convolution of the two functions shown.

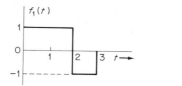

4-14. When dealing with signals having negative-going portions, it is often useful to consider the moments computed using instantaneous power instead of instantaneous amplitude. The moments in this case are defined as

$$M_n = \int_{-\infty}^{\infty} t^n |f(t)|^2 \, dt$$

For the two waveforms of Problem 4-13, compute the centroid (η) and radius of gyration (σ) using the above formula. Compare these results with those obtained using the instantaneous amplitude. What relationship, if any, exists between the moments of the power of the convolution and the moments of the two signals being convolved?

4-15. Show that when two sequences are convolved, the product of the sums of the two sequences equals the sum of the terms of the convolution resultant.

4-16. The inverse of a sequence is defined as that sequence having the property that its convolution with the original sequence leads to a sequence consisting of all zeroes except a single unit element at the origin.

a. Verify the validity of the following causal sequences and their inverse:

$$\{1, -1\} \qquad \{1, 1, 1, 1, \ldots\}$$

$$\{1, 1, 1\} \qquad \{1, -1, 0, 1, -1, 0, \ldots\}$$

b. The following set of data was obtained by averaging over three points at a time. Find the original set of data from which this running average was made.

$$\{1, 2, 4, 5, 4, 3, 1, 1\}$$

CHAPTER

5

Fourier Series and Transforms

5-1 Introduction

Determining the response of a system to a particular excitation when there are several energy storage elements in the system and when the excitation waveform is not simple becomes very involved. A direct attack on the problem through solutions, by classical methods, of the differential equations describing the system is quite difficult when complicated excitation functions are present. In many cases numerical solutions can be obtained readily while analytical solutions cannot. If the system is linear and if its impulse response is available, it is possible to analyze the excitation function into a continuum of impulses, and then superimpose the responses corresponding to each member of this continuum in order to obtain the overall response. This leads to the convolution integral that is discussed in the previous chapter. This method of analysis is very powerful and is particularly useful in theoretical studies. One difficulty with this method, however, is the problem of obtaining the system impulse response. Another difficulty with this method is that for certain types of problems, such as the filtering of signals, only limited physical insight into the problem is gained.

The reasons that the convolution method is theoretically feasible are threefold: First, it is possible to analyze most signals into impulses; second, the response to a single impulse can be determined; and third, the principle of superposition for linear systems allows the total response to be determined from the responses to the elementary signals (impulses) making up the excitation. In principle any other set of basis functions could be used in a manner analogous to that of the impulse functions. The only requirements are that the basis functions be capable of adequately representing the excitation and that the response of the system to the individual basis functions be known. When impulses are used, the responses to the various basis functions are just amplitude-scaled and time-delayed replicas of the response to a unit impulse. With an arbitrary set of basis functions it may turn out that the response to each elementary function is significantly different, thus rendering the procedure useless. In attempting to use this procedure with other basis functions, therefore, the problem becomes one of selecting a set of elementary signals that is appropriate, both in terms of ease of decomposition of the excitation and in terms of calculating the system response to the various members of the set. Fortunately, there are several sets of basis functions that are attractive on both counts. One such signal is the sinusoid, $\cos(\omega_0 t + \phi)$. It will be shown that periodic signals can be decomposed readily into a sum of sinusoids that are harmonically related, and aperiodic signals can be decomposed into a continuum of sinusoids having infinitesimal amplitudes.

The response of a linear system to a steady-state sinusoidal excitation is relatively easy to calculate (or measure). In fact, the response is just an amplitude-scaled and phase-shifted replica of the sinusoidal excitation. Such responses are obtained readily by steady-state ac circuit analysis using the familiar concepts of impedance and admittance. By superimposing the responses to the set of sinusoids representing the excitation, the total response is obtained.

There are also other basis functions, related to the sinusoid, that are useful in signal and system analysis. One of the most frequently encountered basis functions is the imaginary exponential, $\epsilon^{j\omega t}$, which will be used extensively in this chapter because it is easy to manipulate mathematically compared to real sinusoids. Another frequently used basis function is the complex exponential, ϵ^{st}, where $s = \sigma + j\omega$ is a complex number. This elementary signal forms the basis of the Laplace transform, to be discussed in the next chapter, and provides further computational advantages for many problems such as those involving the transient response of a system.

The methods to be discussed in the following sections are based on the use of real and complex sinusoids to represent signals and are properly called Fourier methods after the mathematician who first investigated these techniques. The basic techniques are widely applicable to the solutions of ordinary and partial differential equations. In many such cases the solutions are rather formal and offer little insight into the physical aspects of the problem. In the case of signals and systems, however, the Fourier representations have direct physical interpretation and, in fact, frequently are readily measured quantities. It is for this reason that Fourier analysis plays such a prominent role in signal and system analysis.

5-2   Fourier Series Representation of Time Functions

In Chapter 3 it is shown that an arbitrary function $f(t)$ can be rep· resented over an interval $t_1 < t < t_1 + T$ by an orthogonal expansion in terms of the basis functions $\phi_n(t) = \epsilon^{jn\omega_0 t}$, where $\omega_0 = 2\pi/T$ and $n = 0$, $\pm 1, \pm 2, \ldots$. The expression is

$$f(t) = \sum_{n=-\infty}^{\infty} \alpha_n \epsilon^{jn\omega_0 t} \qquad t_1 < t \leq t_1 + T \tag{5-1}$$

$$\alpha_n = \frac{1}{T} \int_{t_1}^{t_1+T} f(t) \epsilon^{-jn\omega_0 t} dt \tag{5-2}$$

and is referred to as a *Fourier series.*

The coefficients, α_n, chosen according to (5-2), minimize the mean-square error between $f(t)$ and the partial sum $\sum_{n=-N}^{N} \alpha_n \epsilon^{jn\omega_0 t}$ for any N. The fact that the α_n chosen in this manner give the *best* (in a mean-square error sense) fit does not necessarily mean that they give a *good* fit. The question of a good fit is involved with the convergence of the series, (5-1), to the value of the function at each point. Study of the convergence of Fourier series in general is quite complex. However, there is a set of conditions which is sufficient (although not necessary) to assure the desired convergence. Also, these conditions accomodate virtually all cases arising in physical problems. These conditions are known as the Dirichlet conditions and state that a function $f(t)$ has a Fourier series representation if:

1. $\int_{t_1}^{t_1+T} |f(t)| dt$ is finite.

2. There are no more than a finite number of maxima and minima in any finite time period.

3. There are no more than a finite number of discontinuities in any finite time period.

When these conditions are met, the Fourier series representation of $f(t)$ converges at every point, $t = t_0$, in the interval to $f(t_0) = [f(t_0^+) + f(t_0^-)]/2$, where $f(t_0^+)$ and $f(t_0^-)$ are the $\lim_{t \to t_0} f(t)$ as t_0 is approached from the right-and left-hand sides, respectively. If $f(t)$ is continuous at $t = t_0$, then the series converges to $f(t_0)$. However, at a discontinuity the series converges to the arithmetic mean of the left-hand and right-hand limits.

Before illustrating the above properties, it is helpful to obtain an equivalent representation. Making use of the relationship $\epsilon^{j\omega_0 t} =$

$\cos \omega_0 t + j \sin \omega_0 t$, it is possible to rewrite (5-1) as

$$f(t) = \sum_{n=-\infty}^{\infty} \alpha_n (\cos n\omega_0 t + j \sin n\omega_0 t) \qquad (5\text{-}3)$$

From (5-2) it is seen in all cases where $f(t)$ is real that $\alpha_{-n} = \alpha_n^*$, and making use of the even and odd properties of $\cos n\omega_0 t$ and $\sin n\omega_0 t$, (5-3) can be written as

$$f(t) = \alpha_0 + \sum_{n=1}^{\infty} (\alpha_n + \alpha_n^*) \cos n\omega_0 t + j(\alpha_n - \alpha_n^*) \sin n\omega_0 t$$

$$= \frac{a_0}{2} + \sum_{n=1}^{\infty} a_n \cos n\omega_0 t + b_n \sin n\omega_0 t \qquad (5\text{-}4)$$

where

$$a_n = \frac{2}{T} \int_{t_1}^{t_1+T} f(t) \cos n\omega_0 t \, dt \qquad n = 0, 1, 2, \ldots \qquad (5\text{-}5)$$

$$b_n = \frac{2}{T} \int_{t_1}^{t_1+T} f(t) \sin n\omega_0 t \, dt \qquad n = 0, 1, 2, \ldots \qquad (5\text{-}6)$$

Note that a_n and b_n are real numbers if $f(t)$ is real, whereas α_n is, in general, a complex number. In all cases, of course, the series (5-1) and (5-4) must sum to the same value. Therefore, if $f(t)$ is real, the imaginary parts of (5-1) will add to zero.

In general the series (5-1) and (5-4) represent the function $f(t)$ over the interval t_1 to $t_1 + T$, and nothing is specified about $f(t)$ outside this interval. If, however, $f(t)$ is periodic with period T, then the Fourier series representation will be valid for all t.

As an illustration of the Fourier series representation of a time function, consider the periodic rectangular pulse train shown in Fig. 5-1.

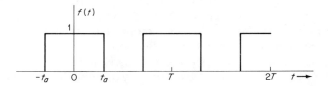

Fig. 5-1 Periodic rectangular pulses.

The Fourier coefficients are found to be

$$a_n = \frac{2}{T} \int_{-t_a}^{t_a} \cos \frac{2\pi nt}{T} \, dt = \frac{2}{T} \frac{\sin 2\pi nt/T}{2\pi n/T} \bigg|_{-t_a}^{t_a}$$

$$= \frac{4t_a}{T} \frac{\sin (2\pi nt_a/T)}{2\pi nt_a/T}$$

$$b_n = \frac{2}{T} \int_{-t_a}^{t_a} \sin \frac{2\pi nt}{T} \, dt = 0$$

and the corresponding Fourier series is

$$f(t) = \frac{2t_a}{T} + \frac{4t_a}{T} \sum_{n=1}^{\infty} \frac{\sin(2\pi n t_a/T)}{2\pi n t_a/T} \cos \frac{2\pi n}{T} t \qquad (5\text{-}7)$$

To make the example somewhat more specific, assume that $T = 10^{-3}$ sec and $t_a/T = \frac{1}{8}$. The frequencies present in the Fourier expansion are then $f_n = \omega_n/2\pi = 1000\,n$ Hz plus a dc term. The frequency corresponding to $n = 1$ is called the *fundamental*, or first, *harmonic*, which in this case is 1000 Hz. The frequency corresponding to $n = 2$ is the *second harmonic*, which in this case is 2000 Hz. Higher harmonics are designated in a similar manner.

The array of amplitudes of the harmonics is called the (discrete) *frequency spectrum*, and for this example is shown in Fig. 5-2. One of the reasons that this type of representation is particularly useful is that it is closely related to directly measurable quantities. In this instance, by using appropriate filters it is possible to isolate each of the frequency components and measure its amplitude, which will be found to be in agreement with (5-7). An alternative physical realization of this representation is obtained by combining separate sinusoids of the correct amplitudes, frequencies, and phases and obtaining a resultant waveform that is the rectangular waveform of Fig. 5-1.

The Fourier series representations of several common periodic waveforms are given in Table 5-1. The representations of numerous other waveforms can be obtained from these by appropriate addition, subtraction, differentiation, or integration of these basic waveforms. Such operations are usually possible, although in the case of differentiation some precautions (as discussed further in Section 5-5) are necessary.

There are a number of reasons why Fourier series analysis is important in the study of signals and systems. One reason is simply the physical insight that it provides concerning the frequency composition of complicated periodic waveforms. A much more important reason, which is discussed in considerable detail in Section 5-17, concerns computer-oriented numerical techniques for analyzing signals that are not periodic per se but necessarily must exist for only a finite time duration. Such techniques turn out to be closely related to Fourier series, and a thorough

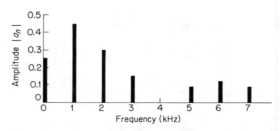

Fig. 5-2 Frequency spectrum of periodic square wave.

Table 5-1   Fourier Series Representation of Common Waveforms

Waveform	Fourier series
General periodic wave $f(t)$	$\sum\limits_{n=-\infty}^{\infty} a_n \epsilon^{[j(2\pi n/T)t]}$ $a_n = \dfrac{1}{T} \int\limits_{t_1}^{t_1+T} f(t) \epsilon^{-[j(2\pi n/T)t]}$ $\dfrac{a_0}{2} + \sum\limits_{n=1}^{\infty} a_n \cos \dfrac{2\pi n}{T} t + b_n \sin \dfrac{2\pi n}{T} t$ $a_n = \dfrac{2}{T} \int\limits_{t_1}^{t_1+T} f(t) \cos \dfrac{2\pi n}{T} t \, dt$ $b_n = \dfrac{2}{T} \int\limits_{t_1}^{t_1+T} f(t) \sin \dfrac{2\pi n}{T} t \, dt$
Odd square wave	$\dfrac{4}{\pi} \sum\limits_{n=1}^{\infty} \dfrac{1}{2n-1} \sin \dfrac{2\pi(2n-1)}{T} t$
Even square wave	$\dfrac{4}{\pi} \sum\limits_{n=1}^{\infty} \dfrac{(-1)^{n+1}}{2n-1} \cos \dfrac{2\pi(2n-1)t}{T}$
Rectangular pulse train	$\dfrac{2t_a}{T} + \dfrac{4t_a}{T} \sum\limits_{n=1}^{\infty} \dfrac{\sin\left(\dfrac{2\pi n t_a}{T}\right)}{\dfrac{2\pi n t_a}{T}} \cos \dfrac{2\pi n t}{T}$
Triangular wave	$\dfrac{8}{\pi^2} \sum\limits_{n=1}^{\infty} \dfrac{1}{(2n-1)^2} \cos \dfrac{2\pi(2n-1)}{T} t$
Sawtooth wave	$\dfrac{2}{\pi} \sum\limits_{n=1}^{\infty} \dfrac{(-1)^{n+1}}{n} \sin \dfrac{2\pi n}{T} t$
Half-wave rectified cosine wave	$\dfrac{2}{\pi} \left(\dfrac{1}{2} + \dfrac{\pi}{4} \cos \dfrac{2\pi t}{T} - \sum\limits_{n=1}^{\infty} \dfrac{(-1)^n}{4n^2-1} \cos \dfrac{4\pi n t}{T} \right)$
Fractional cosine wave	$\dfrac{A}{\pi} \dfrac{\sin(2\pi B) - (2\pi B)\cos(2\pi B)}{[1-\cos(2\pi B)]}$ $+ \dfrac{2AB}{[1-\cos(2\pi B)]} \sum\limits_{n=1}^{\infty} \dfrac{1}{n} \left\{ \dfrac{\sin[2\pi B(n-1)]}{[2\pi B(n-1)]} - \dfrac{\sin[2\pi B(n+1)]}{[2\pi B(n+1)]} \right\} \cos \dfrac{2\pi t}{T}$

understanding of Fourier series is essential in avoiding many problems that might otherwise arise. A third reason for the importance of Fourier series in system analysis is that it provides one way of determining what happens to a periodic waveform when it is passed through a system that alters the relative magnitudes and phases of the various frequency components. Although this same problem can be solved more readily by using

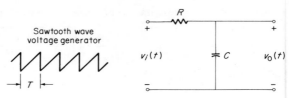

Fig. 5-3 $R–C$ circuit with a sawtooth input.

the Laplace transform technique that is discussed in Chapter 6, it is worth considering a simple illustration of how Fourier series might be used.

The example that is considered concerns the effect of a lowpass $R–C$ circuit on a sawtooth waveform as illustrated in Fig. 5-3. A practical circumstance in which this situation might occur is in the sweep circuit of an oscilloscope. The sawtooth sweep voltage is amplified and applied to the cathode-ray tube deflection plates. Their capacitance is represented by C, and the source resistance of the amplifier is represented by R. For a *sinusoidal* input voltage of frequency ω and amplitude V_i, the amplitude and phase of the output phasor is obtained by steady-state ac circuit analysis as

$$V_0 = \frac{V_i}{1 + j\omega RC} = \frac{V_i(1 - j\omega RC)}{1 + \omega^2 R^2 C^2} \tag{5-8}$$

The Fourier series representation of the sawtooth waveform at the input is given by Table 5-1 as

$$v_i(t) = \frac{2}{\pi} \sum_{n=1}^{\infty} \frac{(-1)^{n+1}}{n} \sin\left(\frac{2\pi nt}{T}\right) \tag{5-9}$$

where T is the duration of one period. Note that only sine components are present and that the radian frequency of each component is $2\pi n/T$, $n = 1, 2, \ldots$. By virtue of the phase shift in the $R–C$ circuit, the output waveform will have both sine and cosine components. Further, each component will be reduced in amplitude by the amount indicated in (5-8). Hence, the Fourier series representation of the output waveform may be written as

$$v_0(t) = \frac{2}{\pi} \sum_{n=1}^{\infty} \frac{(-1)^{n+1}}{n[1 + (2\pi nRC/T)^2]} \left[\sin\left(\frac{2\pi nt}{T}\right) - \left(\frac{2\pi nRC}{T}\right) \cos\left(\frac{2\pi nt}{T}\right)\right] \tag{5-10}$$

This waveform is sketched in Fig. 5-4 for several different values of $R–C$. It is clear that, as $R–C$ becomes larger, the output waveform departs more and more from the desired sawtooth. In order to obtain a more quantitative measure of the amount of distortion, as well as to gain additional insight into the physical interpretation of Fourier series, it is necessary to look further into some of the mathematical relationships involved.

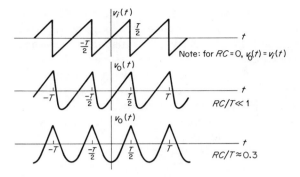

Fig. 5-4 Distorted sawtooth waveform.

Exercise 5-2.1

Find the Fourier series representation for the triangular waveform shown in Table 5-1.

5-3   Average Values and RMS Values

By the average value of a signal is meant the time average or mean height. Mathematically the time average of a signal $x(t)$ over the interval $t_1 \le t \le t_1 + T$ is given by

$$\langle f(t) \rangle_T = \frac{1}{T} \int_{t_1}^{t_1+T} f(t) dt \tag{5-11}$$

For power signals, such as periodic signals, the average is taken over an infinite interval so that (5-11) is written as[1]

$$\langle f(t) \rangle = \lim_{T_1 \to \infty} \frac{1}{2T_1} \int_{-T_1}^{T_1} f(t) dt \tag{5-12}$$

Carrying out the limiting operation in increments of integral numbers of periods results in a considerable simplification. Consider a periodic signal with period T.

$$\langle f(t) \rangle = \lim_{N \to \infty} \frac{1}{NT} \int_{-NT/2}^{NT/2} f(t) dt$$

$$= \lim_{N \to \infty} \frac{1}{NT} \left[N \int_{-T/2}^{T/2} f(t) dt \right] \tag{5-13}$$

$$\langle f(t) \rangle = \frac{1}{T} \int_{-T/2}^{T/2} f(t) dt = \langle f(t) \rangle_T \tag{5-14}$$

---

[1]The subscript on the symbol $\langle \rangle$ indicates the time interval over which the average is taken. When this is infinite, the subscript is omitted.

In obtaining (5-14), use was made of the fact that, for a period signal, the integral over N periods is just N times the integral over one period. From (5-14) it is seen that the average value is just $1/T$ times the area under the waveform in one period. In terms of the Fourier series representation, it is seen readily that $\langle f(t) \rangle_T = a_0/2$. The average value is frequently referred to as the dc value since it represents the value that would be measured by a dc measuring instrument.

A similar analysis can be made for the mean-square value of a periodic waveform, and it is found that

$$\langle f^2(t) \rangle = \frac{1}{T} \int_{-T/2}^{T/2} f^2(t)dt = \langle f^2(t) \rangle_T \tag{5-15}$$

The mean-square value is thus $1/T$ times the area under the square of the waveform in a time equal to one period. When this definition is applied to the Fourier series representation, the result is

$$\langle f^2(t) \rangle_T = \frac{1}{T} \int_{-T/2}^{T/2} \left[\frac{a_o}{2} + \sum_{n=1}^{\infty} (a_n \cos n\omega_o t + b_n \sin n\omega_o t) \right]^2 dt$$

$$= \frac{1}{T} \int_{-T/2}^{T/2} \left[\frac{a_o^2}{4} + a_o \sum_{n=1}^{\infty} a_n \cos n\omega_o t + b_n \sin n\omega_o t \right.$$

$$+ \sum_{n=1}^{\infty} \sum_{m=1}^{\infty} (a_n a_m \cos n\omega_o t \cos m\omega_o t + a_n b_m \cos n\omega_o t \sin m\omega_o t$$

$$+ \left. b_n a_m \sin n\omega_o t \cos m\omega_o t + b_n b_m \sin n\omega_o t \sin m\omega_o t) \right] dt \tag{5-16}$$

The following identities (orthogonality relationships) can be used to simplify (5-16):

$$\int_{-T/2}^{T/2} \cos n\omega_0 t \, dt = \int_{-T/2}^{T/2} \sin n\omega_0 t \, dt = 0 \tag{5-17}$$

$$\int_{-T/2}^{T/2} \sin n\omega_0 t \sin m\omega_0 t \, dt = \int_{-T/2}^{T/2} \cos n\omega_0 t \cos m\omega_0 t \, dt$$

$$= \begin{cases} T/2 & n = m \\ 0 & n \neq m \end{cases} \tag{5-18}$$

$$\int_{-T/2}^{T/2} \sin n\omega_0 t \cos m\omega_0 t \, dt = 0 \tag{5-19}$$

When these identities are inserted in (5-16), the result is

$$\langle f^2(t) \rangle_T = \frac{a_o^2}{4} + \sum_{n=1}^{\infty} \frac{a_n^2}{2} + \frac{b_n^2}{2} \tag{5-20}$$

Thus, the mean-square value is given by the square of the dc component plus one-half the sum of the squares of the amplitudes of the harmonics. The rms (root-mean-square) value is

$$f_{\text{rms}} = \langle f^2(t) \rangle_T^{1/2} = \left[\frac{a_o^2}{4} + \frac{1}{2} \sum_{n=1}^{\infty} (a_n^2 + b_n^2) \right]^{1/2} \tag{5-21}$$

The ac portion of a waveform is usually considered to be all components except the dc portion. Thus, the rms value of the ac component of a periodic waveform is

$$f_{ac} = \sqrt{\frac{1}{2} \sum_{n=1}^{\infty} (a_n^2 + b_n^2)} \qquad (5\text{-}22)$$

The mean-square value of a waveform is directly proportional to the power contained in the waveform. For example, the power dissipated in a resistor of R ohms when a current $i(t)$ flows through it is

$$P = \lim_{T_1 \to \infty} \frac{1}{2T_1} \int_{-T_1}^{T_1} Ri^2(t)dt$$
$$= R \langle i^2(t) \rangle \qquad (5\text{-}23)$$

It is evident from (5-23) that the power contained in any particular frequency component of a periodic waveform is proportional to $\frac{1}{2}(a_k^2 + b_k^2)$, where a_k and b_k are the Fourier coefficients of the frequency component in question.

As an example of the use of the above relationships, consider the problem of determining the amount of distortion that occurs when an audio amplifier does not have the power handling capability required to handle the peak level of a signal. This is illustrated in Fig. 5-5, and an enlarged view of the output signal is shown in Fig. 5-6(a). It is assumed that the limiting occurs only on the positive part of the cycle. In order to make the example specific it is also assumed that the clipping is at 90 percent of the peak value of the sinusoid, which has been normalized to unity for convenience. It is desired to determine the amount of distortion produced by this clipping.

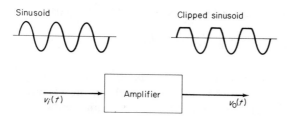

Fig. 5-5 Amplifier with limiting.

Although it is possible to obtain the desired results by expanding $v_o(t)$ in a Fourier series, somewhat greater accuracy and convenience is obtained by working with the portion clipped off, which is designated as $c(t)$ and is shown in Fig. 5-6(b). Thus, $v_o(t)$ can be represented as

$$v_o(t) = \cos\left(\frac{2\pi t}{T}\right) - c(t) \qquad (5\text{-}24)$$

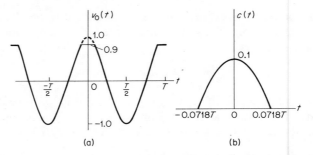

Fig. 5-6 The clipped waveform and the portion removed. (a) The clipped waveform. (b) The portion removed (enlarged).

The Fourier series representation of $c(t)$ is given in Table 5-1, from which the coefficients may be determined immediately. Thus (since $A = 0.1$ and $B = 0.0718$ for this example),

$$\frac{a_o}{2} = \frac{0.1}{\pi} \frac{\sin 0.0718\,(2\pi) - (2\pi)(0.0718) \cos 0.0718(2\pi)}{1 - \cos\,[0.0718(2\pi)]} = 0.00955 \qquad \text{(5-25)}$$

$$a_n = \frac{2(0.1)(0.0718)}{n\{1 - \cos\,[0.0718(2\pi)]\}} \left[\frac{\sin 2\pi(0.0718)(n-1)}{2\pi(0.0718)(n-1)} - \frac{\sin 2\pi(0.0718)(n+1)}{2\pi(0.0718)(n+1)} \right]$$

$$b_n = 0 \qquad \text{(5-26)}$$

For $n = 1$ and $n = 2$,

$$a_1 = 0.01865$$
$$a_2 = 0.0175$$

The amplifier output may now be written as

$$v_o(t) = -0.00955 + 0.98135 \cos\left(\frac{2\pi t}{T}\right) - 0.0175 \cos\left(\frac{4\pi t}{T}\right)$$
$$- \sum_{n=3}^{\infty} a_n \cos\left(\frac{2\pi n t}{T}\right) \qquad \text{(5-27)}$$

in which the coefficient of the fundamental is just $(1 - a_1)$.

It is now possible to relate the powers in the various frequency components. For example, the power in the fundamental is (in a one-ohm resistance)

$$P_1 = \frac{1}{2}(1 - a_1)^2 = \frac{1}{2}(0.98135)^2 = 0.4811$$

and the power in the second harmonic is

$$P_2 = \frac{1}{2}a_2^2 = \frac{1}{2}(0.0175)^2 = 1.542 \times 10^{-4}$$

The total harmonic power, not including the fundamental, can be obtained by subtracting the fundamental and dc power from the power of $c(t)$.

Thus,

$$P_{\text{har}} = \frac{1}{T} \int_{-T/2}^{T/2} c^2(t)\,dt - \left(\frac{a_0}{2}\right)^2 - \frac{1}{2}\,a_1^{\,2}$$

$$= \frac{1}{T} \int_{-0.0718T}^{0.0718T} \left[\cos\left(\frac{2\pi t}{T}\right) - 0.9\right]^2 dt - (0.00955)^2 - \frac{1}{2}(0.01865)^2 \qquad \textbf{(5-28)}$$

$$= 7.96 \times 10^{-4} - 0.912 \times 10^{-4} - 1.74 \times 10^{-4} = 5.31 \times 10^{-4}$$

A common way of expressing distortion power is in terms of its percentage of the fundamental power. Hence, the percent second harmonic distortion is

$$\eta_2 = \frac{1.542 \times 10^{-4}}{0.4811} \times 100 = 0.0321\%$$

and the percent total harmonic distortion is

$$\eta_{\text{har}} = \frac{5.31 \times 10^{-4}}{0.4811} \times 100 = 0.110\%$$

Analysis similar to this can be used in a variety of situations where it is desirable to estimate the relative distribution of power among the harmonics of a periodic waveform.

Exercise 5-3.1

Suppose that the amplifier in the above example is used to amplify a triangular waveform as shown in Table 5-1. Find the magnitudes of the dc component and the second harmonic component appearing in the output of the amplifier.

ANS. 0.0025, 0.005

5-4 Symmetry Properties of Fourier Series

There are a number of easily identifiable properties of waveforms that have very significant effects on the coefficients in a Fourier series expansion. For example, whenever the waveform to be expanded is an even function of t, only cosine terms are present. This is illustrated by the calculations carried out in the previous section. Similarly, whenever the waveform is an odd function of t, that is, when $x(-t) = -x(t)$, then only sine terms will be present in the Fourier series. Also, in both cases, the values of the coefficients can be obtained by integrating over half the period and doubling the result. Thus, the following special results apply to calculating the Fourier coefficient of even and odd time functions:

$f(t)$ *even*

$$a_n = \frac{4}{T} \int_0^{T/2} f(t) \cos\frac{2\pi n t}{T}\,dt$$

$$\qquad\qquad \textbf{(5-29)}$$

$$b_n = 0$$

f(t) odd

$$a_n = 0$$

$$b_n = \frac{4}{T} \int_0^{T/2} f(t) \sin \frac{2\pi n t}{T} dt \qquad (5\text{-}30)$$

Actually, any time function can be resolved into even and odd parts, which lead to cosine and sine terms, respectively. The separation is as follows:

$$f(t) = \underbrace{\frac{f(t) + f(-t)}{2}}_{\text{even part}} + \underbrace{\frac{f(t) - f(-t)}{2}}_{\text{odd part}} \qquad (5\text{-}31)$$

$$= \overbrace{\frac{a_o}{2} + \sum_{n=1}^{\infty} a_n \cos n\omega_0 t + \sum_{n=1}^{\infty} b_n \sin n\omega_0 t}$$

There are other types of symmetry that are equally important. For example, when a signal has the property that $f(t + T/2) = -f(t)$, which is called odd half-wave symmetry, then the Fourier series contains only odd harmonics. This can be seen mathematically from the exponential series. That is,

$$\alpha_n = \frac{1}{T} \int_0^T f(t) \epsilon^{-jn\omega_0 t} dt$$

$$= \frac{1}{T} \int_0^{T/2} f(t) \epsilon^{-jn\omega_0 t} dt + \frac{1}{T} \int_{T/2}^T f(t) \epsilon^{-jn\omega_0 t} dt \qquad (5\text{-}32)$$

$$= \frac{1}{T} \int_0^{T/2} f(t) [\epsilon^{-jn\omega_0 t} - \epsilon^{-jn\omega_0 [t + (T/2)]}] dt$$

Noting that $T/2 = \pi/\omega_0$, (5-32) can be written as

$$\alpha_n = \frac{1}{T} \int_0^{T/2} f(t) \epsilon^{-jn\omega_0 t} (1 - \epsilon^{-jn\pi}) dt$$

$$= \frac{2}{T} \int_0^{T/2} f(t)^{jn\omega_0 t} dt \qquad n \text{ odd} \qquad (5\text{-}33)$$

$$= 0 \qquad n \text{ odd}$$

Thus, only odd harmonics exist. Similarly, it can be shown that for a wave having even half-wave symmetry, that is, $f(t + T/2) = f(t)$, the expansion will contain only even harmonics. The Fourier coefficients become

$$\alpha_n = \frac{2}{T} \int_0^{T/2} f(t) \epsilon^{jn\omega_0 t} dt \qquad n \text{ even}$$

$$= 0 \qquad n \text{ odd} \qquad (5\text{-}34)$$

The expressions for a_n and b_n that correspond to (5-33) and (5-34) involve a factor of $4/T$ in front of the integral. Examples of waveforms having these types of symmetry are shown in Table 5-2.

Table 5-2   Effects of Waveform Symmetry on Fourier Coefficients

Waveform	Symmetry	Effect on coefficient	Harmonics
	None	None	All
	Even	$b_n = 0$	All
	Odd	$a_n = 0$	All
	Half-wave odd	None	Odd only
	Even and half-wave even	$b_n = 0$	Even only

When a waveform is an odd or even time function and also possesses odd or even half-wave symmetry, it is necessary to integrate over only one-fourth of the period, with a corresponding increase by a factor of 4 in the multiplying factor in front of the integral for calculation of the coefficients.

Three additional comments about symmetry are in order. First, translating the origin to the left or the right has no effect on whether odd or even harmonics are present but does alter whether a function is an even or odd time function and, consequently, affects the coefficients a_n and b_n. This corresponds to the physical situation whereby changing the time origin affects the phases of the harmonics of a signal but not their presence or absence (or even their magnitude). Second, changing the average value by moving the abscissa up or down in the waveform does not alter anything in the Fourier series expansion except the dc component, $a_0/2$. Third, a function that has even half-wave symmetry can alternatively be thought of as actually having a period of $T/2$ instead of T. By properly choosing the origin in a waveform, considerable simplification can often be made in the calculation of the Fourier series coefficients. As an example, consider the triangular waveform shown in Fig. 5-7. By choosing the coordinate axes as shown by the dotted lines, the computations are simplified. Thus, when the horizontal axis is placed through the mean value of the waveform, the dc term is eliminated. By choosing the vertical axis, as shown at (a), an even function containing only cosine terms is obtained, and further, since odd half-wave symmetry is present, only odd

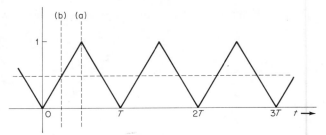

Fig. 5-7 Triangular waveform.

harmonics will be present. By choosing the axis at (b), an odd function is obtained giving only sine terms, and again only odd harmonics are present.

The translation of the axes can be readily compensated for after the expansion is obtained. A shift in the vertical direction is accomplished by adding a constant equal to the amount of shift desired. A shift in the horizontal direction is obtained by replacing t by $t + t_1$, where t_1 is the desired shift of the origin in the positive direction. As an example, consider again the waveform of Fig. 5-7. Using the (a) axes, a zero-average odd function with odd half-wave symmetry is obtained, and thus,

$$a_n = 0 \qquad b_n = 0 \qquad n \text{ even}$$

$$a_n = 0 \qquad n \text{ odd}$$

$$b_n = \frac{8}{T} \int_0^{T/4} \frac{2t}{T} \sin \frac{2\pi n t}{T} \, dt \qquad n \text{ odd}$$

$$= \frac{4}{\pi^2 n^2} \left(\sin \frac{\pi n}{2} - \frac{\pi n}{2} \cos \frac{\pi n}{2} \right)$$

$$= \frac{4}{\pi^2 n^2} (-1)^{(n-1)/2}$$

Hence, the shifted waveform is

$$f_1(t) = \frac{4}{\pi^2} \left(\sin \omega_0 t - \frac{1}{9} \sin 3\omega_0 t + \frac{1}{25} \sin 5\omega_0 t + \cdots \right) \qquad \text{(5-35)}$$

To convert this expression to the correct one for the function as originally shown, it is only necessary to translate the origin downward by one-half and to the left by $T/4 = \pi/2\omega_0$. Thus,

$$f(t) = \frac{1}{2} + \frac{4}{\pi^2} \left[\sin \left(\omega_0 t - \frac{\pi}{2} \right) - \frac{1}{9} \sin \left(3\omega_0 t - \frac{3\pi}{2} \right) + \cdots \right] \qquad \text{(5-36)}$$

Examination of Fig. 5-7 shows that the function is even around its original origin, so the expansion should have only cosine terms. By simple trigonometric identities, (5-36) is changed readily to

$$f(t) = \frac{1}{2} - \frac{4}{\pi^2}\left(\cos \omega_0 t - \frac{1}{9}\cos 3\omega_0 t + \cdots\right) \qquad \text{(5-37)}$$

Translations to other origins can be made as readily.

There are other types of symmetry that affect the coefficients in predictable ways, but they are applicable only in special situations.

Exercise 5-4.1

Determine the kinds of symmetry present in the waveforms of Table 5-1 and check the effects of the symmetry on the coefficients in the expansions given.

5-5 Convergence of Coefficients and Mathematical Operations on Fourier Series

The rate at which the partial sum of a Fourier series approaches the value of the function is an important consideration. If one is approximating a function by a series, the more rapid the convergence of the partial sum, the fewer the number of terms in the approximation required to obtain the desired accuracy. The rate of convergence is directly related to the rate of decrease in the magnitude of the Fourier coefficients. This rate of decrease in magnitude of the coefficients is, in turn, related to the "smoothness" of the original waveform. For any function satisfying the Dirichlet conditions, the coefficients fall off at least as fast as $1/n$, that is,

$$|a_n| < \frac{M}{n} \qquad \text{and} \qquad |b_n| < \frac{M}{n} \qquad \text{(5-38)}$$

where M is some positive integer that is independent of n.

If the function has a derivative that satisfies the Dirichlet conditions, then the Fourier coefficients are of the order of $1/n^2$. In general, if the kth derivative satisfies the Dirichlet conditions, then the Fourier coefficients will be on the order of $1/n^{k+1}$.

For most practical waveforms the primary consideration is the continuity of the function and its derivatives. The order of the derivative that first contains delta functions is the negative order of n representing the rate of decrease of the coefficients. For example, a square wave would have coefficients falling off as $1/n$, while a triangular wave would have coefficients falling off as $1/n^2$.

It is often possible to obtain new Fourier series representations by means of term-by-term differentiation or integration of a known series. It is always permissible to integrate a Fourier series to obtain the series for the integral of the original time function. However, in order for differentiation to be valid, it is necessary that the original time function be continuous and have a derivative with only a finite number of discontinuities in one period. As an example, consider the sawtooth waveform shown in Figure 5-8.

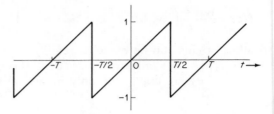

Fig. 5-8 Sawtooth waveform.

This waveform has the equation

$$f(t) = \frac{2t}{T} \qquad -\frac{T}{2} < t < \frac{T}{2}$$

$$f(t + kT) = f(t) \qquad k = 0, \pm 1, \pm 2, \ldots$$

(5-39)

The Fourier series expansion is

$$f(t) = \frac{2}{\pi} \sum_{n=1}^{\infty} \frac{(-1)^{n+1}}{n} \sin \frac{2\pi n t}{T}$$

(5-40)

Integrating this series term-by-term gives the result

$$f_I(t) = \int_{-T/2}^{t} \frac{2\xi}{T} d\xi = \frac{t^2}{T} - \frac{T}{4} \qquad -\frac{T}{2} < t < \frac{T}{2}$$

$$= \int_{-T/2}^{t} \frac{2}{\pi} \sum_{n=1}^{\infty} \frac{(-1)^{n+1}}{n} \sin \frac{2\pi n \xi}{T} d\xi$$

(5-41)

$$= \frac{T}{\pi^2} \sum_{n=1}^{\infty} \frac{(-1)^{n+1}}{n^2} \left[(-1)^n - \cos \frac{2\pi n t}{T} \right]$$

$$f_I(t) = \frac{-T}{6} - \frac{T}{\pi^2} \sum_{n=1}^{\infty} \frac{(-1)^{n+1}}{n^2} \cos \frac{2\pi n t}{T}$$

(5-42)

In arriving at (5-42), use was made of the relationship $\sum_{n=1}^{\infty} 1/n^2 = \pi^2/6$, which led to the constant $T/6$. This constant can be checked by computing the average of $f_I(t)$ from (5-41). The waveform corresponding to (5-42) is shown in Fig. 5-9.

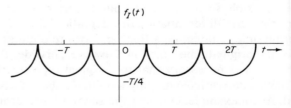

Fig. 5-9 The function $f_I(t)$.

The derivative of $f(t)$ leads to the series

$$\frac{df(t)}{dt} = \frac{4}{T} \sum_{n=1}^{\infty} (-1)^{n+1} \cos \frac{2\pi nt}{T} \qquad (5\text{-}43)$$

This series is seen to diverge, for certain values of t, since its terms do not decrease as $n \to \infty$. The reason for this undesirable behavior is that $f(t)$ is not continuous, and so its derivative contains impulses.

Exercise 5-5.1

At what rate would you expect the Fourier coefficients to decrease for (1) a triangular wave, (2) a square wave, (3) a rectified sine wave?

ANS: See Table 5-1.

5-6 Gibbs Phenomenon

In order to round out the discussion on Fourier series, a certain anomalous behavior must be described. This unusual behavior is observed when the partial sums of a Fourier series are used to represent a waveform at a point of discontinuity. As an example, consider the Fourier series representation of the sawtooth wave given by (5-40). The kth partial sum is

$$f_k(t) = \frac{2}{\pi} \sum_{n=1}^{k} \frac{(-1)^{n+1}}{n} \sin \frac{2\pi nt}{T}$$

Figure 5-10 shows this approximation plotted for several values of k. It is seen from the figure that as k increases the approximation improves everywhere except in the immediate vicinity of a discontinuity. Here the deviation from the true waveform becomes narrower but not any smaller in amplitude. This is indeed the case, and even in the limit as $k \to \infty$ there remains a discrepancy at the point of discontinuity of approximately 9 percent. This behavior is known as the *Gibbs phenomenon,* after its discoverer. For large numbers of terms in the Fourier series the area under the small spike at the discontinuity becomes vanishingly small, so that it does not represent any deviation from the true value in a mean-square sense. Its principal effect is in the *apparent* infidelity of reproduc-

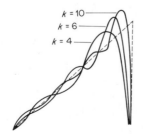

$k = 10$
$k = 6$
$k = 4$

Fig. **5-10** Gibbs phenomenon.

tion at discontinuities, even though it is the best fit of sinuoidal basis functions that is possible. The question of the origin of the Gibbs phenomenon is examined later, after the powerful techniques of the Fourier integral have been considered. At that time, some methods of eliminating this behavior are also discussed.

5-7 Fourier Transform

The Fourier series expansion of a function is a representation in which each basis function has its power concentrated at a single specific frequency, namely, some multiple of the reciprocal of the period of the original waveform. For signals that exist for only a finite time the Fourier series expansion becomes somewhat artificial in that it is valid over some time interval but must be replaced by zero outside that interval. Carrying out calculations on systems with this type of mathematical signal description is frequently very difficult and confusing. One reason that trouble might be anticipated with this representation is that we are trying to represent an energy signal by a set of basis functions, each of which is itself a power signal. Suppose that an attempt is made to compensate for this by limiting the basis functions to a finite time duration, namely, the fundamental period which is set equal to the interval over which the signal exists. Under these conditions the basis functions are finite lengths of sinusoids, such as $\cos n\omega_0 t[u(t) - u(t - T)]$. This modified signal does not have the highly desirable properties of an infinite duration sinusoid. For example, its derivative contains impulses, and the application of this signal to a linear system does not result in an output signal of the same form. Most of the advantages of expanding a signal in a set of basis functions are lost in this procedure. Fortunately, there is a simple way around this problem.

As an illustration of how the problems discussed above can be circumvented, consider the case of a rectangular pulse train in which the pulse shape stays constant, but the repetition period is continuously increased. The complex Fourier series expansion of a rectangular pulse train is

$$f(t) = A \sum_{n=-\infty}^{\infty} \alpha_n \epsilon^{jn\omega_0 t} \tag{5-44}$$

$$\alpha_n = A\left(\frac{t_1}{T}\right) \frac{\sin \dfrac{n\omega_0 t_1}{2}}{\dfrac{n\omega_0 t_1}{2}} \tag{5-45}$$

where A is the pulse amplitude, t_1 is the pulse duration, and $T = 2\pi/\omega_0$ is the repetition period. Figure 5-11 shows the effect on the line spectrum, $|\alpha_n|$, as the period is increased from T to $2T$ to $4T$. There are two important features to be observed in this figure. First, the relative shape

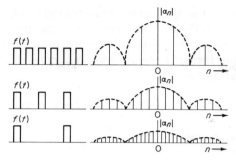

Fig. 5-11 Spectrum of pulse trains with increasing periods.

of the spectrum is not affected by changes in the period but is determined entirely by the pulse shape. The only way the period influences the shape of the spectrum is by altering the amplitude in a manner inversely proportional to the period. In an equally interesting manner, it is clear that the locations of the spectral lines are determined solely by the magnitude of the period. As the period is increased, the spectral lines become closer together, producing a denser spectrum. At the same time, the absolute amplitudes decrease but retain the same relative values among the various components. In the limit as the period approaches infinity the spectrum becomes continuous, with each component having a vanishingly small amplitude. A careful consideration of this limiting case is used to develop a very general way of representing signals, known as the *Fourier integral* theorem, or the *Fourier transform*.

Let the time function $f(t)$ be formally expanded over the interval $-T/2 < t < T/2$ in an exponential Fourier series having a period T. As the period T is increased, more and more of the time function will be included in the series representation. In the limit as $T \to \infty$, the entire function $f(t)$ for all t will be included, at least formally, in the expansion. The validity of the resulting expression will depend on the nature of $f(t)$ and on the validity of the various mathematical operations carried out in the development. This is discussed later. Consider now the representation of $f(t)$ as a complex Fourier series over the interval $-T/2 < t < T/2$. Hence,

$$f(t) = \sum_{n=-\infty}^{\infty} \alpha_n \exp\left(j\frac{2\pi nt}{T}\right) \tag{5-46}$$

where

$$\alpha_n = \frac{1}{T} \int_{-T/2}^{T/2} f(t) \exp\left(-j\frac{2\pi nt}{T}\right) dt \tag{5-47}$$

The fundamental angular frequency is $\omega_0 = 2\pi/T$. In addition to being the lowest frequency component, ω_0 is also the spacing between harmonics. Using this expression for ω_0 and substituting (5-47) into (5-46) gives

$$f(t) = \sum_{n=-\infty}^{\infty} \exp\left(j\frac{2\pi nt}{T}\right)\left[\frac{\omega_0}{2\pi} \int_{-T/2}^{T/2} f(t) \exp\left(-jn\omega_0 t\right) dt\right] \tag{5-48}$$

If we now let $T \to \infty$, the spacing between harmonics becomes a differential; that is, $\omega_0 = 2\pi/T \to d\omega$. The angular frequency of any particular component is given by $n\omega_0$, which becomes the continuous variable ω, and the summation formally passes into an integral. Equation (5-48) may then be written in the form

$$f(t) = \int_{-\infty}^{\infty} \epsilon^{j\omega t}\left[\frac{d\omega}{2\pi}\int_{-\infty}^{\infty} f(t)\epsilon^{-j\omega t}\,dt\right] \tag{5-49}$$

Rearranging terms gives

$$f(t) = \frac{1}{2\pi}\int_{-\infty}^{\infty}\left[\int_{-\infty}^{\infty} f(t)\epsilon^{-j\omega t}\,dt\right]\epsilon^{j\omega t}\,d\omega \tag{5-50}$$

This is the Fourier integral relation. Its significance becomes apparent when we separate the inner and outer integrals. It is evident that the inner integral is only a function of the angular frequency, ω, since time is integrated out. This inner integral is called the *Fourier transform* of $f(t)$ and is designated as

$$\mathcal{F}\{f(t)\} = F(\omega) = \int_{-\infty}^{\infty} f(t)\epsilon^{-j\omega t}\,dt \tag{5-51}$$

The relationship in (5-50) may then be considered as establishing the connection between $F(\omega)$ and $f(t)$. This relationship is called the *inverse* Fourier transform of $F(\omega)$ and is written as

$$\mathcal{F}^{-1}\{F(\omega)\} = f(t) = \frac{1}{2\pi}\int_{-\infty}^{\infty} F(\omega)\epsilon^{j\omega t}\,d\omega \tag{5-52}$$

The functions $f(t)$ and $F(\omega)$ are called *Fourier transform pairs*. It is customary in engineering to use lower-case letters for time functions and capital letters for their transforms, and this practice is followed here. Thus, the Fourier transform of $x(t)$ is $X(\omega)$, and the Fourier transform of $v(t)$ is $V(\omega)$. A word of caution is in order here because this convention of lower-case and capital letters is frequently reversed in mathematical discussions of transform methods.

The factor $1/2\pi$ contained in the inverse transform (5-52) could just as properly have been assigned to the direct transform, or a factor $1/\sqrt{2\pi}$ could have been included in each expression. All of these conventions have been followed by various authors; however, the relationships given in (5-51) and (5-52) are the ones most commonly used in technical literature and will be used throughout this book. They have the distinct advantage of allowing a simple and direct conversion between the Laplace and Fourier transforms in cases where such a conversion is allowable. There is one other convention regarding transform representation that should be mentioned. Frequently, the Fourier transform is written as a function of the variable $j\omega$, so that the transform of $f(t)$ would be $F(j\omega)$. The actual transform expressions are identical to those that are designated here as $F(\omega)$. Either notation is acceptable, provided it is used consistently.

However, the use of $F(j\omega)$ leads to some confusion when such expressions as $F'(j\omega)$ and $F_1(j\omega) * F_2(j\omega)$ occur, because in almost all cases the indicated operations of differentiation and convolution are meant to be carried out with respect to the variable ω and not $j\omega$. In order to avoid these ambiguities and to eliminate the need for special symbols to represent standard operations, the symbolic representation of the transform is given consistently as $F(\omega)$ and not $F(j\omega)$.

Using the analogy between the Fourier series and the Fourier transform, it may be concluded that the function $F(\omega)$ analyzes $f(t)$ into a continuum of complex sinusoids having amplitudes of $(1/2\pi)F(\omega)d\omega$. If $F(\omega)$ is finite, as it is unless discrete frequencies are present, then these amplitudes are infinitesimal. This can be interpreted as the distribution of the signal in the frequency domain. Such a distribution is called a *frequency spectrum*, and in the case of the Fourier transform, $(1/2\pi)|F(\omega)|d\omega$ can be thought of as the amplitude of the signal amplitude in the frequency band of ω to $\omega + d\omega$. Noting that $\omega = 2\pi f$, it is also clear that $|F(2\pi f)|df$ equals the signal amplitude in the frequency band of f to $f + df$ Hz. The Fourier transform can be expressed more clearly in terms of frequency spectra by writing it as

$$F(\omega) = A(\omega)\epsilon^{j\theta(\omega)} \tag{5-53}$$

where

$$A(\omega) = |F(\omega)| \tag{5-54}$$

$$\theta(\omega) = \tan^{-1}\left[\frac{Im\ F(\omega)}{Re\ F(\omega)}\right] \tag{5-55}$$

$A(\omega)$ is then the amplitude spectrum (often just called the frequency spectrum), and $\theta(\omega)$ is the phase spectrum corresponding to the phase (at $t = 0$) of the elementary sinusoid at the angular frequency ω.

Not all time functions can be represented by the Fourier integral. However, when such a representation is possible, there is a unique one-to-one correspondence between a function and its Fourier transform. This means that there is only a single Fourier transform corresponding to a given time function and only a single time function corresponding to a given Fourier transform. The determining factor in the Fourier representation is whether or not the integrals are convergent. One set of conditions that assures convergence is the Dirichlet conditions. These are given in Section 5-2 for the Fourier series and are restated here for the Fourier transform.

1. The function $f(t)$ must be absolutely integrable; that is,

$$\int_{-\infty}^{\infty} |f(t)|dt < \infty$$

2. The function $f(t)$ must have no more than a finite number of maxima and minima in any finite time interval.

3. The function $f(t)$ must have no more than a finite number of finite discontinuities in any finite interval.

Note that the only essential change is that the range of integration in **1** is now infinite instead of a single period. These conditions are sufficient to include virtually all useful finite-energy signals. However, they exclude a number of important signals, such as periodic waveforms and the unit step function, that are not absolutely integrable. By allowing the Fourier transform to include delta functions, it will be found that signals of this type can be handled using essentially the same methods as for finite-energy signals.

5-8 Direct Calculations of Transforms

Rectangular pulse. The Fourier transforms of elementary waveforms can be calculated readily by direct evaluation of the defining integral (5-51). As an example consider the rectangular pulse shown in Fig. 5-11a. This

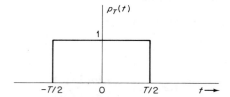

Fig. 5-11a Rectangular pulse.

pulse can be expressed analytically as

$$p_T(t) = 1 \qquad -T/2 < t < T/2$$
$$= 0 \qquad \text{elsewhere} \tag{5-56}$$

The Fourier transform is obtained from (5-51) as

$$P_T(\omega) = \int_{-\infty}^{\infty} p_T(t)\epsilon^{-j\omega t} dt$$

$$= \int_{-T/2}^{T/2} \epsilon^{-j\omega t} dt = \frac{\epsilon^{-j\omega t}}{-j\omega}\bigg|_{-T/2}^{T/2} \tag{5-57}$$

$$= \frac{\epsilon^{+j\omega T/2} - \epsilon^{-j\omega T/2}}{j\omega}$$

Converting the exponentials in (5-57) to the equivalent trigonometric function leads to

$$P_T(\omega) = T\frac{\sin \omega T/2}{\omega T/2} \tag{5-58}$$

The form given in (5-58) makes use of the frequently occurring functional form sin x/x. This function occurs sufficiently often in signal and system analysis that it is convenient to use a shortened notation for it. The most common representation of this function is in terms of the sinc function which is defined as

$$\text{sinc}(x) = \frac{\sin \pi x}{\pi x} \qquad \text{(5-59)}$$

Inclusion of the factor π in the argument of (5-57) leads to considerable simplification when the Fourier transform is expressed in terms of the frequency f instead of the angular frequency ω. This matter is discussed in Section 5-10.

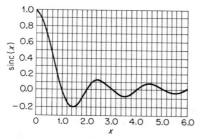

Fig. 5-12 Plot of sinc $x = (\sin \pi x)/\pi x$.

The function sinc x is shown in Fig. 5-12. Approximate values can be read directly from the figure and a short table of sinc (x) is given in Appendix A. The sinc (x) function is an even function of x with its maximum value of unity occurring at the origin and having a damped os-illatory amplitude away from the origin. The zeroes occur at all integral values of x and correspond to the zeroes of sin πx in the numerator of sinc (x).

Returning now to (5-58), we see that $P_T(\omega)$ has an amplitude spectrum $A(\omega) = |T \text{ sinc}(\omega T/2\pi)|$. The phase spectrum, $\theta(\omega)$, oscillates between 0 and $-\pi$ in this instance. Figure 5-13 shows a plot of these spectra. The phase spectrum has been modified by adding or subtracting multiples of 2π from $\theta(\omega)$ to keep it in the range of $\pm \pi$. An important characteristic of signals is evident in the behavior of $A(\omega)$ and $\theta(\omega)$ at multiples of $2\pi/T$. It is observed that at each of these points $A(\omega)$ has a discontinuity in its slope and that this is accompanied by a discontinuity in $\theta(\omega)$. This is a general property of signals and indicates the intimate relationship that exists between the phase and amplitude spectra.

In addition to the details of the spectrum of a specific pulse shape shown in Fig. 5-13, there is also illustrated a very general property of transform pairs. Note that the spectrum is concentrated over a band of frequencies in the vicinity of the origin with the first null of the major lobe of the spectrum occurring at a frequency $f = \omega/2\pi = 1/T$. As the pulse width is decreased, this first null moves to a higher and higher frequency.

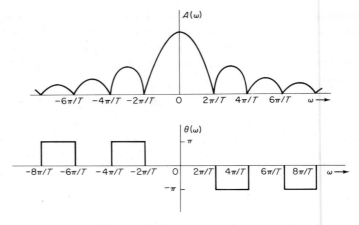

Fig. 5-13 Amplitude and phase spectra of a rectangular pulse.

Conversely, as the pulse width is increased, the first null moves in closer and closer to the origin. The relative amplitudes of the various portions of the spectrum are unchanged by changes in T. It is evident that for this pulse shape there is an inverse relationship between the time duration of the signal and the frequency spread of the spectrum. This is a general property of all signals: The more compact the signal in time, the more spread out it will be in frequency, and vice versa. One consequence of this property is that there is a minimum value of time-duration–bandwidth product that can be obtained with any signal. It is readily shown from the defining equations of the Fourier transform, (5-51) and (5-52), that

$$\int_{-\infty}^{\infty} f(t)dt = F(0) \quad \text{and} \quad \frac{1}{2\pi}\int_{-\infty}^{\infty} F(\omega)d\omega = f(0) \tag{5-60}$$

Multiplying these equations together and rearranging factors leads to the relationship

$$\frac{\int_{-\infty}^{\infty} f(t)dt}{f(0)} \cdot \frac{\int_{-\infty}^{\infty} F(\omega)d\omega}{F(0)} = 2\pi \tag{5-61}$$

The two factors can be thought of as the equivalent duration and equivalent bandwidth of the signals, respectively. In each case it is seen that the equivalent width is the area of the function divided by the ordinate at the origin. This formulation is most useful when the origin is also the centroid of the time signal. If the centroid is at t_0 rather than at the origin, the signal can be translated and the spectrum modified by methods that are discussed in a later section to give

$$\frac{\int_{-\infty}^{\infty} f(t)dt}{f(t_0)} \cdot \frac{\int_{-\infty}^{\infty} F(\omega)\epsilon^{j\omega t_0}d\omega}{F(0)} = 2\pi \tag{5-62}$$

It is seen from these relationships that, when the required equivalent widths exist, that is,

$$f(t_o) \neq 0 \neq F(0)$$

their product is a constant, and an increase in one must, therefore, always result in a compensating decrease in the other. As an example, consider the rectangular pulse, $Ap_T(t)$. The centroid of the pulse is at $t = 0$, and the equivalent duration T_{eq} and equivalent bandwidth B_{eq} are

$$T_{eq} = \frac{A \int_{-\infty}^{\infty} p_T(t)dt}{Ap_T(0)} = \int_{-T/2}^{T/2} dt = T \tag{5-63}$$

$$B_{eq} = \frac{A \int_{-\infty}^{\infty} P_T(\omega)d\omega}{AP_T(0)} = \int_{-\infty}^{\infty} \text{sinc} \frac{\omega T}{2\pi} d\omega = \frac{2\pi}{T} \tag{5-64}$$

The product $T_{eq} \cdot B_{eq} = 2\pi$. When the functions do not have central ordinates, as in the case of bandpass signals, a different formulation is required. In this case the duration and bandwidth are taken as the radii of gyration of the signal and spectrum measured relative to their centroids, and the required relationship states that the duration-bandwidth product must be greater than or equal to π radians.

Exercise 5-8.1
Find the Fourier transform of the triangular pulse signal shown and sketch its amplitude spectrum.

ANS: See Table 5-4.

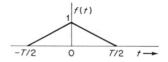

Exercise 5-8.2
Compare the equivalent bandwidths and durations of rectangular and triangular pulses that are nonzero over the same time interval.

ANS: Ratios are $\frac{1}{2}$, 2.

One-sided exponential pulse.   Because of its frequent occurrence in the solution of linear differential equations, the exponential pulse is of great importance in system analysis. Such a pulse is shown in Fig. 5-14 and is represented mathematically by

$$f(t) = \epsilon^{-\alpha t} u(t) \tag{5-65}$$

The Fourier transform is obtained by direct application of the defining

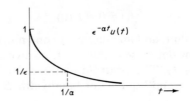

Fig. 5-14 One-sided exponential pulse.

relationship as

$$F(\omega) = \int_{-\infty}^{\infty} \epsilon^{-\alpha t} u(t) \epsilon^{-j\omega t} dt$$

$$= \int_{0}^{\infty} \epsilon^{-\alpha t} \epsilon^{-j\omega t} dt = \frac{\epsilon^{-(\alpha + j\omega)t}}{-(\alpha + j\omega)} \bigg|_{0}^{\infty} \qquad \text{(5-66)}$$

$$= \frac{1}{j\omega + \alpha}$$

The amplitude and phase spectra are given in (5-67) and (5-68) and are illustrated in Fig. 5-15.

$$A(\omega) = \left(\frac{1}{\omega^2 + \alpha^2}\right)^{1/2} \qquad \text{(5-67)}$$

$$\theta(\omega) = -\tan^{-1}\left(\frac{\omega}{\alpha}\right) \qquad \text{(5-68)}$$

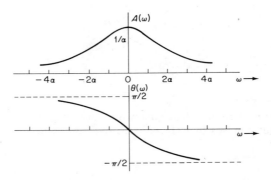

Fig. 5-15 Amplitude and phase spectra of one-sided exponential pulse.

Two-sided exponential pulse.   The two-sided exponential pulse provides a means of introducing a very important concept into the discussion of Fourier transforms. The transform can be calculated by direct application of the defining integral. Referring to Fig. 5-16, we have

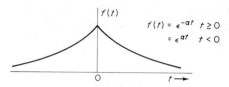

Fig. 5-16 Two-sided exponential pulse.

$$F(\omega) = \int_{-\infty}^{0} \epsilon^{\alpha t} \epsilon^{-j\omega t} dt + \int_{0}^{\infty} \epsilon^{-\alpha t} \epsilon^{-j\omega t} dt$$

$$= \frac{1}{\alpha - j\omega} + \frac{1}{\alpha + j\omega} \tag{5-69}$$

$$= \frac{2\alpha}{\omega^2 + \alpha^2} \tag{5-70}$$

The important concept exhibited here concerns the relationship of the pole location in $F(\omega)$ and the nature of the time function $f(t)$. A *pole* is a value of ω for which $F(\omega)$ goes to infinity. This term is defined more accurately in Appendix B. In (5-70) these poles are the roots of the denominator and are seen from (5-69), which is the partial fraction expansion of (5-70), to be $\omega = \pm j\alpha$. The portion of the time function corresponding to $t < 0$ gives rise to the pole at $\omega = -j\alpha$, and the portion of the time function corresponding to $t > 0$ gives rise to the pole at $\omega = +j\alpha$. This is a general property of the Fourier transform; that is, poles at negative imaginary frequencies correspond to functions existing in negative time, and poles at positive imaginary frequencies correspond to functions existing in positive time. This property can be used to break up the transform into portions corresponding to positive and negative time functions, and it can be used to determine if a given transform corresponds to a time function existing only for positive or negative time.

Derivative of a time function.   The Fourier transform of the derivative of a time function, when that derivative exists, can be expressed in terms of the transform of the original function. The relationship can be derived in a number of ways, but a particularly simple method is the following one, which starts from the definition of the inverse transform.

$$f(t) = \frac{1}{2\pi} \int_{-\infty}^{\infty} F(\omega) \epsilon^{j\omega t} d\omega$$

$$\frac{df(t)}{dt} = \frac{d}{dt} \left[\frac{1}{2\pi} \int_{-\infty}^{\infty} F(\omega) \epsilon^{j\omega t} d\omega \right] \tag{5-71}$$

$$= \frac{1}{2\pi} \int_{-\infty}^{\infty} j\omega F(\omega) \epsilon^{j\omega t} d\omega$$

Since there is a unique one-to-one correspondence between a Fourier transform and its inverse and since the right-hand side of (5-71) is the inverse Fourier transform of $j\omega F(\omega)$, it immediately follows that

$$\mathcal{F}^{-1}\{j\omega F(\omega)\} = \frac{df(t)}{dt} \tag{5-72}$$

and, conversely,

$$\mathcal{F}\left\{\frac{df(t)}{dt}\right\} = j\omega F(\omega) \tag{5-73}$$

Thus, it is seen that the transform of a derivative is just $j\omega$ times the transform of the original function.

Exercise 5-8.3

Use (5-73) and the transform of a triangular pulse given in Table 5-4 to find the transform of the function shown.

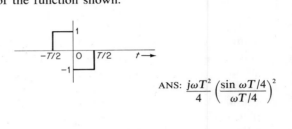

ANS: $\dfrac{j\omega T^2}{4}\left(\dfrac{\sin \omega T/4}{\omega T/4}\right)^2$

Circuit analysis.   As an example of the use of the Fourier transform in circuit analysis, consider the R-C circuit shown in Fig. 5-17, in which an exponential current pulse is applied and it is desired to determine the resulting voltage across the capacitor. Using Kirchhoff's current law, the differential equation for the circuit is found to be

$$v(t) + 2\frac{dv(t)}{dt} = i(t) \tag{5-74}$$

Designating $V(\omega)$ as the transform of $v(t)$, the transform of (5-74) is found to be

$$V(\omega) + j2\omega V(\omega) = \mathcal{F}\{12\epsilon^{-t}u(t)\} = \frac{12}{j\omega + 1} \tag{5-75}$$

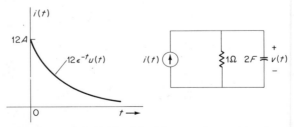

Fig. 5-17 R–C circuit excited by an exponential pulse.

Solving for $V(\omega)$ gives

$$V(\omega) = \frac{6}{(j\omega + 1)(j\omega + \frac{1}{2})} \tag{5-76}$$

The right-hand side of (5-76) can be written as the sum of two terms as follows:[2]

$$V(\omega) = \frac{12}{j\omega + \frac{1}{2}} - \frac{12}{j\omega + 1} \tag{5-77}$$

The voltage is obtained by taking the inverse Fourier transform of (5-77). Thus, using (5-66), the following result is obtained:

$$v(t) = \mathcal{F}^{-1} \left\{ \frac{12}{j\omega + \frac{1}{2}} - \frac{12}{j\omega + 1} \right\} \tag{5-78}$$

$$= \mathcal{F}^{-1} \left\{ \frac{12}{j\omega + \frac{1}{2}} \right\} - \mathcal{F}^{-1} \left\{ \frac{12}{j\omega + 1} \right\} \tag{5-79}$$

$$= \begin{cases} 12\epsilon^{-1/2t} - 12\epsilon^{-t} & t \geq 0 \\ 0 & t < 0 \end{cases} \tag{5-80}$$

The resulting voltage is sketched in Fig. 5-18.

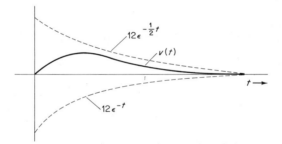

Fig. 5-18 Voltage in circuit of Fig. 5-17.

A little study of the process just carried out reveals that it closely resembles steady-state ac analysis. The Fourier transforms of the terms in the differential equation lead to new terms that are identical to the impedances encountered in steady-state ac circuit analysis. In the example just considered, these impedances are $j\omega C$ and R. The principal difference is that instead of a phasor representing a single sinusoidal driving function, we use the Fourier transform of the driving function. It will be recalled, however, that the Fourier transform can be considered as a continuum of sinusoidal functions. The analysis of circuits can be carried out readily by representing the circuit elements by their corresponding impedances (or

---

[2]A rational function such as (5-76) can always be expressed as the sum of simple fractions whose denominators are the factors of the denominator of the original function. This is called a *partial fraction expansion* and is discussed in detail in Chapter 6.

admittances) and representing the driving functions by their Fourier transforms. The circuit is then solved, as in the case of steady-state ac circuit analysis, for the unknown response in the frequency domain. The response in the time domain is obtained by taking the inverse Fourier transform. This technique is very powerful and is widely applicable. The concept of the Fourier transform approach to circuit analysis is of the utmost importance and will be used frequently in later discussions. In actually carrying out computation of circuit response, however, it will be found that the closely related Laplace transform (to be discussed in Chapter 6) leads to somewhat simpler algebraic expressions and is, therefore, generally preferred. Although the Laplace transform is of great utility in obtaining time-domain solutions, there is also much that can be learned about systems and signals by considering their frequency-domain characteristics. This requires use of the Fourier transform or interpretation of the Laplace transform as a Fourier transform. Because of this, it is important to have a good understanding of the physical significance of the Fourier transform and to be able to manipulate and interpret such transforms. When the initial conditions in the circuit are not zero, the computations are somewhat more involved. This matter is discussed in Chapter 6.

Exercise 5-8.4
Write the differential equation corresponding to the following transform equation:

$$I(\omega) = \frac{5}{j\omega + 1} - \frac{5}{j\omega + 2}$$

$$\text{ANS: } \frac{d^2 i(t)}{dt^2} + 3\frac{di(t)}{dt} + i(t) = 5\delta(t)$$

5-9 Elementary Properties of the Fourier Transform

Before considering application of the Fourier transform to specific problems, it is worthwhile examining some properties inherent in the transform itself.

Symmetry of f(t) about the time origin. The real and imaginary parts of the Fourier transform are dependent on the oddness and evenness of the corresponding time function. This relationship can be examined in detail by making use of the fact that any signal can be considered as being made up of an even part and an odd part relative to the time origin as discussed in Section 5-4. Thus, if $f(t)$ is a given time function, and $f_e(t)$ is the even part, and $f_o(t)$ is the odd part,

$$f(t) = f_e(t) + f_o(t) \tag{5-81}$$

The even function, $f_e(t)$, has the property that $f_e(-t) = f_e(t)$, and the odd function, $f_o(t)$, has the property that $f_o(-t) = -f_o(t)$. Substituting $t = -t$

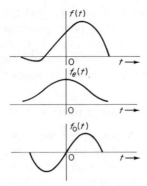

Fig. 5-19 A time function and its odd and even parts.

into (5-81) and using these properties gives

$$f_e(t) = \frac{f(t) + f(-t)}{2} \qquad \text{(5-82)}$$

$$f_o(t) = \frac{f(t) - f(-t)}{2} \qquad \text{(5-83)}$$

An example of a function and its odd and even parts is shown in Fig. 5-19.

By use of these relationships, the Fourier transform of an arbitrary time function can be found as follows:

$$F(\omega) = \int_{-\infty}^{\infty} f(t)\epsilon^{-j\omega t}\,dt$$

$$= \int_{-\infty}^{\infty} [f_e(t) + f_o(t)](\cos \omega t - j \sin \omega t)\,dt \qquad \text{(5-84)}$$

$$= \int_{-\infty}^{\infty} [f_e(t) \cos \omega t - jf_e(t) \sin \omega t + f_o(t) \cos \omega t - jf_o(t) \sin \omega t]\,dt$$

Since the second and third terms in the integrand are odd functions of t, their integrals over symmetrical limits are zero. Therefore,

$$F(\omega) = \int_{-\infty}^{\infty} f_e(t) \cos \omega t\, dt - j \int_{-\infty}^{\infty} f_o(t) \sin \omega t\, dt \qquad \text{(5-85)}$$

Since the integrands are even, $F(\omega)$ can also be written as

$$F(\omega) = 2 \int_{0}^{\infty} f_e(t) \cos \omega t\, dt - j\,2 \int_{0}^{\infty} f_o(t) \sin \omega t\, dt \qquad \text{(5-86)}$$

$$\underbrace{\qquad\qquad\qquad}_{\text{Real part}} \qquad \underbrace{\qquad\qquad\qquad}_{\text{Imaginary part}}$$

From (5-86) the following properties may be deduced for real time functions:

1. Even-time functions lead to pure real transforms.
2. Odd-time functions lead to pure imaginary transforms.
3. The real part of $F(\omega)$ is an even function of ω.
4. The imaginary part of $F(\omega)$ is an odd function of ω.
5. The amplitude spectrum $|F(\omega)|$ is an even function of ω. This follows from:

$$|F(\omega)| = \sqrt{\{\text{Re}\,[F(\omega)]\}^2 + \{\text{Im}\,[F(\omega)]\}^2}$$

in which all of the terms under the radical are even because the square of either an odd or an even function is even.

6. The phase spectrum $\theta(\omega)$ is an odd function of ω since the ratio of an odd to an even function is odd and the arctan is an odd function.

These properties are useful in sketching transforms, in checking mathematical operations, and in deducing properties of time functions from their spectra, and vice versa. From (5-85) we can write $F(\omega)$ in terms of an odd and an even part, as follows:

$$F(\omega) = F_e(\omega) + F_o(\omega) \tag{5-87}$$

$$F_e(\omega) = \int_{-\infty}^{\infty} f_e(t)\cos\omega t\,dt \tag{5-88}$$

$$F_o(\omega) = -j\int_{-\infty}^{\infty} f_o(t)\sin\omega t\,dt \tag{5-89}$$

Exercise 5-9.1
From their oddness or evenness what can be said about the real and imaginary parts of the Fourier transforms of the functions shown?

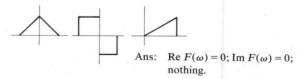

Ans: Re $F(\omega) = 0$; Im $F(\omega) = 0$; nothing.

Symmetry of $F(\omega)$ about the frequency origin.   Just as in the case of symmetrical time functions, there are also special properties associated with symmetry in the frequency domain. Expressing $F(\omega)$ as the sum of an odd and an even part, we have

$$f(t) = \frac{1}{2\pi}\int_{-\infty}^{\infty} [F_e(\omega) + F_o(\omega)]\epsilon^{j\omega t}\,d\omega \tag{5-90}$$

Expanding $\epsilon^{j\omega t}$ in terms of $\cos\omega t$ and $\sin\omega t$ and using the oddness and evenness of the resulting integrands gives

$$f(t) = \frac{1}{2\pi}\int_{-\infty}^{\infty} F_e(\omega)\cos\omega t\,d\omega + j\frac{1}{2\pi}\int_{-\infty}^{\infty} F_o(\omega)\sin\omega t\,d\omega \tag{5-91}$$

Note that there can be no imaginary part to $f(t)$, since we are restricting consideration to real time functions. The even and odd parts of the time function associated with a given transform are, from (5-91),

$$f_e(t) = \frac{1}{2\pi} \int_{-\infty}^{\infty} F_e(\omega) \cos \omega t d\omega$$

$$= \frac{1}{2\pi} \int_{-\infty}^{\infty} \operatorname{Re} F(\omega) \cos \omega t d\omega \tag{5-92}$$

$$f_o(t) = j\frac{1}{2\pi} \int_{-\infty}^{\infty} F_o(\omega) \sin \omega t d\omega$$

$$= -\frac{1}{2\pi} \int_{-\infty}^{\infty} \operatorname{Im} F(\omega) \sin \omega t d\omega \tag{5-93}$$

These relationships could also have been found directly from (5-88) and (5-89) because of the uniqueness between transform pairs.

Causal time functions.   Time functions that are identically zero prior to some specified time are called *causal* time functions. Unless otherwise specified, it is assumed that a causal time function is zero for $t < 0$, since a simple translation of the time scale can always be applied to the time function to make its occurrence correspond to $t = 0$. Causal time functions have great importance in system analysis, since all impulse responses of physical systems have this characteristic. The Fourier transform of a causal function has the important property that either its real or its imaginary part is sufficient to specify the time function completely. Consider the causal function $f(t)$, which has the property that $f(t) = 0$ for $t < 0$. The even and odd parts of $f(t)$ are for $t > 0$,

$$f_e(t) = \frac{f(t) + f(-t)}{2} = \frac{1}{2} f(t) \tag{5-94}$$

$$f_o(t) = \frac{f(t) - f(-t)}{2} = \frac{1}{2} f(t) \tag{5-95}$$

Therefore,

$$f(t) = 2f_e(t) = 2f_o(t) \qquad t > 0 \tag{5-96}$$

Now since $f_e(t)$ can be found from $F_e(\omega)$, and $f_o(t)$ can be found from $F_o(\omega)$, it follows that $f(t)$ can be found either from the real part, $F_e(\omega)$, or from the imaginary part, $-jF_o(\omega)$. The required equations are

$$f(t) = \frac{1}{\pi} \int_{-\infty}^{\infty} F_e(\omega) \cos \omega t d\omega$$

$$= \frac{1}{\pi} \int_{-\infty}^{\infty} \operatorname{Re} F(\omega) \cos \omega t d\omega \tag{5-97}$$

$$= j\frac{1}{\pi} \int_{-\infty}^{\infty} F_o(\omega) \sin \omega t d\omega$$

$$= -\frac{1}{\pi} \int_{-\infty}^{\infty} \operatorname{Im} F(\omega) \sin \omega t d\omega \tag{5-98}$$

It must be remembered that (5-97) and (5-98) are only applicable to time functions that are zero for negative values of time.

Just as the even and odd parts of $F(\omega)$ are uniquely related to each other, it is found that the amplitude spectrum, $A(\omega)$, and the phase spectrum, $\theta(\omega)$, are also uniquely related to each other for causal functions. In particular, it is found that only functions meeting certain requirements belong to the class of functions that could be transforms of causal time functions. One of the most powerful requirements of this kind relating to the amplitude spectrum of a causal function is the *Paley–Wiener criterion*,[3] which may be stated as follows:

$$\int_{-\infty}^{\infty} \frac{|\ln A(\omega)|}{1 + \omega^2} d\omega < \infty \qquad \text{(5-99)}$$

Two requirements on $A(\omega)$ are immediately evident from (5-99). First, $A(\omega)$ cannot go to zero over any band of frequencies, for then $|\ln A(\omega)|$ would become infinite, and the integral would also become infinite. Second, the amplitude spectrum cannot fall off to zero more rapidly than exponential order, or again the integral will become infinite. These results are of great significance in relationship to the physical realizability of systems that are specified in terms of the Fourier transform of their impulse response, a subject that is discussed in the next section.

5-10 System Function

A general linear system is shown in block diagram form in Fig. 5-20.

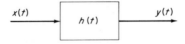

Fig. 5-20 General linear system.

As was discussed in Chapter 4, the output, $y(t)$, can be expressed in terms of the input, $x(t)$, and the impulse response, $h(t)$, by means of the convolution integral as follows:

$$y(t) = \int_{-\infty}^{\infty} h(\lambda)x(t - \lambda)d\lambda \qquad \text{(5-100)}$$

This integral applies when the initial energy storage in the system is zero. When the effects of nonzero initial energy storage must be taken into account, the situation becomes somewhat more involved, and discussion

[3]R. Paley and N. Wiener, "Fourier Transforms in the Complex Domain," Am. Math. Soc. Colloq. Rule 19, 1934. Also see G. Valley and N. Wallman, *Vacuum Tube Amplifiers*, McGraw-Hill Book Company, Inc. New York, 1948, pp. 721–727.

of that problem is deferred until Chapters 6 and 8. The simpler problem of zero initial energy storage is, however, very important because it provides much insight into system operation and system properties, as well as providing the solution to many practical problems. The Fourier transform provides a particularly simple method of solving (5-100). Taking the Fourier transform of both sides of (5-100) gives

$$Y(\omega) = \int_{-\infty}^{\infty} \left[\int_{-\infty}^{\infty} h(\lambda)x(t - \lambda)d\lambda \right] \epsilon^{-j\omega t} dt$$

Interchanging the order of integration on the right-hand side, changing the variable of integration, and carrying out the indicated operation leads to the following result:

$$Y(\omega) = \int_{-\infty}^{\infty} h(\lambda) \left[\int_{-\infty}^{\infty} x(t - \lambda)\epsilon^{-j\omega t} dt \right] d\lambda$$

$$= \int_{-\infty}^{\infty} h(\lambda) \left[\int_{-\infty}^{\infty} x(\xi)\epsilon^{-j\omega(\xi + \lambda)} d\xi \right] d\lambda$$

$$= \int_{-\infty}^{\infty} h(\lambda)\epsilon^{-j\omega\lambda} X(\omega) d\lambda \qquad\qquad \text{(5-101)}$$

$$= X(\omega) \int_{-\infty}^{\infty} h(\lambda)\epsilon^{-j\omega\lambda} d\lambda$$

$$= X(\omega)H(\omega)$$

Two important results are implied in (5-101). First, the Fourier transform of the convolution of two functions is equal to the product of the Fourier transforms of the functions taken separately. Symbolically, this may be stated as

$$\mathcal{F}\{f_1(t) * f_2(t)\} = F_1(\omega)F_2(\omega) \qquad\qquad \text{(5-102)}$$

Second, the Fourier transform of the output of a linear system is given by the Fourier transform of the input multiplied by the Fourier transform of the system impulse response. Because of its frequent use, the Fourier transform of the impulse response, $H(\omega) = \mathcal{F}\{h(t)\}$, is called the *system function*, or transfer function, and represents another mathematical model for the system when there is no initial stored energy. The system function is generally found as the ratio $H(\omega) = Y(\omega)/X(\omega)$ by solution of the circuit equations of the system. Use of the system function concept often greatly simplifies computation of system response and is of great value in the theoretical analysis of systems.

The system function $H(\omega)$ is the ratio of the component of the output at the frequency ω to the component of the input at the same frequency. This ratio is commonly called the *frequency response* of a network, or system. It should be noted that $H(\omega)$ contains both amplitude and phase information. The value of $H(\omega)$ at some particular frequency can be measured by applying a signal of known frequency and amplitude and measuring the output signal. The ratio of the phasors representing the input and output sinusoids gives the value of $H(\omega)$. The physical explanation of why $H(\omega)$, the Fourier transform of the system impulse response, is the frequency response of the system is readily obtained by formally carrying out the computation of the response of a system in the frequency domain when a unit impulse is applied. The output will be the product of $H(\omega)$ and the Fourier transform of the unit impulse. The Fourier transform of the unit impulse is

$$\mathcal{F}\{\delta(t)\} = \int_{-\infty}^{\infty} \delta(t)\epsilon^{-j\omega t}dt = 1 \qquad \delta(t) \Leftrightarrow 1 \qquad \textbf{(5-103)}$$

From (5-103)[4] it is seen that the spectrum of the unit impulse is uniform; that is, all frequency components are present with equal amplitudes and zero initial phase. When a signal having these characteristics is applied to a system, the output is a direct measure of the transmission of each frequency component, and this in turn is just the frequency response of the system. The uniform spectrum exhibited by the unit impulse is the primary characteristic which makes it such a valuable test signal. Unfortunately, an impulse is not a physically realizable signal because of its infinite amplitude and infinite frequency extent; however, by using signals having uniform spectra over the transmission band of a system, it is possible to obtain system responses that, for all practical purposes, are indistinguishable from the true impulse response. Thus, pulses make suitable test signals, provided they are sufficiently narrow.

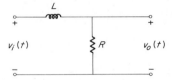

Fig. 5-21 Simple $R–L$ circuit.

As an example of a system function, consider the simple R-L circuit shown in Fig. 5-21. As discussed previously, such circuits can be analyzed by replacing the circuit elements with their equivalent impedances using the frequency variable ω and then proceeding in the same manner as in steady-state analysis using the Fourier transforms of the input and output signals. The circuit obtained using this procedure is shown in Fig. 5-22. The system function can now be written down by inspection as

$$H(\omega) = \frac{V_o(\omega)}{V_i(\omega)} = \frac{R}{j\omega L + R} = \frac{R/L}{j\omega + R/L} \qquad \textbf{(5-104)}$$

---

[4]The double arrow is a shorthand notation relating a function and its transform. The same representation will also be used in connection with the Laplace transform.

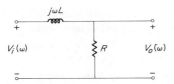

Fig. 5-22 Transformed circuit of Fig. 5-21.

The impulse response corresponding to $H(\omega)$ is

$$h(t) = \mathscr{F}^{-1}\{H(\omega)\} = \mathscr{F}^{-1}\left\{\frac{R/L}{j\omega + R/L}\right\}$$

$$= \frac{R}{L}\epsilon^{-Rt/L}u(t)$$

(5-105)

The system function and impulse response corresponding to (5-104) and (5-105) are shown in Fig. 5-23. Several important points should be considered in connection with this example. A frequent method of describing a circuit is in terms of its time constant. By this is meant that interval of time during which an exponential term in the response changes by a factor of $1/\epsilon$. In the present instance it is seen that, when t goes from $t = 0$ to $t = L/R$, the exponent goes from 0 to 1. Therefore, the time constant is L/R. In general the time constant is the reciprocal of the factor multiplying t in the exponential terms. Many circuits have more than one time constant; for example, $h(t) = (\epsilon^{-t} + \epsilon^{-2t})u(t)$ has time constants of 1 sec and 0.5 sec. For a single time constant circuit the time constant is simply related to the frequency response. In fact, at an angular frequency equal to the reciprocal of the time constant, the amplitude spectrum has dropped to $1/\sqrt{2}$ of its value at the origin. This is the half-power point on the frequency response curve, since a sinusoid having its amplitude reduced by a factor of $1/\sqrt{2}$ has its power reduced by one-half.

It frequently is desirable to be able to sketch rapidly the impulse response or system function associated with a particular system. In the time

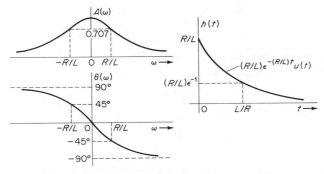

Fig. 5-23 System function and impulse response for circuit in Fig. 5-22.

domain an exponential can be easily sketched by making use of the fact that, at the end of an interval equal to one time constant, the amplitude decreases to $\epsilon^{-1} = 0.368$ of its value at the beginning of that interval. For crude sketches this factor can be approximated as $\frac{1}{3}$, or 0.4, and good qualitative results obtained. To make the sketch, intervals of duration equal to 1 time constant are marked off. Starting with the value at $t = 0$, values at each succeeding interval are plotted using the above factor. After 4 time constants, the amplitude is down to 1.8 percent of its initial value and cannot be distinguished from 0 on a linear scale plot.

Rapid plotting of the frequency response is also very useful and can often be done in a simple manner with reasonable accuracy. The most frequently used plot, called a *Bode diagram*, after an early researcher on feedback systems, employs a logarithmic frequency scale. The amplitude spectrum is plotted in decibels,[5] and the phase spectrum is plotted in degrees. The use of logarithmic scales allows a very wide range of amplitudes and frequencies to be accommodated on a compact graph. The frequency response corresponding to Fig. 5-23, assuming a time constant of 1, is shown in Fig. 5-24, using a logarithmic frequency scale and a decibel amplitude scale.

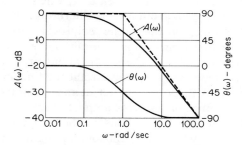

Fig. 5-24 Frequency response on a logarithmic scale.

The amplitude spectrum is particularly simple to sketch if the asymptotes are sketched first. The asymptotes are straight lines intersecting at the half-power frequency. The low-frequency asymptote is determined by the variation in $A(\omega)$ as $\omega \to 0$. From (5-102) it is seen that in the present case $A(\omega) = (\omega^2 + 1)^{-1/2} \to 1$ for small ω. For large ω, $A(\omega) \to 1/\omega$. In terms of decibels, these become the lines $A(\omega) = 0$ and $A(\omega) = -20 \log \omega$, and they are shown as dotted lines in Fig. 5-24.

---

[5]The decibel is a logarithmic (to the base 10) measure of the ratio of the powers in two signals x_1 and x_2 and is defined as

$$\left.\frac{P_{x_1}}{P_{x_2}}\right|_{dB} = 10 \log \frac{P_{x_1}}{P_{x_2}} = 10 \log \frac{x_1^2}{x_2^2} = 20 \log \frac{x_1}{x_2}$$

The last two expressions are based on the proportionality between the power in a signal and the square of the signal amplitude.

This same sketching procedure can be extended to more complicated system functions. In general, for lumped parameter, time-invariant, linear systems, the system functions occur only as ratios of polynomials that can be factored into linear and quadratic factors in ω. The contribution of each factor is computed separately. The procedure is best explained by an example. Consider the following system function:

$$H(\omega) = \frac{10(j\omega + 4)}{(j\omega + 1)(-\omega^2 + 5j\omega + 100)} \tag{5-106}$$

For purposes of computation it is convenient to replace $j\omega$ by s and to put all factors into the form $(Ts + 1)$ or $(T^2 s^2 + 2\xi Ts + 1)$. Carrying out this procedure on (5-106) leads to the expression

$$H(s) = \frac{0.4(0.25s + 1)}{(s + 1)(0.01s^2 + 0.05s + 1)} \tag{5-107}$$

This can be further simplified by assigning an amplitude and phase to each factor, giving

$$H(s) = \frac{0.4 A_4(s)\underline{/\theta_4(s)}}{A_1(s)\underline{/\theta_1(s)}A_{10}(s)\underline{/\theta_{10}(s)}} \tag{5-108}$$

where the subscripts refer to the *critical frequency* corresponding to that factor, that is, the angular frequency ω, at which the magnitude of $T\omega = 1$. The amplitude spectrum can now be written as

$$A(\omega)_{dB} = 20 \log \frac{0.4 A_4(\omega)}{A_1(\omega)A_{10}(\omega)} \tag{5-109}$$

$$= 20 \log 0.4 + 20 \log A_4(\omega) - 20 \log A_1(\omega) - 20 \log A_{10}(\omega) \tag{5-110}$$

Thus, by expressing the amplitude of each factor in decibels, the overall amplitude is found from their algebraic sum. The constant factor is the low frequency gain and in this case is $20 \log 0.4 = -8$ dB. The linear factors contribute in a manner similar to that of the previous example. A normalized plot of $A(\omega)$ and $\theta(\omega)$ for $H(s) = (Ts + 1)^{-1}$ is shown in Fig. 5-25. If the factor is in the numerator instead of the denominator, all that

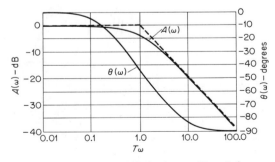

Fig. 5-25 Response of the factor $(Ts + 1)^{-1}$.

is required is to change the signs of $A(\omega)$ and $\theta(\omega)$ from negative to positive and use the same magnitude as read from Fig. 5-25. Some easy-to-remember numbers are that at the critical frequency the departure from the asymptotes is 3 dB, at one-half and twice the critical frequency the departure is 1 dB, and at 0.1 and 10 times the critical frequency the departure is essentially 0. Below the critical frequency the slope of the asymptote is zero, and above the critical frequency it is $-20\,\mathrm{dB}$ per decade, or 6 dB per octave.

The factor $(T^2 s^2 + 2\xi Ts + 1)^{-1}$ is handled similarly, but it represents a resonant response in which the damping is determined by ξ. The normalized response for this factor is shown in Fig. 5-26. As before, the contribu-

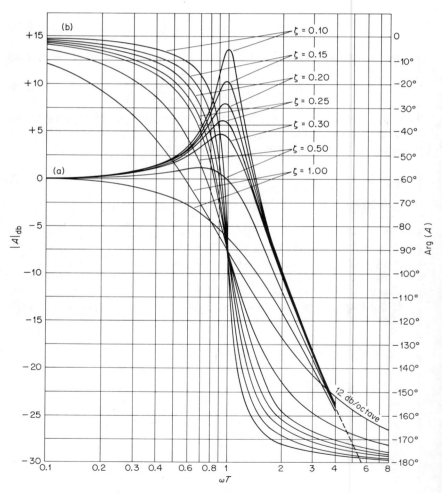

Fig. 5-26 Frequency response for the factor $(T^2 s^2 + 2\zeta Ts + 1)^{-1}$. (a) Gain curves. (b) Phase-shift curves.

tion of this term at low frequencies is 0 dB; however, at high frequencies the asymptote falls off at 40 dB per decade, or 12 dB per octave, as would be expected from the asymptotic form of $A(\omega)$. When such a term is in the numerator, the signs of $A(\omega)$ and $\theta(\omega)$ are reversed as before.

The procedure for plotting $A(\omega)$ is as follows: Determine the critical frequencies, sketch in the asymptotes, and correct the curve away from the asymptotes by the deviations of each factor from its asymptotic value as determined from Fig. 5-25 or Fig. 5-26. The result of carrying out this procedure for the system function given in (5-106) is shown in Fig. 5-27.

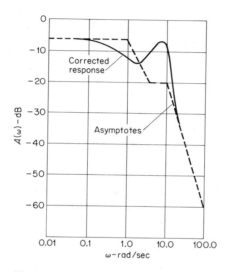

Fig. 5-27 Amplitude spectrum of (5-106).

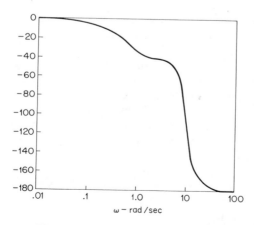

Fig. 5-28 Phase spectrum of (5-106).

The phase spectrum $\theta(\omega)$ is determined by adding the phases of each factor at each frequency. The validity of this procedure is evident from (5-108), when use is made of the fact that when multiplying phasors the phase angles add, and when dividing phasors the phase angles subtract. The subtraction of phase angles is accomplished in the above procedure by using negative phase angles for denominator factors. Figures 5-25 and 5-26 give the phase values for linear and quadratic factors. Using data from these curves, the overall phase of $H(\omega)$ in the example just considered is shown in Fig. 5-28.

Exercise 5-10.1
Find the transfer functions $H_1(\omega) = V_0(\omega)/V_i(\omega)$ and $H_2(\omega) = V_R(\omega)/V_i(\omega)$ for the circuit shown.

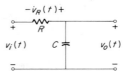

Exercise 5-10.2
Sketch the amplitude and phase spectrum of $H(\omega) = j\omega/(j\omega + 1)(j\omega + 4)$.

5-11 Ideal Filters

As is evident from Fig. 5-25, a system function of the form $H(\omega) = a/j\omega + a$ has a constant amplitude at low frequencies and falls off at higher frequencies. Another way of saying this is that the *attenuation* of signals at low frequencies is low (0 in this case) and increases at high frequencies. The attenuation is the fractional loss in signal amplitude that occurs when a signal is transmitted through a system and is usually measured in decibels. A circuit having the above type of frequency response is called a *lowpass filter*. There are many other kinds of filters, including bandpass, bandstop, highpass, and equalizing filters. Often it is convenient to consider certain idealized forms of filter characteristics. The amplitude responses of two frequently used idealizations are shown in Fig. 5-29. The ideal lowpass filter is characterized by a system function whose magnitude is constant over the angular frequency band, $-2\pi W \leq \omega \leq 2\pi W$, where W is the bandwidth of the filter measured in Hz and $2\pi W$ is the bandwidth of the filter measured in radians per second. The amplitude of the system function is referred to as the *gain* of the system and is unity for the ideal filter shown.

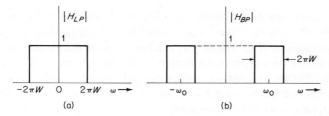

Fig. 5-29 Ideal filter characteristics. (a) Lowpass filter. (b) Bandpass filter.

The *ideal bandpass filter* is characterized by unity gain over a band of frequencies of width $2\pi W$ radians per second symmetrically located about the center angular frequencies $\pm\omega_0$ as shown.

Both the ideal lowpass and ideal bandpass filters have the property of passing all frequency components lying within their passbands and completely attenuating all other frequency components. It is worth noting that systems having characteristics like an ideal filter are not physically realizable. That this is the case may be seen readily by applying the Paley–Wiener criterion to $H(j\omega)$. Inserting $H(j\omega)$ into (5-99) gives the requirement

$$\int_{-\infty}^{\infty} \frac{|\ln|H(\omega)||}{1+\omega^2} \, d\omega < \infty$$

Since all ideal filters have frequency bands over which $|H(j\omega)|$ is 0, the logarithm is infinite, and the criterion is not met. In a later section the theoretical responses of ideal filters to various input signals are considered, and it is shown that for such systems there is an output prior to application of the input. As a practical matter, a given amplitude spectrum can be approximated arbitrarily closely over any finite band of frequencies with a real filter provided sufficient time delay (linear phase shift) is permitted. If the Paley–Wiener criterion is satisfied, an exact synthesis can be accomplished (theoretically) with a finite time delay. If the Paley–Wiener criterion is not satisfied, an exact synthesis requires an infinite time delay, but an approximation can be obtained with finite time delay.

For purposes of describing filter characteristics, and in fact, Fourier transforms in general, it is often convenient to talk in terms of frequency, f, measured in Hz, rather than angular frequency, ω, measured in radians per second. This can be accomplished readily merely by substituting $2\pi f$ for ω in the Fourier transforms being considered. For example, consider the system function of the simple R–L lowpass filter shown in Fig. 5-21. We can write this in equivalent forms as

$$H(\omega) = \frac{R/L}{j\omega + R/L}$$

$$H(f) = \frac{R/L}{j2\pi f + (R/L)} = \frac{R/2\pi L}{jf + (R/2\pi L)}$$

Here we have used a somewhat simplified notation. To be correct mathematically, the left-hand side should be $H(2\pi f)$, since it is obtained by substituting $\omega = 2\pi f$. However, for the sake of convenience and simplified notation, we suppress the factor 2π in the argument of the symbolic form of the Fourier transform when using f as the variable. This should not result in confusion, once it is realized what is being done; in fact, this type of representation is commonly encountered in the technical literature. When computations are carried out, there will be no possibility of confusion if either $F(\omega)$ or $F(f)$ is used consistently. If it is necessary to change from one variable to the other, then all that is required is to make the substitution $f = \omega/2\pi$ into the algebraic expression of $F(f)$ to get $F(\omega)$ or to make the substitution $\omega = 2\pi f$ into the algebraic expression for $F(\omega)$ to get $F(f)$.

Using this notation, the ideal lowpass and bandpass filters corresponding to Fig. 5-29 can be specified in terms of frequency f, as shown in Fig. 5-30. Further examples of this notation are given in later sections.

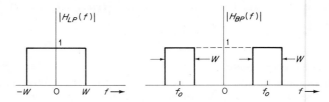

Fig. 5-30 Ideal filter characteristic as functions of f.

5-12 Energy Spectrum

In developing the Fourier transform concept, we have discussed the idea of amplitude and phase spectra, which relate to the relative magnitudes and phases of the infinitesimal signals making up a continuum of complex sinusoids representing the original signal. The idea of signal spectra can be put on a more intuitively satisfying basis by considering the distribution of signal energy as a function of frequency. The required relationship is found by expressing the energy in the time domain and then writing an equivalent expression using inverse transforms of the frequency-domain representation of the time function. As discussed in Chapter 3, the instantaneous power in a signal is generally taken as the square of the signal amplitude. Thus, for the signal, $f(t)$, the instantaneous signal power is $[f(t)]^2$. Actually, this is a true power for only certain situations. For example, if the signal were a voltage across a resistance R, the true instantaneous power would be $(1/R)[f(t)]^2$. If $R = 1\,\Omega$, then $[f(t)]^2$ is the true power. However, if R is something different from one ohm, then $[f(t)]^2$ is only proportional to true power, with the constant of proportionality being $1/R$. A similar situation occurs for the case in which $f(t)$ is a current. The power is always proportional to the square of the signal amplitude, but the constant of proportionality varies with the particular resistance level in the circuit. In order to avoid carrying along extra constants in computation, it is customary in signal analysis to consider that the units of power are the squares of the units of the signals involved. For a signal voltage the power is, therefore, measured in volts squared and energy measured in volts squared-seconds. This is equivalent to assuming that the signal voltage is measured across a one-ohm resistance. A similar relationship is used for signals measured in other units. Little confusion results if the units are included, and talking about power is greatly simplified when this convention is employed.

Using this convention, the energy in a signal $f(t)$ is given as

$$E = \int_{-\infty}^{\infty} [f(t)]^2 \, dt$$

Letting $F(\omega)$ be the Fourier transform of $f(t)$, this can be written as

$$\int_{-\infty}^{\infty} [f(t)]^2 \, dt = \int_{-\infty}^{\infty} f(t) \left[\frac{1}{2\pi} \int_{-\infty}^{\infty} F(\omega) \epsilon^{j\omega t} d\omega \right] dt \qquad \text{(5-111)}$$

Interchanging the order of integration on the right-hand side of (5-111) and rearranging gives

$$\int_{-\infty}^{\infty} [f(t)]^2 \, dt = \frac{1}{2\pi} \int_{-\infty}^{\infty} F(\omega) \left[\int_{-\infty}^{\infty} f(t) \epsilon^{j\omega t} dt \right] d\omega \qquad \text{(5-112)}$$

The factor in this integrand that is enclosed in brackets is seen to be $F(-\omega)$. Therefore, (5-112) can be written as

$$\int_{-\infty}^{\infty} [f(t)]^2 dt = \frac{1}{2\pi} \int_{-\infty}^{\infty} F(\omega) F(-\omega) \, d\omega \qquad \text{(5-113)}$$

This can be put into a somewhat simpler form by noting from the definition of the Fourier transform that if $f(t)$ is real (that is, if it has no imaginary part), then $F(-\omega) = F^*(\omega)$, the complex conjugate of $F(\omega)$. Using this relationship, (5-113) can be written as

$$\int_{-\infty}^{\infty} [f(t)]^2 dt = \frac{1}{2\pi} \int_{-\infty}^{\infty} F(\omega) F^*(\omega) d\omega$$

$$= \frac{1}{2\pi} \int_{-\infty}^{\infty} |F(\omega)|^2 d\omega \qquad \text{(5-114)}$$

Expressed in terms of the frequency f, this expression becomes

$$\int_{-\infty}^{\infty} [f(t)]^2 dt = \int_{-\infty}^{\infty} |F(f)|^2 df \qquad \text{(5-115)}$$

The relationship in (5-114), which is a fundamental property of Fourier transforms, is called *Parseval's theorem*. In words, Parseval's theorem states that the energy in the signal $f(t)$ is equal to $1/2\pi$ times the area under the square of the magnitude of the Fourier transform of $f(t)$. The quantity $|F(\omega)|^2$ is called the *energy spectrum*, or energy-density spectrum, of $f(t)$ since, from (5-115), it can be interpreted as the distribution of energy with frequency.[6] The units of $|F(\omega)|^2$ are dependent on the units of $f(t)$; for example, if $f(t)$ were a voltage, then $|F(\omega)|^2$ would have units of volts squared-seconds per Hz.

A greater understanding of the significance of the energy spectrum is obtained by considering how the system function of a linear system alters the spectrum of a signal transmitted through the system. It was previously shown that, for the case of a simple system having a system function $H(\omega)$, the output and input are related by

$$Y(\omega) = H(\omega)X(\omega)$$

The energy spectrum of the output is then found to be

$$|Y(\omega)|^2 = Y(\omega)Y^*(\omega)$$

$$= [H(\omega)X(\omega)][H^*(\omega)X^*(\omega)] \tag{5-116}$$

$$= |H(\omega)|^2|X(\omega)|^2$$

From (5-116) it is seen that the output signal energy spectrum is related to the input signal energy spectrum by the quantity $|H(\omega)|^2$. Because of this relationship, $|H(\omega)|^2$ is sometimes called the *energy transfer function* of the system.

Using the energy transfer function concept, it is possible to obtain a physical interpretation of the energy spectrum. Suppose that a signal having an arbitrary energy spectrum is passed through an ideal bandpass filter having a narrow passband centered at a frequency f_1. The energy transfer function of the filter will be unity for those components lying in the filter passband and will be zero for all other components. The energy spectrum of the output will therefore be just that portion of the energy spectrum of the input corresponding to the frequencies in the filter passband. Figure 5-31 shows a typical example of such an arrangement. The total energy of the output will be given by

$$E_o = \int_{-\infty}^{\infty} |V_o(f)|^2 df = 2 \int_{f_1-W/2}^{f_1+W/2} |V_i(f)|^2 df \tag{5-117}$$

For a sufficiently narrow filter bandpass (narrow enough so that the input spectrum is essentially constant over the band), the output can be approximated as

$$E_o \doteq 2W|V_i(f_1)|^2$$

---

[6]More precisely, $|F(\omega)|^2(\Delta\omega/2\pi)$ is the signal energy contained in the differential frequency band from ω to $\omega + \Delta\omega$, and $|F(f)|^2\Delta f$ is the energy contained in the frequency band from f to $f + \Delta f$.

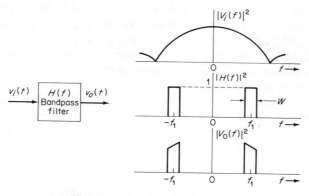

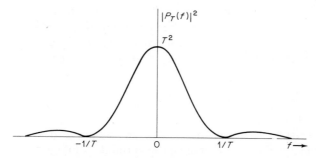

Fig. 5-31 Measurement of energy spectrum.

Solving for $|V_i(f_1)|^2$ gives

$$|V_i(f_1)|^2 \doteq \frac{E_o}{2W}$$

From this expression it is evident that $|V_i(f)|^2$ can be interpreted as the energy per unit bandwidth.

As an example of the energy spectrum of a signal, consider the pulse signal, $p_T(t)$, which was discussed earlier. From (5-58) we have

$$|P_T(\omega)|^2 = T^2 \left(\frac{\sin \omega T/2}{\omega T/2} \right)^2 \tag{5-118}$$

Changing from ω to $f = \omega/2\pi$ gives the energy spectrum as

$$|P_T(f)|^2 = T^2 \left(\frac{\sin \pi Tf}{\pi Tf} \right)^2 = T^2 \text{sinc}^2 (fT) \tag{5-119}$$

The energy spectrum, $|P_T(f)|^2$, of the rectangular pulse is shown in Fig. 5-32. It is seen that the energy is concentrated in the low-frequency portion of the spectrum. The extent of this concentration can be found by

Fig. 5-32 Energy spectrum of rectangular pulse.

computing the energy in the first loop (that is, for $|f| < 1/T$) and comparing this to the total energy. The ratio, found by graphical integration, is 0.902. Thus, 90.2 percent of the energy in a rectangular pulse is contained in the band of frequencies below a frequency equal to reciprocal of the pulse length. As a useful rule of thumb, it is often assumed that a pulse transmission system having a bandwidth equal to the reciprocal of the pulse width will perform satisfactorily. Actually, if high-fidelity reproduction of the pulse shape is required, a much greater bandwidth will be necessary. However, it can be seen that a system with this bandwidth will transmit most of the pulse energy.

Exercise 5-12.1
A time function of the form $v(t) = \epsilon^{-t}u(t)$ is applied to an ideal lowpass filter having a cutoff frequency of $1/2\pi$ Hz. Find the fraction of the signal energy that is transmitted by the filter.

ANS: 50 percent.

5-13   Fourier Transforms corresponding to Mathematical Operations

Certain mathematical operations performed on time functions give rise to modified functions whose Fourier transforms can be obtained by performing suitable mathematical operations on the Fourier transform of the original function. By studying some of the more frequently occurring operations and establishing the corresponding operations in the domain of the image function, great flexibility and power are added to the Fourier transform technique. An example that has already been considered is differentiation in the time domain, which was found to correspond to multiplication by $j\omega$ in the frequency domain. A number of additional operations will now be considered. In all cases it is assumed that $F(\omega)$ is the Fourier transform of the unmodified function $f(t)$.

Scaling.   By scaling is meant multiplication of the variable in the time function by a constant. This has the effect of expanding or contracting the time scale, depending on whether the magnitude of the constant is less than or greater than unity. If the constant is negative, the time scale is reversed. The Fourier transform of a scaled-time function can be obtained as follows:

$$\mathscr{F}\{f(at)\} = \int_{-\infty}^{\infty} f(at)\epsilon^{-j\omega t}dt$$

We must consider the case of positive and negative a separately. Consider first the positive a and change the variable of integration to $\lambda = at$.

$$\mathcal{F}\{f(at)\} = \frac{1}{a}\int_{-\infty}^{\infty} f(\lambda)\epsilon^{-j\omega\lambda/a}d\lambda$$

$$= \frac{1}{a}F\left(\frac{\omega}{a}\right) \qquad a > 0 \tag{5-120}$$

When a is negative, the limits on the integral will be reversed when the variable of integration is changed; the final result is

$$\mathcal{F}\{f(at)\} = -\frac{1}{a}F\left(\frac{\omega}{a}\right) \qquad a < 0 \tag{5-121}$$

These two results can be combined to give

$$f(at) \Leftrightarrow \frac{1}{|a|}F\left(\frac{\omega}{a}\right) \tag{5-122}$$

The relationship is useful in extending a table of transforms of normalized functions to cover more general cases. It also illustrates once more the property that, as we expand the time scale of a function, its frequency spectrum is contracted. In addition, it is seen that the amplitude of the spectrum also changes. This latter effect is necessary to maintain an energy balance between the two domains.

Delay. When a new variable $t - t_o$ is substituted for the original variable t in a time function, the resulting function is an exact replica of the original function delayed by an amount t_o. The Fourier transform of the modified function is

$$\mathcal{F}\{f(t - t_o)\} = \int_{-\infty}^{\infty} f(t - t_o)\epsilon^{-j\omega t}dt$$

Changing the variable of integration and carrying out the indicated operations gives

$$\mathcal{F}\{f(t - t_o)\} = \int_{-\infty}^{\infty} f(\lambda)\epsilon^{-j\omega(\lambda + t_o)}d\lambda$$

$$= \epsilon^{-j\omega t_o}\int_{-\infty}^{\infty} f(\lambda)\epsilon^{-j\omega\lambda}d\lambda \tag{5-123}$$

$$= \epsilon^{-j\omega t_o}F(\omega)$$

$$f(t - t_o) \Leftrightarrow \epsilon^{-j\omega t_o}F(\omega)$$

Delay in the time domain is, thus, seen to correspond to introduction of a phase shift in the frequency domain that varies linearly with frequency. As an example of the use of (5-123), consider the rectangular pulse shown in Fig. 5-33. This signal is identical to the pulse signal, $p_T(t)$, previously considered, except that it is delayed by an amount $T/2$. Therefore, $f(t) = p_T(t - T/2)$.

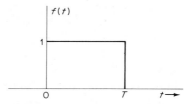

Fig. 5-33 Symmetrical pulse signal.

The corresponding transform is then found from (5-123) and (5-58) to be

$$F(\omega) = \epsilon^{-j\omega T/2} P_T(\omega)$$
$$= T\epsilon^{-j\omega T/2}\frac{\sin \omega T/2}{\omega T/2} \tag{5-124}$$

Expressed in terms of the frequency f, this can be stated as

$$F(f) = T\epsilon^{-j\pi Tf}\frac{\sin \pi Tf}{\pi Tf} = T\epsilon^{-j\pi fT} \text{ sinc } (Tf) \tag{5-125}$$

Modulation. Multiplication of a time function by the imaginary exponential $\epsilon^{j\omega_o t}$ causes a translation in the frequency domain. Thus,

$$\mathscr{F}\{\epsilon^{j\omega_o t}f(t)\} = \int_{-\infty}^{\infty} f(t)\epsilon^{-j(\omega-\omega_0)t}dt$$
$$= F(\omega - \omega_o) \tag{5-126}$$
$$\epsilon^{j\omega_o t}f(t) \Leftrightarrow F(\omega - \omega_o) \qquad \text{or} \qquad F(f - f_o)$$

The relationship in (5-126) can be thought of as a process in which the complex sinusoid is modulated by the time function, $f(t)$. If instead of $\epsilon^{j\omega_o t}$, we consider the real function, $\cos \omega_o t$, we obtain the following relationship:

$$\mathscr{F}\{f(t) \cos \omega_o t\} = \int_{-\infty}^{\infty} f(t)\frac{\epsilon^{j\omega_o t} + \epsilon^{-j\omega_o t}}{2}\epsilon^{-j\omega t}dt$$
$$f(t) \cos \omega_o t \Leftrightarrow \tfrac{1}{2}F(\omega - \omega_o) + \tfrac{1}{2}F(\omega + \omega_o) \tag{5-127}$$
$$\Leftrightarrow \tfrac{1}{2}F(f - f_o) + \tfrac{1}{2}F(f + f_o)$$

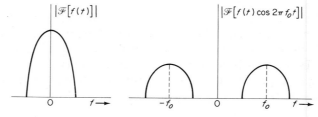

Fig. 5-34 Spectrum of modulated cosine wave.

Thus, modulation of a cosine wave by a time function $f(t)$ leads to a new function having a spectrum consisting of half the original spectrum translated along the positive frequency axis by an amount f_o and half the original spectrum translated along the negative frequency axis by the amount $-f_o$. An example of this process is shown in Fig. 5-34.

Reversal.   When a time function is reflected about the origin, the corresponding spectrum is also reflected about the origin. In equation form this is

$$f(-t) \Leftrightarrow F(-\omega) \qquad (5\text{-}128)$$

This relationship follows immediately from the expression derived for scaling, where the scale factor is -1. As an example of (5-128), consider the exponential pulse $f_1(t)$, shown in Fig. 5-35(a). This pulse can be thought of as the reflection of the exponential pulse in Fig. 5-35(b). The transform of the pulse in Fig. 5-35(b) is simply $\epsilon^{j\omega t_o}$ times the transform of a unit amplitude exponential pulse at the origin which is given in (5-66).

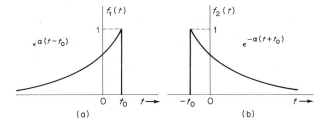

Fig. 5-35 Reflection of time functions.

The required transform is, therefore,

$$\mathscr{F}\{f_1(t)\} = \epsilon^{j\omega t_0} \frac{1}{j\omega + \alpha} \Big|_{\omega = -\omega}$$

$$= \frac{-\epsilon^{-j\omega t_0}}{j\omega - \alpha} \qquad (5\text{-}129)$$

Symmetry.   Because of the similarity between the integrals defining the Fourier transform and the inverse Fourier transform, there is a very close relationship between the transform of a particular function of t and the inverse transform of that same function of ω. The precise relationship is

$$F(t) \Leftrightarrow 2\pi f(-\omega) \qquad (5\text{-}130)$$

$$\frac{1}{2\pi} f(-t) \Leftrightarrow f(\omega) \qquad (5\text{-}131)$$

The corresponding relationship when the Fourier transform variable is f is somewhat simpler and is given by

$$F(t) \Leftrightarrow f(-f) \qquad (5\text{-}132)$$

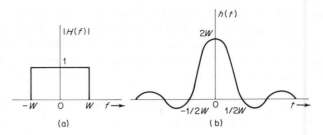

Fig. 5-36 System function and impulse response of ideal lowpass filter.

These relationships can be used to extend tables of transforms and also to gain further insight into corresponding time and frequency representation of signals. As an example of this relationship, consider the time function corresponding to a transform that is constant over a specified frequency band and zero elsewhere. This situation is illustrated in Fig. 5-36(a), in which the shape of the transform is assumed to be a rectangular spectrum of width $2W$ Hz and unity amplitude. The phase function is assumed to be zero. Such a transform corresponds to the transfer function, $H(f)$, of an ideal lowpass filter. Accordingly, the inverse transform will be the impulse response, $h(t)$, of such a filter. The inverse transform can be found directly from (5-131) by using the previously established transform for a pulse signal; that is,

$$p_T(t) \Leftrightarrow T \frac{\sin \pi Tf}{\pi Tf} = T \text{ sinc } Tf \tag{5-133}$$

In the present instance we can write the frequency function as

$$H(f) = p_{2w}(f)$$

Therefore, the corresponding inverse transform is

$$h(t) = 2W \text{ sinc } (2Wt) \tag{5-134}$$

This impulse response is shown in Fig. 5-36(b). It is clear from the figure that this is not a physically realizable system, because the output occurs prior to application of the input. One way of approximating the ideal filter response is to employ a system having a response similar in form to that of Fig. 5-36(b) but delayed in time. The greater the delay, the more nearly the shape of Fig. 5-36(b) can be reproduced by a physically realizable filter. The effect of the delay in the frequency domain is to produce a phase shift which varies linearly with frequency.

Convolution in the frequency domain.   It is shown in Section 5-10 that convolution in the time domain corresponds to multiplication in the frequency domain. From the symmetry properties of the Fourier transform, it follows that convolution of two transforms in the frequency domain corresponds to multiplication of the original functions in the time domain.

The exact relationship is as follows:

$$f_1(t)f_2(t) \Leftrightarrow \frac{1}{2\pi} \int_{-\infty}^{\infty} F_1(\xi)F_2(\omega - \xi)d\xi = \frac{1}{2\pi} F_1(\omega) * F_2(\omega) \qquad \text{(5-135)}$$

$$f_1(t)f_2(t) \Leftrightarrow F_1(f) * F_2(f) \qquad\qquad\qquad\qquad\qquad \text{(5-136)}$$

These expressions can be verified formally by inverting the right-hand side. The relationships given in (5-135) and (5-136) are particularly useful in the study of modulation processes when one of the time functions has a transform consisting of impulses. Examples of this will be considered in later sections.

Table 5-3   Fourier Transforms of Mathematical Operations

Operation	$f(t)$	$F(\omega)$
Transformation	$f(t)$	$\int_{-\infty}^{\infty} f(t)\epsilon^{-j\omega t}dt$
Inversion	$\frac{1}{2\pi} \int_{-\infty}^{\infty} F(\omega)\epsilon^{j\omega t}d\omega$	$F(\omega)$
Superposition	$a_1f_1(t) + a_2f_2(t)$	$a_1F_1(\omega) + a_2F_2(\omega)$
Reversal	$f(-t)$	$F(-\omega)$
Symmetry	$F(t)$	$2\pi f(-\omega)$
Scaling	$f(at)$	$\frac{1}{|a|}F\left(\frac{\omega}{a}\right)$
Delay	$f(t - t_0)$	$\epsilon^{-j\omega t_0}F(\omega)$
Modulation	$\epsilon^{j\omega_0 t}f(t)$	$F(\omega - \omega_0)$
Time differentiation	$\frac{d^n}{dt^n}f(t)$	$(j\omega)^n F(\omega)$
Frequency differentiation	$t^n f(t)$	$(j)^n \frac{d^n}{d\omega^n} F(\omega)$
Integration	$\int_0^t f_e(t)dt + \int_{-\infty}^t f_o(t)dt$	$\frac{1}{j\omega}F(\omega)$
Integration	$\int_{-\infty}^t \cdot f(t)dt$	$\frac{1}{j\omega}F(\omega) + \pi F(0)\delta(\omega)$
Convolution	$f_1 * f_2 = \int_{-\infty}^{\infty} f_1(\lambda)f_2(t - \lambda)d\lambda$	$F_1(\omega)F_2(\omega)$
Multiplication	$f_1(t)f_2(t)$	$\frac{1}{2\pi} \int_{-\infty}^{\infty} F_1(\xi)F_2(\omega - \xi)d\xi$

A collection of the most frequently used Fourier transforms of operations is given in Table 5-3. In Table 5-4 are listed a number of elementary transform pairs. Additional transform pairs are given in Table 5-5 and relate to signals having nonzero average power. More extensive tabulations of Fourier transforms are available in the references listed at the end of the chapter. A short table of Fourier transforms using the variable f instead of the variable ω is included in Appendix A.

Table 5-4 Fourier Transforms of Energy Signals

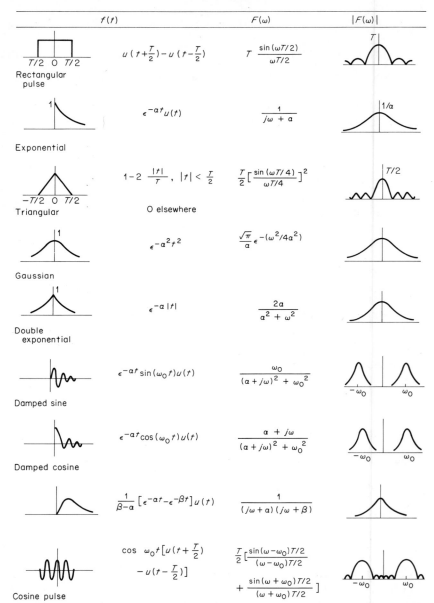

$f(t)$		$F(\omega)$	$|F(\omega)|$
Rectangular pulse	$u\left(t+\dfrac{T}{2}\right)-u\left(t-\dfrac{T}{2}\right)$	$T\,\dfrac{\sin(\omega T/2)}{\omega T/2}$	
Exponential	$\epsilon^{-at}u(t)$	$\dfrac{1}{j\omega+a}$	
Triangular	$1-2\,\dfrac{|t|}{T},\quad |t|<\dfrac{T}{2}$ O elsewhere	$\dfrac{T}{2}\left[\dfrac{\sin(\omega T/4)}{\omega T/4}\right]^2$	
Gaussian	$\epsilon^{-a^2t^2}$	$\dfrac{\sqrt{\pi}}{a}\epsilon^{-(\omega^2/4a^2)}$	
Double exponential	$\epsilon^{-a|t|}$	$\dfrac{2a}{a^2+\omega^2}$	
Damped sine	$\epsilon^{-at}\sin(\omega_0 t)u(t)$	$\dfrac{\omega_0}{(a+j\omega)^2+\omega_0^2}$	
Damped cosine	$\epsilon^{-at}\cos(\omega_0 t)u(t)$	$\dfrac{a+j\omega}{(a+j\omega)^2+\omega_0^2}$	
	$\dfrac{1}{\beta-a}\left[\epsilon^{-at}-\epsilon^{-\beta t}\right]u(t)$	$\dfrac{1}{(j\omega+a)(j\omega+\beta)}$	
Cosine pulse	$\cos\,\omega_0 t\left[u\left(t+\dfrac{T}{2}\right)-u\left(t-\dfrac{T}{2}\right)\right]$	$\dfrac{T}{2}\left[\dfrac{\sin(\omega-\omega_0)T/2}{(\omega-\omega_0)T/2}+\dfrac{\sin(\omega+\omega_0)T/2}{(\omega+\omega_0)T/2}\right]$	

Table 5-5   Fourier Transforms of Power Signals

$f(t)$		$F(\omega)$	$[F(\omega)]$
Unit impulse	$\delta(t)$	1	
Unit step	$u(t)$	$\pi\delta(\omega) + \dfrac{1}{j\omega}$	
Signum function	$\operatorname{sgn} t = \dfrac{t}{|t|}$	$\dfrac{2}{j\omega}$	
Constant	K	$2\pi K\delta(\omega)$	$2\pi K$
Cosine wave	$\cos \omega_0 t$	$\pi[\delta(\omega-\omega_0) + \delta(\omega+\omega_0)]$	
Sine wave	$\sin \omega_0 t$	$-j\pi[\delta(\omega-\omega_0) - \delta(\omega+\omega_0)]$	
Periodic wave	$\displaystyle\sum_{n=-\infty}^{\infty} f_T(t-nT)$	$\dfrac{2\pi}{T}\displaystyle\sum_{n=-\infty}^{\infty} F_T\left(\dfrac{2\pi n}{T}\right)\delta\left(\omega - \dfrac{2\pi n}{T}\right)$	
Impulse train	$\sum\delta(t-nT)$	$\dfrac{2\pi}{T}\sum\delta\left(\omega - \dfrac{2\pi n}{T}\right)$	
Complex sinusoid	$\epsilon^{j\omega_0 t}$	$2\pi\delta(\omega-\omega_0)$	2π
Unit ramp	$tu(t)$	$j\pi\delta'(\omega) - \dfrac{1}{\omega^2}$	

5-14   Fourier Transforms of Power Signals

The ordinary Fourier transform is limited to the transformation of functions that are absolutely integrable—that is, functions that obey the inequality

$$\int_{-\infty}^{\infty} |f(t)|\,dt < \infty$$

A number of functions having great usefulness do not meet this requirement; for example, a sine wave or a step function does not. Many such functions can, nevertheless, be handled by allowing the Fourier transform to contain impulses or, in some cases, higher-order singularity functions. This procedure can be put on a rigorous mathematical basis by means of the theory of generalized functions[7]; however, it will be sufficient for our purposes to justify this approach by considering the impulse as a limiting form of a proper function and by showing that correct results are obtained when this method is used.

Consider the function sgn (t), called signum t, which is defined as

$$\begin{aligned}
\text{sgn}\,(t) &= -1 & t &< 0 \\
&= 0 & t &= 0 \\
&= +1 & t &> 0
\end{aligned}$$

(5-137)

This function is shown in Fig. 5-37. It is evident that it has a zero average value and is not absolutely integrable. The Fourier transform of this function cannot be computed in the formal manner, since the integral does not exist. Consider instead a sequence of functions that approaches sgn (t) as a limit. Such a sequence can be obtained by introducing a suitable convergence factor multiplying sgn (t). A suitable function is $\epsilon^{-\alpha|t|}$ sgn (t).

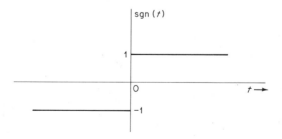

Fig. 5-37 Signum function, sgn (t).

The transform may now be computed as

$$\mathscr{F}\{\text{sgn}\,(t)\} = \mathscr{F}\{\lim_{\alpha \to 0} \epsilon^{-\alpha|t|}\,\text{sgn}\,(t)\}$$

[7]A. H. Zemanian, *Distribution Theory and Transform Analysis*, McGraw-Hill Book Company, Inc., New York, 1965.

Interchanging the limiting and integration operations gives[8]

$$\mathcal{F}\{\text{sgn}\,(t)\} = \lim_{\alpha \to 0} \int_{-\infty}^{\infty} \epsilon^{-\alpha|t|}\,\text{sgn}\,(t)\epsilon^{-j\omega t}dt$$

$$= \lim_{\alpha \to 0} \left[\int_{-\infty}^{0} -\epsilon^{(\alpha - j\omega)t}dt + \int_{0}^{\infty} \epsilon^{-(\alpha + j\omega)t}dt \right]$$

$$= \lim_{\alpha \to 0} \left[\left. \frac{-\epsilon^{(\alpha - j\omega)t}}{\alpha - j\omega} \right|_{-\infty}^{0} + \left. \frac{\epsilon^{-(\alpha + j\omega)t}}{-(\alpha + j\omega)} \right|_{0}^{\infty} \right] \qquad \text{(5-138)}$$

$$= \lim_{\alpha \to 0} \left(\frac{-1}{\alpha - j\omega} + \frac{1}{\alpha + j\omega} \right)$$

$$\text{sgn}\,(t) \Leftrightarrow \frac{2}{j\omega}$$

The amplitude of the spectrum is shown plotted in Fig. 5-38.

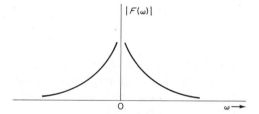

Fig. 5-38 Amplitude spectrum of sgn (t).

The spectrum of a constant can be found in a similar fashion using the same type of convergence factor:

$$\mathcal{F}\{1\} = \lim_{\alpha \to 0} \int_{-\infty}^{\infty} \epsilon^{-\alpha|t|}\epsilon^{-j\omega t}dt$$

$$= \lim_{\alpha \to 0} \left[\int_{-\infty}^{0} \epsilon^{(\alpha - j\omega)t}dt + \int_{0}^{\infty} \epsilon^{-(\alpha + j\omega)t}dt \right]$$

$$= \lim_{\alpha \to 0} \left[\left. \frac{\epsilon^{(\alpha - j\omega)t}}{\alpha - j\omega} \right|_{-\infty}^{0} - \left. \frac{\epsilon^{-(\alpha + j\omega)t}}{\alpha + j\omega} \right|_{0}^{\infty} \right] \qquad \text{(5-139)}$$

$$= \lim_{\alpha \to 0} \left(\frac{1}{\alpha - j\omega} + \frac{1}{\alpha + j\omega} \right)$$

$$= \lim_{\alpha \to 0} \left(\frac{2\alpha}{\alpha^2 + \omega^2} \right)$$

---

[8]This interchange of operations is the source of trouble in justifying mathematically a derivation of this type. Mathematical proof of the validity of the results obtained by this method of derivation is well beyond the assumed mathematical level of the readers. The best justification that can be offered to the reader is that correct results are obtained when transforms obtained in this manner are used.

It is seen that the limit as $\alpha \to 0$ in (5-139) is 0 except when $\omega = 0$, for which case an indeterminate form is obtained. The indeterminate form can be evaluated by L'Hospital's rule to give

$$\lim_{\alpha \to 0} \left(\frac{2}{2\alpha}\right) = \infty \tag{5-140}$$

The area under the function is found to be

$$\text{area} = \int_{-\infty}^{\infty} \frac{2\alpha}{\alpha^2 + \omega^2} d\omega = 2 \tan^{-1}\left(\frac{\omega}{\alpha}\right)\Big|_{-\infty}^{\infty}$$

$$= 2\left(\frac{\pi}{2} + \frac{\pi}{2}\right) = 2\pi \tag{5-141}$$

The results of (5-140) and (5-141) can be interpreted as corresponding to an impulse at the origin having a strength of 2π. Therefore, the Fourier transform of a constant can be written as

$$\mathscr{F}\{1\} = 2\pi\delta(\omega) \tag{5-142}$$

$$1 \Leftrightarrow 2\pi\delta(\omega) \qquad \text{or} \qquad 1 \Leftrightarrow \delta(f) \tag{5-143}$$

This result can be extended to include sinusoidal functions in the following manner: By considering the imaginary exponential $\epsilon^{j\omega_0 t}$ to be the product of $(1)\epsilon^{j\omega_0 t}$ and invoking the modulation theorem of the Fourier transform, we can obtain its transform as

$$\mathscr{F}\{\epsilon^{j\omega_0 t}\} = \mathscr{F}\{1 \cdot \epsilon^{j\omega_0 t}\} = 2\pi\delta(\omega - \omega_o) \tag{5-144}$$

From (5-144) the transform of the sine and cosine functions are found to be

$$\mathscr{F}\{\cos \omega_o t\} = \mathscr{F}\left\{\frac{\epsilon^{j\omega_o t} + \epsilon^{-j\omega_o t}}{2}\right\}$$

$$\cos \omega_o t \Leftrightarrow \pi[\delta(\omega - \omega_o) + \delta(\omega + \omega_o)] \tag{5-145}$$

or

$$\cos \omega_0 t \Leftrightarrow \frac{1}{2}[\delta(f - f_o) + \delta(f + f_o)]$$

$$\sin \omega_o t \Leftrightarrow -j\pi[\delta(\omega - \omega_o) - \delta(\omega + \omega_o)] \tag{5-146}$$

or

$$\sin \omega_o t \Leftrightarrow -j\frac{1}{2}[\delta(f - f_o) - \delta(f + f_o)]$$

From these relationships and the various transform operations, it is possible to derive formally (although not rigorously) most of the transforms needed. For example, the transform of the step function can be found using the equality $u(t) = \frac{1}{2} + \frac{1}{2}\operatorname{sgn}(t)$ as follows:

$$\mathscr{F}\{u(t)\} = \mathscr{F}\left\{\frac{1}{2} + \frac{1}{2}\operatorname{sgn}(t)\right\} = \pi\delta(\omega) + \frac{1}{j\omega}$$

$$u(t) \Leftrightarrow \pi\delta(\omega) + \frac{1}{j\omega} \tag{5-147}$$

Because the transforms of sinusoids consist only of impulses, it is very easy to carry out convolution operations involving these transforms. As an example, consider the determination of the spectrum of an amplitude-modulated signal. Such a signal can be represented as

$$f(t) = A[1 + m(t)]\cos(2\pi f_o t) \tag{5-148}$$

In (5-148) the modulating function $m(t)$ is usually normalized to have a maximum value of unity in order to prevent the coefficient $[1 + m(t)]$ from going negative when $m(t)$ is negative. This requirement is a practical one in the sense that, if it is not met, distortion will occur when the signal is detected by ordinary methods. Mathematically, it makes no difference how large $m(t)$ becomes. The function $m(t)$ can have either a continuous or a discrete spectrum. The spectrum of the total signal is the Fourier transform of (5-148). This can be obtained by means of the convolution theorem as follows:

$$\begin{aligned} F(f) &= \mathscr{F}\{A[1 + m(t)]\cos 2\pi f_o t\} \\ &= \frac{A}{2}\{[\delta(f) + M(f)] * [\delta(f + f_o) + \delta(f - f_o)]\} \\ &= \frac{A}{2}[\delta(f + f_o) + \delta(f - f_o)] + \frac{A}{2}M(f + f_o) + \frac{A}{2}M(f - f_o) \end{aligned} \tag{5-149}$$

It is seen that the spectrum consists of a discrete component at the carrier frequency f_o and a reproduction of $M(f)$ around $\pm f_o$. Figure 5-39 shows the amplitude spectrum for a typical example.

A useful expression for the Fourier transform of an indefinite integral can be obtained by considering the integral as the convolution of a

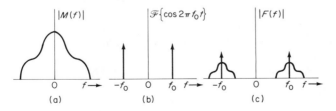

Fig. 5-39 Amplitude modulation. (a) Modulation spectrum. (b) Carrier spectrum. (c) AM signal spectrum.

function, $f_1(t)$, and the unit step. Thus,

$$f_2(t) = \int_{-\infty}^{t} f_1(\lambda)d\lambda = \int_{-\infty}^{\infty} f_1(\lambda)u(t-\lambda)d\lambda$$
$$= f_1(t) * u(t)$$

From (5-102), which states that the transform of the convolution of two functions is equal to the product of the transforms of the functions taken separately, we have

$$F_2(\omega) = F_1(\omega)\left[\frac{1}{j\omega} + \pi\delta(\omega)\right]$$

Therefore,

$$\int_{-\infty}^{t} f_1(\lambda)d\lambda \Leftrightarrow \frac{1}{j\omega}F_1(\omega) + \pi F_1(0)\delta(\omega) \tag{5-150}$$

$$\int_{-\infty}^{t} f_1(\lambda)d\lambda \Leftrightarrow \frac{1}{j2\pi f}F_1(f) + \frac{1}{2}F_1(0)\delta(f) \tag{5-151}$$

This relationship is useful in transforming integral equations and in various analytical studies.

Since a periodic function can be expressed as a sum of imaginary exponentials, which can be Fourier transformed, it is possible to compute the Fourier transform of a periodic function by taking the transforms of each term in the expansion. Suppose that we are given a periodic function $f(t)$ with period T. We can proceed formally to obtain the Fourier transform of $f(t)$ by first writing the Fourier series for $f(t)$ as

$$f(t) = \sum_{n=-\infty}^{\infty} \alpha_n \epsilon^{jn\omega_o t}$$

The Fourier transform then becomes

$$F(\omega) = \mathscr{F}\left\{ \sum_{n=-\infty}^{\infty} \alpha_n \epsilon^{jn\omega_o t}\right\}$$
$$= \sum_{n=-\infty}^{\infty} \alpha_n \mathscr{F}\{\epsilon^{jn\omega_o t}\}$$
$$= 2\pi \sum_{n=-\infty}^{\infty} \alpha_n \delta(\omega - n\omega_o)$$

The constants α_n can be expressed in terms of the Fourier transform of $f_T(t)$, the waveform over one period, as follows:

$$\alpha_n = \frac{1}{T}\int_{-T/2}^{T/2} f(t)\epsilon^{-jn\omega_o t}dt$$
$$= \frac{1}{T}\int_{-\infty}^{\infty} f_T(t)\epsilon^{-jn\omega_o t}dt$$
$$= \frac{1}{T}F_T(n\omega_o)$$

Therefore,

$$F(\omega) = \omega_o \sum_{n=-\infty}^{\infty} F_T(n\omega_o)\delta(\omega - n\omega_o) \qquad (5\text{-}152)$$

$$f(t) \Leftrightarrow \frac{2\pi}{T} \sum_{n=-\infty}^{\infty} F_T\left(\frac{2\pi n}{T}\right)\delta\left(\omega - \frac{2\pi n}{T}\right) \qquad (5\text{-}153)$$

Converting to the variable f instead of ω, the relationship becomes

$$F(f) = \sum_{n=-\infty}^{\infty} \alpha_n \delta(f - nf_o) \qquad (5\text{-}154)$$

$$f(t) \Leftrightarrow \frac{1}{T} \sum_{n=-\infty}^{\infty} F_T\left(\frac{n}{T}\right)\delta\left(f - \frac{n}{T}\right) \qquad (5\text{-}155)$$

The Fourier transform is thus seen to be a series of impulses at the harmonics of the repetition period with strengths determined by the shape of the waveform in one period.

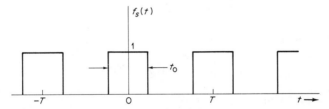

Fig. 5-40 The sampling function, $f_s(t)$.

As an example of the Fourier transform of a periodic function, consider the *sampling function $f_s(t)$*, shown in Fig. 5-40. The Fourier transform of the truncated function $f_{sT}(t)$ is

$$f_{sT}(t) = f_s(t) \qquad -\frac{T}{2} < t < \frac{T}{2}$$

$$= 0 \qquad \text{elsewhere} \qquad (5\text{-}156)$$

$$F_{sT}(\omega) = t_o \frac{\sin \omega t_o/2}{\omega t_o/2}$$

Therefore, the Fourier transform of $f_s(t)$ is

$$F_s(\omega) = \frac{2\pi t_o}{T} \sum_{n=-\infty}^{\infty} \frac{\sin \pi n t_o/T}{\pi n t_o/T}\delta\left(\omega - \frac{2\pi n}{T}\right) \qquad (5\text{-}157)$$

$$F_s(f) = \frac{t_o}{T} \sum_{n=-\infty}^{\infty} \frac{\sin \pi n t_o/T}{\pi n t_o/T}\delta\left(f - \frac{n}{T}\right) \qquad (5\text{-}158)$$

The amplitude spectrum of $f_s(t)$ is shown plotted in Fig. 5-41 for $t_o/T = \frac{1}{2}$.

Another important signal is the unit impulse train. Consider a function $f(t)$ composed of an infinite train of unit impulses having a repetition

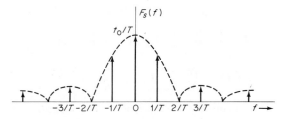

Fig. 5-41 Amplitude spectrum of the sampling function.

period T. Such a function is shown in Fig. 5-42. The Fourier transform of this function is very important in sampling theory and can be determined quite readily by first expanding $f(t)$ in a Fourier series and then taking the Fourier transform of the terms in the expansion. Following this procedure we obtain

$$f(t) = \sum_{n=-\infty}^{\infty} \delta(t - nT) \tag{5-159}$$

$$= \sum_{n=-\infty}^{\infty} \alpha_n \epsilon^{jn\omega_o t} \qquad \omega_o = \frac{2\pi}{T} \tag{5-160}$$

The coefficient α_n is given by

$$\alpha_n = \frac{1}{T} \int_{-T/2}^{T/2} \delta(t)\epsilon^{-jn\omega_o t} dt = \frac{1}{T} \tag{5-161}$$

Therefore, $f(t)$ can be written as

$$f(t) = \frac{1}{T} \sum_{n=-\infty}^{\infty} \epsilon^{jn\omega_o t} \tag{5-162}$$

Taking the Fourier transform gives

$$F(\omega) = \frac{1}{T} \sum_{n=-\infty}^{\infty} \mathcal{F}\{\epsilon^{jn\omega_o t}\}$$

$$= \frac{2\pi}{T} \sum_{n=-\infty}^{\infty} \delta(\omega - n\omega_o) \tag{5-163}$$

$$\sum_{n=-\infty}^{\infty} \delta(t - nT) \Leftrightarrow \frac{2\pi}{T} \sum_{n=-\infty}^{\infty} \delta\left(\omega - \frac{2\pi n}{T}\right)$$

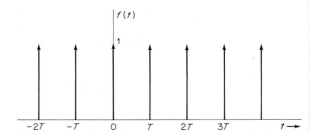

Fig. 5-42 Unit impulse train.

In terms of the frequency f, (5-163) can be written as

$$\sum_{n=-\infty}^{\infty} \delta(t - nT) \Leftrightarrow \frac{1}{T} \sum_{n=-\infty}^{\infty} \delta\left(f - \frac{n}{T}\right) \qquad \text{(5-164)}$$

Thus, it is seen that an impulse train in the time domain has as its Fourier transform an impulse train in the frequency domain. The amplitude spectrum is shown in Fig. 5-43.

Some of the more frequently used transforms of signals having non-zero average power are given in Table 5-5.

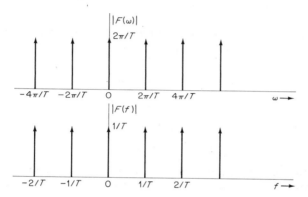

Fig. 5-43 Amplitude spectrum of a unit impulse train.

5-15 Gibbs Phenomenon

When an attempt is made to recover a signal from its Fourier transform, a very striking kind of behavior is encountered whenever the original function contains discontinuities. At each discontinuity the reconstructed function invariably displays overshoots, as shown in Fig. 5-44(a),

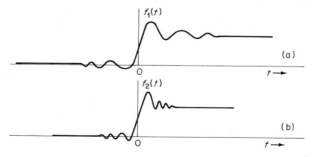

Fig. 5-44 Reconstructed step function. (a) Spectrum truncated at f_1, (b) Spectrum truncated at $2f_1$.

which represents the reconstruction of a unit step function using only a limited portion of the spectrum. This same behavior, known as the *Gibbs phenomenon*, is mentioned in connection with the reconstruction of a discontinuous signal from its Fourier series components. One's initial reaction to such a phenomenon is to conclude that it is the result of not using enough of the spectrum in the recovery process. However, increasing the spectrum utilized does not cause the overshoot to go away. All it does is to compress it along the time axis in the vicinity of the discontinuity. This is illustrated in Fig. 5-44(b), in which a spectral width twice as large as in Fig. 5-44(a) is used in reconstructing the original time function.

The cause of the Gibbs phenomenon can be seen readily by considering the reconstruction of a function $f(t)$ from its truncated spectrum which is given by

$$F_{\omega_o}(\omega) = F(\omega) \qquad |\omega| \le \omega_o$$
$$= 0 \qquad \text{elsewhere} \tag{5-165}$$

The truncated spectrum can be rewritten as

$$F_{\omega_o}(\omega) = F(\omega)p_{\omega_o}(\omega) \tag{5-166}$$

where $p_{\omega_o}(\omega)$ is a unit amplitude rectangle in the frequency domain extending from $-\omega_o < \omega < \omega_o$. The reconstructed function is then

$$\hat{f}(t) = f(t) * \mathcal{F}^{-1}\{p_{\omega_o}(\omega)\} \tag{5-167}$$

From (5-134) the inverse transform of $p_{\omega_o}(\omega)$ is given by

$$\mathcal{F}^{-1}\{p_{\omega_o}(\omega)\} = \frac{\omega_o}{\pi} \frac{\sin \omega_o t}{\omega_o t} \tag{5-168}$$

This function is sketched in Fig. 5-45. From (5-167) it is clear that the reconstructed function $\hat{f}(t)$ is the convolution of the true function, $f(t)$, and the "window" function shown in Fig. 5-45. It is easy to see that in the vicinity of a discontinuity this convolution will lead to anticipatory "wiggles" as it slides toward the discontinuity and echoing wiggles as it leaves the discontinuity. The overshoot is best handled by considering the problem analytically. In the case of a step function the convolution is

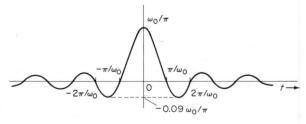

Fig. 5-45 Inverse transform of $p_{\omega_0}(\omega)$.

given by

$$\hat{u}(t) = u(t) * \frac{\omega_o}{\pi} \frac{\sin \omega_o t}{\omega_o t}$$

$$\frac{\omega_o}{\pi} \int_{-\infty}^{t} \frac{\sin \omega_o \lambda}{\omega_o \lambda} d\lambda \qquad\qquad \textbf{(5-169)}$$

Letting $\omega_o \lambda = \xi$, this becomes

$$\hat{u}(t) = \frac{1}{\pi} \int_{-\infty}^{\omega_o t} \frac{\sin \xi}{\xi} d\xi \qquad\qquad \textbf{(5-170)}$$

This integral cannot be expressed in terms of elementary functions, but can be directly written in terms of the sine integral function, $\text{Si}(x)$, which is available in tabular form. The sine integral is defined as

$$\text{Si}(x) = \int_0^x \frac{\sin \xi}{\xi} d\xi \qquad\qquad \textbf{(5-171)}$$

Making use of the fact that $\text{Si}(-x) = \text{Si}(x)$ and $\text{Si}(\infty) = \pi/2$, the expression (5-170) can be written as

$$\hat{u}(t) = \frac{1}{2} + \frac{1}{\pi} \text{Si}(\omega_o t) \qquad\qquad \textbf{(5-172)}$$

The quantity $\frac{1}{2} + (1/\pi)\text{Si}(x)$ is shown plotted in Fig. 5-46. From this figure and (5-169) it is clear that the discontinuity at $t = 0$ will be preceded by the oscillations corresponding to the portion of Fig. 5-46 to the left of the origin and followed by the oscillations to the right of the origin. The minimum of the anticipatory oscillations has a magnitude of 9 percent of the discontinuity and occurs at $\omega_o t = -\pi$ or $t = -\pi/\omega_o$ relative to the true position of the discontinuity. The maximum of the overshoot is also 9 percent of the magnitude of the discontinuity and occurs at $t = \pi/\omega_o$ beyond the discontinuity. If the spectrum cutoff is measured in Hz, $W = \omega_o/2\pi$, then the peaks of the oscillations are at $t = \pm n/2W$, and their frequency of oscillation is approximately W. It is clear from this analysis that increasing the width of the spectrum only changes the time scale of the oscillations and does not alter their amplitude in any way.

The type of analysis carried out here can be used to explain many unusual results that are obtained when signals are reconstructed from

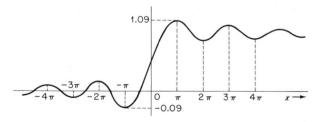

Fig. 5-46 The function $\frac{1}{2} + 1/\pi\, \text{Si}(x)$.

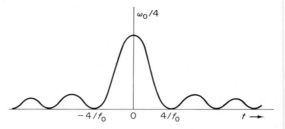

Fig. 5-47 Inverse transform of a triangular window.

truncated spectra. It can also be used to devise methods of eliminating overshoot or the anticipatory oscillations. For example, by choosing a "window function" in the frequency domain whose transform is better behaved than that of the rectangular window, the overshoot can be completely removed. One such function (not by any means the best) is a triangular window, $\Lambda_{\omega_0}(\omega)$, of unit amplitude at $\omega = 0$ extending to $\pm \omega_0$. The corresponding time function is

$$\mathcal{F}^{-1}\{\Lambda_{\omega_0}(\omega)\} = \frac{\omega_0}{4} \left(\frac{\sin \omega_0 t/8}{\omega_0 t/8} \right)^2 \tag{5-173}$$

The general shape of this function is shown in Fig. 5-47. Since the function is always positive, its integral will be monotonic and so cannot have any overshoot. The result of convolving this function with a step is shown in Fig. 5-48. It is seen that, although there is no overshoot, there is a substantial loss of rise time. A somewhat better window function is the so-called *Hamming window*, which is defined by

$$F_H(\omega) = 0.54 + 0.46 \cos \frac{\pi\omega}{\omega_0} \qquad |\omega| < \omega_0$$
$$= 0 \qquad\qquad\qquad \text{elsewhere} \tag{5-174}$$

The result of reproducing a step with a spectrum truncated by this function is shown in Fig. 5-49. It is seen that in this case a more pleasing reproduction is obtained. The mean-square error between the reproduced signal and the original is greater when the tapered window functions are

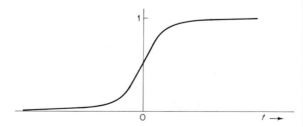

Fig. 5-48 Reproduction of step with a triangular window function.

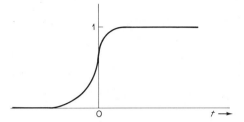

Fig. 5-49 Reproduction of a step using a Hamming window function.

used, but the result is generally more realistic in appearance. It is usually worth employing spectral window functions when signals containing discontinuities are being reconstructed from their spectra.

The same arguments apply directly to Fourier series, and similar results can be obtained by tapering the coefficients in accordance with the corresponding values of the window function to be used. The particular amount of taper for any particular component will be dependent on the extent of the spectrum to be included.

Exercise 5-15.1
Show that the effects of using a Hamming window can be obtained by first reconstructing the signal on the basis of a rectangular window of width $\pm f_0$ to give $\hat{x}_1(t)$ and then combining quantities according to the formula

$$\hat{x}_2(t) = 0.23\hat{x}_1[t - (\pi/\omega_0)] + 0.54\hat{x}_1(t) + 0.23\hat{x}_1[t + (\pi/\omega_0)]$$

5-16   Practical Filters

It was pointed out in Section 5-11 that ideal filter characteristics having rectangular passbands do not correspond to physically realizable structures. As a consequence, the best that can be done is to approximate the ideal response by some nonideal but physically realizable network, or system. Many approaches to this problem have been developed and are considered in detail in books on network synthesis. In order to illustrate the kind of approximations that are made, as well as to present a very widely used filter structure, one particular type of filter will be considered in some detail. The case of a lowpass filter will be considered first, and the results then extended to cover other filter types.

Three lowpass filter characteristics that are physically realizable are shown in Fig. 5-50. In Fig. 5-50(a) the filter characteristic is such that it is the flattest possible curve at the origin that can be obtained for a given number of filter elements. This is called a *maximally flat* filter and will be considered in detail shortly. The filter characteristic of Fig. 5-50(b) is not as smooth as that of Fig. 5-50(a), in that it has ripples in the passband.

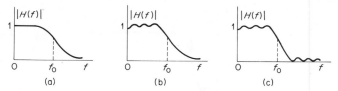

Fig. 5-50 Three lowpass filter characteristics.

However, by allowing these ripples to be present, it is found that a much higher slope of the filter characteristic can be obtained in the vicinity of the cutoff frequency. Figure 5-50(c) is a further step in this direction in which ripples in both the passband and the stopband are present. These various filter characteristics result from selection of particular sets of coefficients for the powers of ω occurring in the numerator and denominator of the transfer function, $H(\omega)$, and the choice among them is made on the basis of the specific application.

The case of the maximally flat transfer function (also called a *Butterworth* filter) is particularly simple in that $H(\omega)$ is specified in terms of its energy transfer function as

$$|H(\omega)|^2 = \frac{1}{1 + \left(\dfrac{\omega}{\omega_0}\right)^{2n}} \tag{5-175}$$

where ω_0 is the cutoff frequency or half-power frequency, and n is the order of the filter and corresponds to the number of energy storage elements present in the network. For $\omega \ll \omega_o$ the transfer function is unity, and for $\omega \gg \omega_0$ it is seen that $|H(\omega)|$ falls off at a rate of $20n$ dB per decade. Some further insight into the behavior of $|H(\omega)|$ is obtained by expanding it in a power series as follows:

$$|H(\omega)| = \left[1 + \left(\frac{\omega}{\omega_0}\right)^{2n}\right]^{-1/2}$$

$$= 1 - \frac{1}{2}\left(\frac{\omega}{\omega_0}\right)^{2n} + \frac{1 \cdot 3}{2 \cdot 4}\left(\frac{\omega}{\omega_0}\right)^{4n} - \frac{1 \cdot 3 \cdot 5}{2 \cdot 4 \cdot 6}\left(\frac{\omega}{\omega_0}\right)^{6n} + \cdots \tag{5-176}$$

It is evident from (5-176) that the first $2n - 1$ derivatives of $|H(\omega)|$ are 0 at the origin, and this is the reason for calling this a maximally flat transfer function. A family of curves representing normalized responses for various orders of maximally flat transfer function is shown in Fig. 5-51. The maximally flat transfer function is useful for theoretical studies as well as for the design of practical filters. The general process of arriving at the correct values for the components required to realize a particular response function can become quite involved and will not be considered here. However, some typical lowpass filter configurations are shown in Table 5-6. In all of these examples the gain at the origin is unity, and the halfpower frequency is one radian per second. The impedance level of these filters can

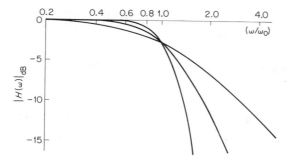

Fig. 5-51 Maximally flat transfer function.

be changed by a factor k by multiplying the resistances and inductances by k and the capacitances by $1/k$. Such scaling leaves the transfer function unchanged.

Transfer functions for other filter types can be obtained readily from those of the lowpass filters shown in Table 5-6 by means of appropriate transformations of the frequency variable. In doing this, it is convenient to designate the transfer function of the normalized lowpass filter as $H(\lambda)$, where λ is a normalized frequency variable. The transformations are shown below.

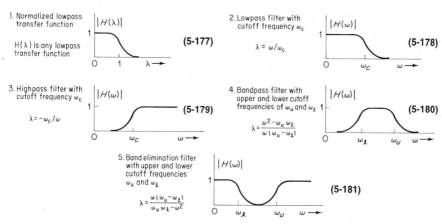

1. Normalized lowpass transfer function

$H(\lambda)$ is any lowpass transfer function

$(5\text{-}177)$

2. Lowpass filter with cutoff frequency ω_c

$\lambda = \omega/\omega_c$

$(5\text{-}178)$

3. Highpass filter with cutoff frequency ω_c

$\lambda = -\omega_c/\omega$

$(5\text{-}179)$

4. Bandpass filter with upper and lower cutoff frequencies of ω_u and ω_ℓ

$\lambda = \dfrac{\omega^2 - \omega_u \omega_\ell}{\omega(\omega_u - \omega_\ell)}$

$(5\text{-}180)$

5. Band elimination filter with upper and lower cutoff frequencies ω_u and ω_ℓ

$\lambda = \dfrac{\omega(\omega_u - \omega_\ell)}{\omega_u \omega_\ell - \omega^2}$

$(5\text{-}181)$

A method for computing the component values for a bandpass filter from those of a lowpass filter is discussed in the problems at the end of this chapter.

Exercise 5-16.1

If a lowpass filter has a transfer function $H(\omega) = 1/1 + j\omega$, find the transfer function of a filter that will pass the band of frequencies from 100 to 1000 rad/sec.

ANS: $H(\omega) = \dfrac{-j900\omega}{\omega^2 - j900\omega + 10^5}$

Table 5-6 Lowpass Butterworth Filters Normalized to $\omega_c = 1$*

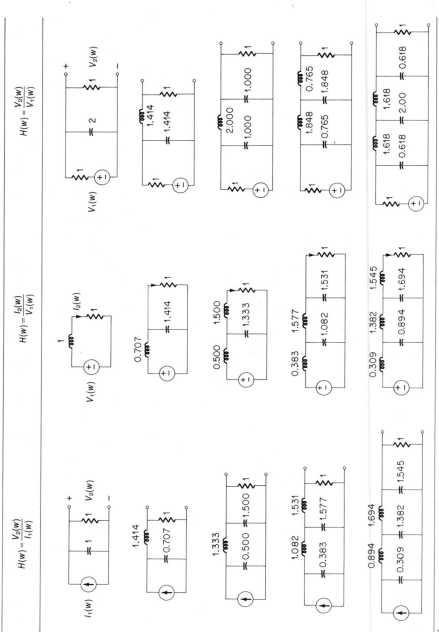

* Element values for L, C, and R are expressed in henrys, farads, and ohms, respectively.

5-17 Sampling Theorem

An important and useful result in signal analysis is the property that a band-limited signal can be uniquely represented by a set of samples taken at time intervals spaced $1/2W$ seconds apart, where W is the signal bandwidth in Hz.[9] By the terminology "uniquely represented" is meant that the original signal can be exactly recovered from the samples. The validity of this statement can be demonstrated readily for the lowpass case by considering the system shown in Fig. 5-52. Here the signal is sampled periodically by the switch, and the period of the sampler is assumed to be $1/2W$. The sampling operation can be mathematically represented as the product of the input signal, $f_1(t)$, and the sampling function, $f_s(t)$, which was discussed in Section 5-14, and corresponds to a unit amplitude periodic rectangular wave. The output of the sampler, $f_2(t)$, is, therefore, given by

$$f_2(t) = f_1(t)f_s(t) \tag{5-182}$$

The spectrum of $f_2(t)$ is found from the convolution theorem to be

$$F_2(f) = F_1(f) * F_s(f) \tag{5-183}$$

The spectrum of $f_s(t)$ is a series of impulses at the harmonics of the sampling frequency $(2W)$, and the convolution of (5-183) leads to a replication of the spectrum of $f_1(t)$ around each of these harmonics. The magnitude of the spectra reproduced at each harmonic depends on the magnitude of the δ function in $F_s(f)$ at that frequency and can be computed readily from (5-158). A set of typical spectra is shown in Fig. 5-53. It is evident that if the spacing between the impulses equals or exceeds twice the bandwidth of $f_1(t)$, there will be no overlapping of the spectra centered at adjacent harmonics. This is the reason for the requirement that the sampling frequency be $2W$. In order to recover the original signal, it is necessary only to separate the spectral component centered at

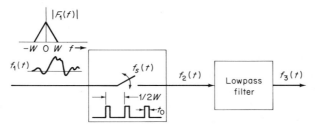

Fig. 5-52 Signal sampling and reconstruction.

[9]A band-limited signal of width W is one whose Fourier transform is identically 0 everywhere except in a frequency region of width W when $f > 0$ and a corresponding region when $f < 0$. From the Paley–Wiener criterion of (5-99), it is clear that such a signal can never be causal.

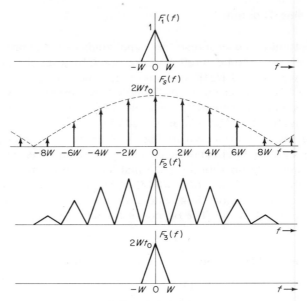

Fig. 5-53 Sampled signal spectra.

the origin from the remainder of the spectrum. This can be accomplished readily by means of a lowpass filter with bandwidth W, and the result is shown as $F_3(f)$ in Fig. 5-53. The waveform $f_3(t)$ will be a replica of $f_1(t)$, since its spectrum is identical except for a scale factor of $2Wt_0$, where t_0 is the "on time" of the sampling switch. By including an amplifier having a gain, $G = 1/2Wt_0$, the original signal would be recovered exactly. It is important to note that the duration and shape of the pulses in $f_s(t)$ do not affect the process of recovering $f_1(t)$ from its samples. All that is required is that $f_s(t)$ be periodic with a fundamental frequency at least twice as great as the highest frequency component in $f_1(t)$.

The lowpass filtering operation described above for recovery of the original signal from its samples can be carried out in the time domain rather than the frequency domain. Such an operation might be required, for example, in carrying out an analysis or making a simulation on a digital computer. Assume that the sample values of a band-limited function, $f(t)$, are known at discrete-time instants $t_n = n/2W$. The value of $f(t)$ for any value of t is related to the values at the sampling times by the following expression:

$$f(t) = \sum_{n=-\infty}^{\infty} f(t_n) \frac{\sin 2\pi W(t - n/2W)}{2\pi W(t - n/2W)} \tag{5-184}$$

$$= \sum_{n=-\infty}^{\infty} f\left(\frac{n}{2W}\right) \operatorname{sinc}(2Wt - n) \tag{5-185}$$

From (5-185) it is seen that the value of $f(t)$ at any particular point is a weighted sum of values at earlier and later sampling instants, with the weighting factor being sinc $(2Wt - n)$. The amplitude of the sinc function decreases inversely with the first power of n, so it may be necessary to use many terms in (5-185) to obtain a desired degree of precision. By sacrificing some of the sharpness in cutoff of the filter characteristic, it is possible to obtain more rapid convergence.

The derivation of (5-185) can be made in the following manner: Since it is known that $F(f)$ is 0 outside the interval $-W < f < W$, it can be expanded in a Fourier series in f, giving

$$F(f) = \sum_{n=-\infty}^{\infty} \alpha_n \epsilon^{j(\pi n f / W)} \qquad (5\text{-}186)$$

$$\alpha_n = \frac{1}{2W} \int_{-W}^{W} F(f) \epsilon^{-j(\pi n f / W)} df \qquad (5\text{-}187)$$

But it is evident that α_n is just the inverse transform of $F(f)$ evaluated at $t = -n/2W$, so that

$$F(f) = \frac{1}{2W} \sum_{n=-\infty}^{\infty} f\left(\frac{-n}{2W}\right) \epsilon^{j\pi n f / W} \qquad (5\text{-}188)$$

Now taking the inverse transform of $F(f)$ and recalling that it is 0 outside of $\pm W$, gives

$$f(t) = \mathscr{F}^{-1}\{F(f)\} = \frac{1}{2W} \int_{-W}^{W} \sum_{n=-\infty}^{\infty} f\left(\frac{-n}{2W}\right) \epsilon^{j\pi n f / W} \epsilon^{j 2\pi f t} df \qquad (5\text{-}189)$$

$$= \frac{1}{2W} \sum_{n=-\infty}^{\infty} f\left(\frac{-n}{2W}\right) \int_{-W}^{W} \epsilon^{j[(\pi n / W) + 2\pi t] f} df$$

$$= \frac{1}{2W} \sum_{n=-\infty}^{\infty} f\left(\frac{-n}{2W}\right) \frac{\epsilon^{j 2\pi W [t + (n/2W)]} - \epsilon^{-j 2\pi W [t + (n/2W)]}}{j 2\pi [t + (n/2W)]}$$

$$= \sum_{n=-\infty}^{\infty} f\left(\frac{n}{2W}\right) \text{sinc}\, (2Wt - n) \qquad (5\text{-}190)$$

where the last equation is obtained by replacing $-n$ with n in the summation.

An important consideration in the use of sampled signals is the question of what happens if the signal is not limited in frequency content to a value less than one-half the sampling frequency. Similarly, the question arises as to what distortion occurs in the reconstruction process if the filter cutoff is not sharp enough to exclude portions of the spectrum coming from higher harmonics of the sampling frequency. These are very real problems and must be considered in any practical use of the sampling theorem. Figure 5-54 illustrates the case in which the sampled signal has frequency components greater than one-half the sampling frequency. This is referred to as *undersampling*. It is seen that the spectral component centered around the first harmonic overlaps that centered around

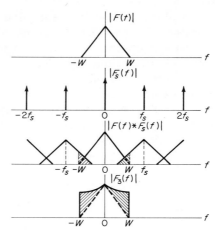

Fig. 5-54 Effects of undersampling.

zero frequency. Thus, there is no portion of the spectrum of the sampled signal that corresponds to the spectrum of the original signal, and it is, therefore, not possible to recover the original signal when undersampling has occurred. This overlapping of spectra is referred to as *aliasing* and is a major source of distortion in the reconstruction of signals from their samples. Aliasing can be avoided by sampling at very high frequencies or by filtering the signal before sampling to properly restrict its bandwidth. In any case it is possible to compute readily the amount of distortion that will occur for a given situation. The general effect of undersampling is to reinforce the even part of the spectrum and to diminish the odd part of the spectrum, as can be seen from Fig. 5-54.

Exercise 5-17.1
A singer's performance is to be recorded by sampling and storing the sample values. Assuming the highest frequency tone to be recorded is 15,800 Hz, what is the minimum sampling frequency that can be used? How many samples would be required to store a three-minute performance? If each sample is quantized into 128 levels, how many binary digits (bits) would be required to store the three-minute performance?

ANS: 5.688×10^6; 3.16×10^4; 3.9816×10^7.

5-18   Discrete Fourier Transform

In the preceding section the question of recovering a signal from samples taken at periodic intervals is discussed, and the sampling theorem for band-limited signals is derived.

With the widespread use of digital computers for analysis, simulation, and actual signal processing, there has developed a need for more general methods of working with sampled signals. Frequently, such signals correspond to the values of a continuous waveform sampled at periodic intervals. In other cases, however, the signals are part of a completely digital processor and are discrete-time signals that are defined only at the sampling instants. In either case the signal consists of a discrete sequence of numbers corresponding to the samples. The Fourier transform, as previously defined, of such a sequence of discrete-time samples does not have any significance, if in fact it can be considered to exist at all. This is because any attempt to compute such a transform would give a value of zero, since the discrete-time samples have no area associated with them. In order to circumvent this problem, it is possible to consider each sample as being associated with a suitable "sample representing" function, such that a transform can be computed. A desired property of sample representing functions is that their influence on transforms computed in this manner can be taken readily into account and removed if desired. It is possible to carry out the development of an expression for a *discrete Fourier transform* (DFT) along these lines. However, it turns out that in order to obtain the proper form, a number of *a priori* assumptions must be made, and certain of the intermediate steps include some rather involved expressions. An alternative approach, and one that turns out to be considerably simpler, is to define the DFT of a sequence and then show how it is related to the more familiar Fourier series and Fourier transform. It is this latter procedure that will be followed here.

Let a sequence of samples be represented by

$$\{f(nT)\} = f(0), f(T), f(2T), \ldots, f[(N-1)T]$$

The discrete Fourier transform is a sequence of (complex) samples $\{F_D(k\Omega)\}$ in the frequency domain defined by[10]

$$F_D(k\Omega) = \sum_{n=0}^{N-1} f(nT)\epsilon^{-j\Omega Tnk}, \qquad k = 0, 1, \ldots, N-1 \qquad \text{(5-191)}$$

where N is the number of samples, and $\Omega = 2\pi/NT$ is the separation of the components in the frequency domain. With this definition there are only N distinct values of $F_D(k\Omega)$ that can be computed, since $F_D(k\Omega)$ is periodic in Ω with a period of $N\Omega$. In order to see this, consider calculation of $F_D(k\Omega)$ for k greater than N by a small integer k_1; i.e., $k = N + k_1$,

---

[10]Various definitions appear in the literature. This one is consistent with the transform of an impulse-sampled function and the z transform discussed in Chapter 7. It differs by a factor N from the Fourier series coefficients.

$k_1 = 0, 1, 2, \ldots$. Equation (5-191) then gives

$$F_D(k\Omega) = \sum_{n=0}^{N-1} f(nT)\epsilon^{-j\Omega T(N+k_1)n}$$

But

$$\epsilon^{-j\Omega NTn} = \epsilon^{-(j2\pi NTn/NT)} = \epsilon^{-j2\pi n} = 1$$

and therefore,

$$F_D[(N+k_1)\Omega] = \sum_{n=0}^{N-1} f(nT)\epsilon^{-j\Omega k_1 n} = F_D(k_1\Omega) \tag{5-192}$$

Thus, it is seen that any value of $F_D(k\Omega)$ for $k > N$ can be expressed in terms of a smaller argument, $F_D(k_1\Omega)$, where $k_1 = k$ modulo N.[11] In view of the above, it is clear that $F(k\Omega)$ can be thought of as a sequence of N numbers or as a periodic sequence of numbers that repeats after N.

There is also an inverse discrete Fourier transformation (IDFT), $\mathscr{F}_D^{-1}$, whereby $\{f(nT)\}$ can be recovered exactly from $\{F_D(k\Omega)\}$. This transformation is given by

$$f(nT) = \mathscr{F}_D^{-1}\{F_D(k\Omega)\} = \frac{1}{N}\sum_{k=0}^{N-1} F_D(k\Omega)\epsilon^{j\Omega Tkn} \tag{5-193}$$

The validity of the inverse transformation can be established by substituting the value for $F_D(k\Omega)$ from (5-191) into (5-193), giving

$$f(nT) = \frac{1}{N}\sum_{k=0}^{N-1}\sum_{m=0}^{N-1} f(mT)\epsilon^{-j\Omega Tmk}\epsilon^{j\Omega Tkn} \tag{5-194}$$

Combining the exponents of the exponential terms and reversing the order of summation gives

$$f(nT) = \frac{1}{N}\sum_{m=0}^{N-1} f(mT)\sum_{k=0}^{N-1}\epsilon^{j\Omega Tk(n-m)} \tag{5-195}$$

Recalling that $\Omega = (2\pi/NT)$, the inner sum can be written as

$$S(n-m) = \sum_{k=0}^{N-1}\epsilon^{j[(2\pi k/N)(n-m)]} \tag{5-196}$$

This sum is obviously N if $n = m$, modulo N. For $n \neq m$, modulo N, the sum will be zero. This is most easily seen by writing (5-196) as the summation of a geometric progression with the multiplier, $\epsilon^{j[2\pi(n-m)/N]}$. Thus,

$$S(n-m) = \sum_{k=0}^{N-1}\epsilon^{j(2\pi k/N)(n-m)} = \frac{1 - \epsilon^{j2\pi(n-m)}}{1 - \epsilon^{j(2\pi/N)(n-m)}} = 0 \tag{5-197}$$

since the numerator is identically zero, and the denominator is not. Using this result, (5-195) can be written as

---

[11] A number is written modulo N by expressing only the remainder after all integral multiples of N have been subtracted. The number 15, modulo 4, is 3; or modulo 10, it is 5.

$$f(nT) = \frac{1}{N} \sum_{m=0}^{N-1} S(n-m)f(mT) = \frac{1}{N}(N)f(nT), \qquad m=n$$

$$= 0, \qquad\qquad m \neq n$$

$$= f(nT)$$

It is interesting to note that the IDFT is identical to the DFT except for the multiplicative factor $1/N$ and the reversal of sign in the exponent. The similarity to the continuous transform is also clearly evident.

Exercise 5-18.1
Find the DFT of $\{1, 1, 0, 0\}$ and the IDFT of $\{1, 0, 1, 0\}$.

ANS: $\{\frac{1}{2}, 0, \frac{1}{2}, 0\}$; $\{2, 1-j1, 0, 1+j1\}$

Properties of the Discrete Fourier Transform.   As will be discussed shortly, the DFT is useful in approximating the Fourier transform of a continous function. However, it should be borne in mind that the DFT is actually an *exact* relationship between an original sequence $f(nT)$ and an *image*[12] sequence $F_D(k\Omega)$. The IDFT is capable of producing samples $f(nT)$ for values of n outside the range $0 \leq n < N-1$. However, these values are simply repetitions of values of $f(nT)$ taken on within this range. Such values can be thought of as corresponding to a periodic extension of $\{f(nT)\}$, or the entire sequence can be considered as being arranged around a circle with N divisions, one for each sample.

There are several properties of the DFT that are useful in carrying out or interpreting calculations involving these transforms. First of all, the DFT is linear; that is,

$$\mathscr{F}_D\{af_1(nT) + bf_2(nT)\} = a\mathscr{F}_D\{f_1(nT)\} + b\mathscr{F}_D\{f_2(nT)\} \qquad \text{(5-198)}$$

where $\mathscr{F}_D\{f(nT)\}$ is taken to mean the DFT of the sequence $\{f(nT)\}$.

The various operations involving the DFT are analogous (but not identical) to those for the continuous Fourier transform. For example, the IDFT of the product of two DFTs is the convolution of the original sequences. However, the convolution is a *periodic, or circular, convolution*. To see this, consider two sequences $\{(f_1(nT)\}$ and $\{f_2(nT)\}$. In order to simplify the notation, let the sequences be represented by $f_1(nT)$ and $f_2(nT)$ and let their DFTs be $F_{D1}(k\Omega)$ and $F_{D2}(k\Omega)$. In order for this type of operation to have meaning, it is necessary, of course, that both sequences have the same sampling interval and the same number of samples. If one sequence has fewer sample values than the other, enough zero-value sample values can be added to make the numbers of samples equal.

---

[12] A fourier transform is frequently referred to in the older literature as an *image*. This is particularly appropriate in the case of the DFT.

Consider now the IDFT, $f_3(nT)$, of the product of the two DFTs; that is,

$$f_3(nT) = \frac{1}{N} \sum_{k=0}^{N-1} F_{D1}(k\Omega) F_{D2}(k\Omega) \epsilon^{i\Omega Tkn} \tag{5-199}$$

Substituting the defining relations for $F_1(k\Omega)$ and $F_2(k\Omega)$ from (5-191) gives

$$f_3(nT) = \frac{1}{N} \sum_{k=0}^{N-1} \left[\sum_{i=0}^{N-1} f_1(iT)\epsilon^{-j\Omega Tki} \right]\left[\sum_{m=0}^{N-1} f_2(mT)\epsilon^{-j\Omega Tkm} \right] \cdot \epsilon^{j\Omega Tkn} \tag{5-200}$$

This can be written as a triple summation over the indices k, i, m. By interchanging the order of summation so that the summation over k is innermost and factoring out $f_1(iT)$ and $f_2(mT)$, the following expression is obtained:

$$f_3(nT) = \frac{1}{N} \sum_{i=0}^{N-1} \sum_{m=0}^{N-1} f_1(iT)f_2(mT) \sum_{k=0}^{N-1} \epsilon^{j\Omega Tk(n-i-m)} \tag{5-201}$$

The inner summation is of the same form as encountered in connection with (5-196) and is zero except for combinations of $i = (n - m)$, modulo N, for which case it equals N. Thus, (5-201) can be written as

$$f_3(nT) = \sum_{i=0}^{N-1} f_1(iT)f_2[(n - i)T] \tag{5-202}$$

$$= \sum_{i=0}^{N-1} f_1[(n - i)T]f_2(iT) \tag{5-203}$$

where the quantity $(n - i)$ in (5-202) and (5-203) is understood to be modulo N. The interpretation of (5-202) or (5-203) is most easily seen by considering $f_1(iT)$ and $f_2(iT)$ to be arranged in circles as shown in Fig. 5-55 for $N = 6$. As in ordinary convolution, one sequence is "folded" (that is, reversed) and indexed around by an amount n relative to the other. The

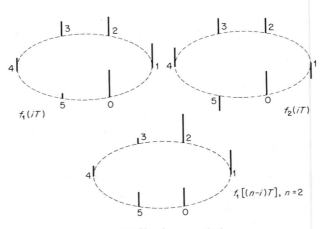

Fig. 5-55 Circular convolution.

terms are then multiplied serially, and the products are summed. An exactly equivalent result is obtained by considering the original sequences to be periodic, with period NT, and then carrying out the normal folding and shifting operation employed in ordinary convolution. In this case the summation is carried out over a single period. This is shown in Fig. 5-56. For obvious reasons this type of convolution is called *circular, or periodic.* The analogous operation for continuous Fourier transforms can be thought of as *aperiodic convolution.* It is usually the aperiodic type of convolution that is encountered in system analysis problems, and in order to simulate it using the DFT it is necessary to extend the interval with zero-value samples so that there is no overlap when periodic convolution is employed. This process is illustrated in Fig. 5-57.

A result analogous to that of time delay in continuous functions is obtained with *circular shifting* of discrete sequences. Thus,

$$\text{DFT}\{f[[(n-i) \bmod N]T]\} = F_D(k\Omega)\epsilon^{-j\Omega Tki} \qquad \textbf{(5-204)}$$

It is seen that moving i samples from the end of a sequence to the beginning is equivalent to multiplying the DFT by a linear phase function.

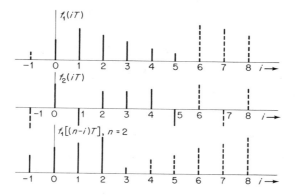

Fig. 5-56 Periodic convolution.

There are some symmetry properties of the DFT that are useful. A sequence $\{f(nT)\}$ can be considered as *even* if $f(nT) = f(NT - nT)$ and *odd* if $f(nT) = -f(NT - nT)$. When these sequences are plotted around a circle or periodically extended, these definitions are seen to correspond to conventional interpretations of even and odd functions. With regard to the DFT, it is shown readily that an even sequence of real values has a DFT that is real and even, and an odd sequence of real values has a DFT that is odd and pure imaginary. These and other related properties are given in Table 5-7. Because of the symmetry of the DFT and the IDFT, the headings of the two columns can be interchanged without altering the validity of the relationships.

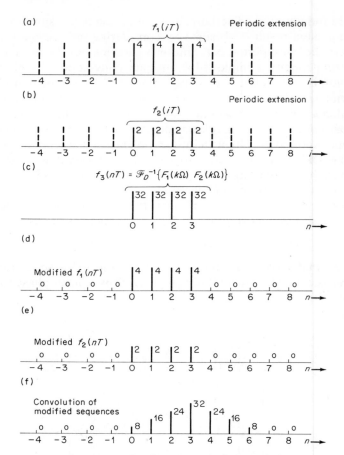

Fig. 5-57 Approximation of a periodic convolution.

Table 5-7   Symmetry Properties of the DFT

$f(nT)$	$F(k\Omega)$
Even	Even
Odd	Odd
Real	Real part even; imaginary part odd
Imaginary	Real part odd; imaginary part even

Exercise 5-18.2

Convolve the following sequences by multiplying their DFTs and inverting. Check by convolving in the time domain.

$$\{1, 1, 0, 0\} * \{1, 1, 0, 0\}$$

ANS: $\{1, 2, 1, 0\}$

The Discrete Fourier Transform in System Analysis. With these rather extensive preliminaries out of the way, it is now possible to consider how the DFT can be used in signal and system analysis. First of all, consider a continuous-time function, $f(t)$, that exists over the time interval, $0 \le t \le T_0$. As shown in Chapter 4, such a function can be represented exactly by a continuum of impulses. A less exact but more computationally tractable representation can be obtained by representing the function as a sequence of equally spaced impulses, where the strength of the impulse at $t = nT$ is the area under the function, $f(t)$, in the interval $T(n - \frac{1}{2}) \le t < T(n + \frac{1}{2})$. Such a representation is shown in Fig. 5-58. If the sampling interval, T, is small enough so that the function does not change appreciably over one interval, then the strength of the δ function can be taken as $Tf(nT)$. Thus, the representation can be written as

$$f(t) \doteq \hat{f}(t) = Tf(t) \sum_{n=0}^{N-1} \delta(t - nT) \qquad \text{(5-205)}$$

$$= T \sum_{n=0}^{N-1} f(nT)\delta(t - nT) \qquad \text{(5-206)}$$

This type of representation is very useful (and quite accurate) when the computations to be made involve convolutions (or integrations) of $f(t)$ with other functions that do not change very much over the interval T. Such applications include computing the spectrum up to a frequency on the order of $f = (1/2T)$ or computing the output of a linear system having a relatively smooth impulse response. By making the interval T smaller,

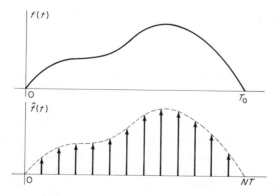

Fig. 5-58 Impulse approximation of functions.

the fidelity of the representation can be extended to higher frequencies. As an example, let it be assumed that it is desired to compute the Fourier transform of the function represented by (5-206). This leads to

$$\hat{F}(\omega) = T \int_{-\infty}^{\infty} \epsilon^{j\omega t} \sum_{n=0}^{N-1} f(nT)\delta(t - nT)$$

$$= T \sum_{n=0}^{N-1} f(nT)\epsilon^{-j\omega nT} \tag{5-207}$$

When $\omega = 2\pi k/NT = k\Omega$, it is seen that (5-207) gives the DFT of $\{f(nT)\}$; that is,

$$\hat{F}(\omega)|_{\omega=k\Omega} = T\mathscr{F}_D\{f(nT)\}, \qquad 0 \le k \le N - 1 \tag{5-208}$$

From (5-208) it is seen that the DFT is equal to $1/T$ times the sequence of samples taken at intervals of $\Omega = 2\pi/NT$ from the Fourier transform of the δ-function approximation to the continuous-time function.

There is another way to express the continuous Fourier transform of $\hat{f}(t)$ that provides considerable insight into the proper interpretation of the DFT as an approximation to the continuous *FT*. Since it is assumed that $f(t)$ exists only over the interval $0 < t < T_0 = NT$, it is evident that extending the summation in (5-205) beyond $N - 1$ will not alter the result; that is,

$$\hat{f}(t) = Tf(t) \sum_{n=0}^{N-1} \delta(t - nT) = Tf(t) \sum_{n=-\infty}^{\infty} \delta(t - nT) \tag{5-209}$$

The Fourier transform of $f(t)$ may, therefore, be written in the equivalent form

$$\mathscr{F}\{\hat{f}(t)\} = \hat{F}(\omega) = \mathscr{F}\left\{ Tf(t) \sum_{n=-\infty}^{\infty} \delta(t - nT)\right\} \tag{5-210}$$

$$= TF(\omega) * \frac{2\pi}{T} \sum_{n=-\infty}^{\infty} \delta\left(\omega - \frac{2\pi n}{T}\right) \tag{5-211}$$

$$= \sum_{n=-\infty}^{\infty} F\left(\omega - \frac{2\pi n}{T}\right) \tag{5-212}$$

Evaluating (5-212) at $\omega = 2\pi k/NT$ gives

$$\hat{F}(\omega)|_{\omega=(2\pi k/NT)} = \sum_{n=-\infty}^{\infty} F\left(\frac{2\pi k}{NT} - \frac{2\pi n}{T}\right) \tag{5-213}$$

and equating this result to the right-hand side of (5-208) gives

$$\mathscr{F}_D\{f(nT)\} = \frac{1}{T} \sum_{n=-\infty}^{\infty} F\left(\frac{2\pi k}{NT} - \frac{2\pi n}{T}\right) \tag{5-214}$$

From (5-214) it is seen that the discrete Fourier transform of the sequence of samples $\{f(nT)\}$ is equal to $1/T$ times samples at $\omega = 2\pi k/NT$ of the infinite replication of $F(\omega)$ along the ω-axis at intervals of $\Lambda\omega = 2\pi/T$.

When the discrete Fourier transform is used to approximate samples of the continouus Fourier transform, it is generally assumed that contributions from $F(2\pi k/NT - 2\pi n/T)$ for $n \neq 0$ are negligible, and this leads to the simple expression

$$F(\omega)|_{\omega = (2\pi k/NT)} \doteq T\mathscr{F}_D\{f(nT)\} \tag{5-215}$$

However, it is evident from (5-214) that the error of this approximation is

$$\epsilon = \sum_{n=-\infty}^{\infty} F\left(\frac{2\pi k}{NT} - \frac{2\pi n}{T}\right) \qquad n \neq 0 \tag{5-216}$$

Therefore, in order to obtain a satisfactory approximation, it is necessary that (5-216) be made sufficiently small. This is normally accomplished by choosing T small enough to assure that the replications of $F(\omega)$ are sufficiently far removed from the origin so that their contribution will be negligible. An illustration will help to clarify this problem.

Let it be assumed that it is desired to compute an approximation to the continuous Fourier transform of a rectangular pulse $u(t) - u(t - 1)$ by means of the discrete Fourier transform. For purposes of discussion let $T = 1/5$, thus dividing the signal into five segments. The sample sequence will consist of five ones; that is, $f(nT) = 1$, $n = 0, 1, 2, 3, 4$. This is illustrated in Fig. 5-59. The discrete Fourier transform is computed to be

$$\mathscr{F}_D\{f(nT)\} = \sum_{n=0}^{4} f(nT)\epsilon^{-j(2\pi nk/N)}$$

$$= \sum_{n=0}^{4} \epsilon^{-j(2\pi nk/5)} = \begin{matrix} 5, & k = 0 \\ 0, & k \neq 0, \end{matrix} \tag{5-217}$$

Samples of the continuous Fourier transform can now be computed from (5-217) and (5-215) to be

$$\mathscr{F}\{u(t) - u(t-1)\}|_{\omega = 2\pi k} \doteq T\mathscr{F}_D\{f(nT)\} = \frac{1}{5}\begin{cases} 5 \\ 0 \end{cases}$$

$$= \begin{cases} 1 & k = 0, \\ 0 & k \neq 0, \end{cases} \qquad k = 0, 1, 2, 3, 4 \tag{5-218}$$

Thus, the approximation to the continuous Fourier transform consists of a single pulse of unit magnitude at the origin and is zero elsewhere.

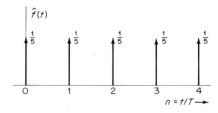

Fig. 5-59 Samples of a rectangular pulse.

On the surface this is a very unsettling result, since we know that the Fourier transform of a rectangular pulse is of the form $\sin x/x$. Fortunately, there is a reasonable explanation for this unexpected result. The exact expression for $\mathcal{F}_D\{f(nT)\}$ is given by (5-214) in terms of the continuous Fourier transform. For the waveform under consideration

$$F(\omega) = \epsilon^{-j(1/2)\omega} \frac{\sin\frac{1}{2}\omega}{\frac{1}{2}\omega}$$

and therefore, $\mathcal{F}_D\{f(nT)\}$ is

$$\mathcal{F}_D\{f(nT)\} = \sum_{n=-\infty}^{\infty} \epsilon^{-j\pi(k-5n)} \frac{\sin \pi(k-5n)}{\pi(k-5n)} \tag{5-219}$$

That is clearly zero except for $k = 0$. The reason is that the spacing between components is such as to cause the samples to fall at the nulls in the spectrum, as shown in Fig. 5-60. This difficulty can be overcome by placing some augmenting zeros after the initial sequence of samples. The number of zeros used depends on the resolution desired in the frequency domain. The spacing between harmonics is $\Omega = 2\pi/NT$, and therefore, increasing N, while keeping T constant, will bring the frequency components closer together. For example, if 5 zeros are placed after the 5 samples, the sequence becomes $\{f_1(nT)\} = \{1,1,1,1,1,0,0,0,0,0\}$ and the DFT becomes

$$\mathcal{F}_D\{f_1(nT)\} = \sum_{n=0}^{9} f\left(\frac{n}{5}\right)\epsilon^{-j(2\pi nk/10)} \tag{5-220}$$

The spacing of the frequency components is now $\Omega = 2\pi/NT = \pi$, and the components corresponding to $k = 0, 1, 3, 5, 7, 9$ fall between the nulls, while the components corresponding to $k = 2, 4, 6, 8$ are at the nulls. This is illustrated in Fig. 5-61. Consider now the component corresponding to $k = 1$. This component is computed from (5-220) to be

$$F_D(\pi) = \sum_{n=0}^{9} f\left(\frac{n}{5}\right)\epsilon^{-j(2\pi n/10)}$$

$$= \epsilon^0 + \epsilon^{-j(2\pi/10)} + \epsilon^{-j(4\pi/10)} + \epsilon^{-j(6\pi/10)} + \epsilon^{-j(8\pi/10)} \tag{5-221}$$

$$= 1.00000 - j\,3.07768$$

$$|F(\pi)| \doteq |TF_D(\pi)| = 0.64721$$

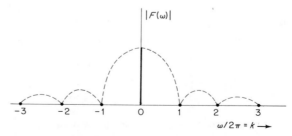

Fig. 5-60 Sampled spectrum of unit pulse.

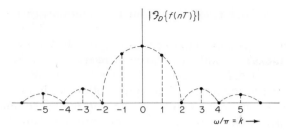

Fig. 5-61 Spectrum of augmented signal.

The correct value computed from the analytical expression is $F(\pi) = -j0.63662$. The error in the magnitude of $F(\omega)$ is 0.01059, or 1.7 percent. As discussed in connection with (5-216), this error can be thought of as resulting from contributions from the replications of the spectrum at all multiples of $N\Omega = 2\pi/T$. This phenomenon is illustrated in Fig. 5-62 for the present example. Errors of this kind are frequently referred to as aliasing and result from the presence of significant spectral components in the signal at frequencies in excess of $1/2T$, that is, greater than half the sampling frequency. This type of error can be reduced by using more closely spaced samples. For instance, in the present example, if T is made 1/10 there will be 10 samples of unity and 10 zeros, provided $\Omega = 2\pi/NT$ is held constant. This should cause the replicated spectra to be twice as far away as in the case just considered, and therefore, the error should be reduced. To test this we will recompute the approximation to the continuous Fourier transform at $\omega = \pi$ using these parameters:

$$F(\omega)|_{\omega=\pi} \doteq \frac{1}{10} \sum_{n=0}^{20} f(nT)\epsilon^{-j(2\pi n/20)} = \frac{1}{10} \sum_{n=0}^{10} \epsilon^{-j(2\pi n/20)}$$

$$= 0.1 - j0.63138 \qquad \text{(5-222)}$$

$$|F(\pi)| \doteq 0.63925$$

The error in the magnitude is 0.00262, or 0.41 percent, which represents an appreciable improvement.

In practice, it is not generally possible to compute the error analytically, and the usual procedure is to keep reducing the sampling interval and recomputing the transform until no change in the desired number of significant figures occurs. This value of T is then used in subsequent calculations. The number of samples (including augmenting zeros) is deter-

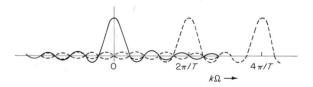

Fig. 5-62 Spectrum replication in the DFT.

mined by the desired resolution from the relationship $\Omega = 2\pi/NT$, or $N = 2\pi/\Omega T$.

The algorithms usually employed to compute the DFT assume that the data starts at the origin, and the points run from $n = 1, 2, 3, \ldots, N$, corresponding to $t = 0, T, 2T, \ldots, (N-1)T$. When the functions involved have values to the left of the origin some modification of either the data or the result is required in order to obtain the correct approximation to the Fourier transform. One procedure is to compute the transform as if the first point were at the origin and then, using (5-204), compute the spectrum of the data translated to the left by multiplying by $\epsilon^{j\Omega Tkm}$, where m is the number of samples of $f(nT)$ that lie to the left of the origin. An alternative, and often more convenient, procedure is to make use of the fact that the DFT and its inverse are periodic with periods of $2\pi/T$ radians per second and NT seconds, respectively. Because of this periodicity it is possible to correctly compute the transform by using the points in sequence starting with the point at $t = 0$ after the waveform has been periodically extended, including any augmenting zeros required. Some self-explanatory examples are shown in Fig. 5-63.

Another requirement on the number of zeros to be included in the approximating sequence comes about when convolution operations in the time domain are being carried out by multiplying the transforms of the functions. The circular convolution operation that occurs when the DFT is used will fold over and produce distortion unless a sufficient number of zeros is added. Specifically, if one function has a duration of $N_1 T$ and the other $N_2 T$, then the period of the approximating sequences should be equal to or greater than $(N_1 + N_2)T$ to avoid fold over.

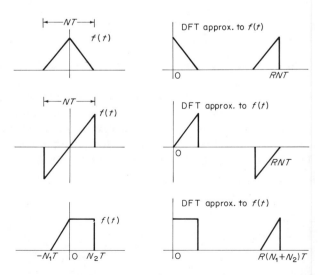

Fig. 5-63 Translation of negative time portions of waveforms.

Fast Fourier Transform.   The *fast Fourier transform* is the name applied to various algorithms that rapidly make the computations required to obtain the DFT of a sequence. A dramatic reduction in computation time is obtained by making use of the fact that, when the number of points in a sequence is the product of several factors, the sequence can be broken up into smaller sequences whose transforms can be determined with fewer computations and from which the transform of the original sequence can be obtained readily. For example, consider the sequence $x_0, x_1, \ldots, x_{N-1}$, where N is an even number. The DFT of this sequence can be broken down into separate computations of the DFT of the odd and even elements in the following manner:

$$X_D(k\Omega) = \sum_{n=0}^{(N/2)-1} x(2nT)\epsilon^{-j[2\pi k(2n)/N]} + \sum_{n=0}^{(n/2)-1} x[(2n+1)T]\epsilon^{-j[2\pi k(2n+1)/N]}$$

$$\tag{5-223}$$

$$= \sum_{n=0}^{(n/2)-1} x(2nT)\epsilon^{-j[2\pi kn/(N/2)]} + \epsilon^{-j[2\pi k/N]}\sum_{n=0}^{(N/2)-1} x[(2n+1)T]\epsilon^{-j[2\pi kn/(N/2)]}$$

The two summations are seen to be the DFTs of sequences of $N/2$ components and to be periodic in k with a period of $N/2$. Thus, even though k goes through a range of 0 to $N-1$, only $N/2$ values of the two DFTs need be computed. There are an additional $N/2$ multiplications of the second summation by the phase factor $\epsilon^{(-j2\pi k/N)}$ necessary to give the required values for $N/2 \leq k < N-1$. There are, therefore, a total of $2[(N/2)^2] + N/2 = N^2/2 + N/2$ complex multiplications required to compute all values of $X_D(k\Omega)$ using the two half sequences. This compares to N^2 complex multiplications working directly with the full sequence. For large N this would amount to a 50-percent reduction in computation time. This process can be continued as long as the sequences can be further divided, and each reduction to the computation of DFTs having half as many terms gives a reduction in required multiplications by a factor of 2. By choosing N to be a power of 2, it is always possible to carry out this reduction completely, and in this case the number of multiplications required by the FFT algorithm is $2\,N\log_2 N$. In the case of large N the savings in time are enormous. Consider the computation of a 4096-point data set. By direct computation this would require 16,777, 216 multiplications, while by the FFT algorithm only 98,304 would be required. This reduces the computation time by a factor of 174. The number of multiplications required for several frequently used types of computation are given in Table 5-8.

The exact relationship of the FFT values to the corresponding samples of the continuous Fourier transform depends on the multiplicative constants assigned to the transform and the inverse transform. For a particular algorithm, the relationship can be found by examining the formulas directly. For example, in the widely used algorithm. FORT,[13] a quantity

---

[13]Share Program Library SDA 3465, Finite Complex Fourier Transform, Sept. 8, 1966.

Table 5-8 Comparison of Direct and FFT Computation

Operation	Formula	Approximate No. Multiplications Direct	Approximate No. Multiplications FFT
DFT	$\displaystyle\sum_{k=0}^{N-1} x_k \epsilon^{-2\pi jkr/N}, \qquad r = 1, 2, \ldots, N-1$	N^2	$2N \log_2 N$
Convolution	$\displaystyle\sum_{k=0}^{N-1} x_k y_{u-k}, \qquad u = 0, 1, \ldots, N-1$	N^2	$3N \log_2 N$
Autocorrelation	$\displaystyle\sum_{k=0}^{N-1-r} x_k x_{r+k}, \qquad r = 0, 1, \ldots, N-1$	$\dfrac{N}{4}\left(\dfrac{N}{2}+3\right)$	$3N \log_2 N$

called the *finite Fourier transform* (FFT) is computed that is equal to $1/N$ times the DFT. From (5-209) it is evident that values of the continuous Fourier transform are given by NT times the finite Fourier transform. The same computer program computes a quantity designated the complex Fourier series (CFS) which equals N times the IDFT. Thus, samples of the continuous-time function are given by $1/NT$ times the complex Fourier series of the samples of the continuous Fourier transform. If this algorithm is used to carry out the convolution of two continuous-time functions by multiplication of their transforms, the required relationship is as follows:

$$
\begin{aligned}
f_3(nT) &= \frac{1}{T}\,\text{IDFT}\{TF_{D1}(k\Omega)\cdot TF_{D2}(k\Omega)\}\\[6pt]
&= T\,\text{IDFT}\{N\,\text{FFT}[f_1(nT)]\cdot N\,\text{FFT}[f_2(nT)]\}\\[6pt]
&= \frac{T}{N}\,\text{CFS}\{N^2\,\text{FFT}[f_1(nT)]\,\text{FFT}[f_2(nT)]\}\\[6pt]
&= NT\,\text{CFS}\{\text{FFT}[f_1(nT)]\,\text{FFT}[f_2(nT)]\}
\end{aligned}
$$

(5-224)

One further word of caution is in order with regard to using the DFT to approximate samples of the continuous transform. If the time function is sampled every T seconds and has the duration of NT, then the spacing of samples in the frequency domain will be $1/NT$ Hz, and the highest frequency will be $1/2T$ Hz. This latter result is in accord with the sampling theorem and occurs in the present instance as a result of the fact that, although N samples in the frequency domain are computed, the first $N/2$ of them are the positive frequency components and the last $N/2$ are the negative frequency components. This is illustrated in Fig. 5-64. In this figure the absolute value of the DFT and samples of the continuous FT are used to simplify the picture—in general, these samples are complex numbers.

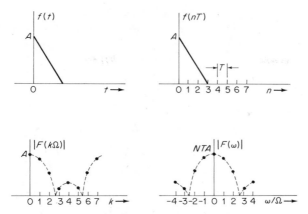

Fig. 5-64 FFT approximation to Fourier spectrum.

The following points are helpful in using the FFT in the processing of continuous time functions: (1) The FFT of equally spaced samples of a function must be multiplied by NT (signal time duration) to obtain an approximation to samples of the continuous FT. (2) In order to gain maximum computational efficiency, the total number of samples is normally required to be a power of 2; that is, $N = 2^m$, where m is an integer. (3) If there are a total of N samples in the time domain, then there will be N samples in the frequency domain. (4) The first $N/2 + 1$ samples (that is, $k = 0 \rightarrow N/2$) in the frequency domain may be considered as the positive frequency components having frequencies of 0 to $1/2T$ Hz. The last $N/2$ samples may be considered as the negative frequency components having frequencies of $(k - N)/NT$ Hz. Note that the sample at $k = N/2$ is used for both the positive and negative frequency component. This occurs automatically for periodic functions but is introduced artificially here from known symmetry conditions. Thus, there will be $N + 1$ known frequency components, counting both positive and negative frequencies. (5) The spacing between frequencies will be $1/NT$ Hz, and the highest frequency will be $1/2T$ Hz. (6) Augmenting zeros may be added to the sample sequence to increase the frequency resolution, and closer spacing of time samples may be used to increase the highest frequency present in the spectrum. (7) In using the CFS to compute the corresponding time function from *samples of a continuous FT*, the resulting time samples must be multiplied by $1/NT$. (8) In using the CFS to compute samples of the time function *from values of the DFT*, the result of the CFS operation must be multiplied by $1/N$.

As an illustration and summary of the above procedures we shall compute the spectrum of a triangular pulse and will also compute the error of the approximation by comparing the result with values computed from the analytical expression for the transform. The signal to be considered is

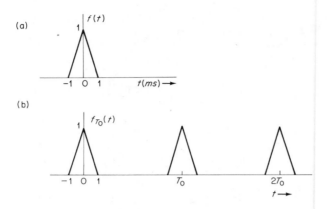

Fig. 5-65 Time function to be transformed.

a unit-amplitude triangular pulse of duration 2 ms. A resolution of 100 Hz is to be obtained to a frequency of at least 25 kHz. The time function is shown in Fig. 5-65(a). In order to use the FORT subroutine to compute the FFT, it is convenient to convert the waveform to a periodic function. This is done by replicating the function with period T_0, giving

$$f_{T_0}(t) = \sum_{k=-\infty}^{\infty} f(t + kT_0) \tag{5-225}$$

If the function is a pulse type of function whose duration is less than T_0, then the periodic function will consist of distinct replications of $f(t)$ at intervals of T_0. However, if the duration of $f(t)$ is greater than T_0, then the periodic function will have aliasing, giving rise to distorted reproductions of $f(t)$ at intervals of T_0. In either case it is this periodic function that should be used to assign the data points $f(nT)$. If one of the points falls at a point of discontinuity in the waveform, some improvement in accuracy will result from assigning to that point the average value of the left and right limits; that is, $f(nT) = \frac{1}{2}[f(nT^+) + f(nT^-)]$. It is to this value that the Fourier transform converges. Figure 5-65(b) shows the periodic extension of $f(t)$.

Since the highest frequency that will be obtained from the DFT is $f_{max} = 1/2T$, where T is the sampling interval, it follows that

$$\frac{1}{2T} \geq 25 \times 10^3 \text{ Hz}$$
$$T \leq 2 \times 10^{-5} \text{ sec} \tag{5-226}$$

A resolution of 100 Hz is to be obtained, and therefore,

$$\frac{1}{NT} = 100$$
$$N = \frac{1}{100T} \geq \frac{1}{100 \times 2 \times 10^{-5}} = 500 \tag{5-227}$$

Since N must be a power of 2, it will be taken to be the first number greater than 500, or $N = 2^9 = 512$. If it is desired to have a spacing of the frequency samples of exactly 100 Hz, then T must be selected such that

$$\frac{1}{100T} = 512$$

$$T = \frac{1}{51,200} = 19.53125 \ \mu \sec$$

(5-228)

The total duration of the augmented signal will be $T_0 = NT = 512(1/51,200) = 10$ msec, and the signal to be transformed is as shown in Fig. 5-66. Using the FORT computer program to compute the FFT, the samples must be numbered from 1 to 2^N and correspond to the time samples $f(0)$ to $f[(N-1)T]$ of the periodically extended waveform. The values of the original function corresponding to negative time occur in the periodically extended waveform at the end of the period, as shown in Fig. 5-66. The point corresponding to $n = 1$ must be the point at $t = 0$, or there will be errors in the phase of the spectrum computed, since this is equivalent to translation of the function.

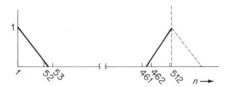

Fig. 5-66 Augmented time function.

In the particular example considered here, the point at which $f(t)$ goes to zero falls between samples 52 and 53, which corresponds to times $t_{52} = 51T = 996 \ \mu \sec$ and $t_{53} = 52T = 1016 \ \mu \sec$. Also note that the last sample corresponding to $n = N = 512$ occurs one sample interval short of the complete period.

The finite Fourier transform computed by the FORT subroutine must be multiplied by $NT = 0.01$ in order to obtain an approximation to the continuous spectrum.

For purposes of this example only, the magnitude of the spectrum was computed, and the results for the first 50 points are shown in Table 5-9. It is evident that the error is small and oscillatory. Figure 5-67 shows a plot of the peak values of the error oscillations as a function of frequency. For many purposes this would be an excellent approximation to the amplitude spectrum of a triangular pulse.

The increasing availability of large digital computers for engineering analysis has made Fourier techniques into a powerful and convenient tool in signal and system analysis, and there seems little doubt that their use will continue to increase for some time to come.

Table 5-9   Computed Spectrum of 2 msec Triangular Pulse

f	FFT	F(f)	Error
0.00	1.00006E-03	1.00000E-03	6.10352E-08
100.00	9.67593E-04	9.67531E-04	6.14332E-08
200.00	8.75203E-04	8.75150E-04	6.25092E-08
300.00	7.36904E-04	7.36840E-04	6.39413E-08
400.00	5.72852E-04	5.72787E-04	6.52926E-08
500.00	4.05351E-04	4.05285E-04	6.61362E-08
600.00	2.54638E-04	2.54572E-04	6.61847E-08
700.00	1.35403E-04	1.35338E-04	6.53870E-08
800.00	5.47602E-05	5.46963E-05	6.39612E-08
900.00	1.20072E-05	1.19448E-05	6.23453E-08
1000.00	6.10719E-08	1.17757E-32	6.10719E-08
1100.00	8.05672E-06	7.99613E-06	6.05985E-08
1200.00	2.43706E-05	2.43095E-05	6.11450E-08
1300.00	3.93026E-05	3.92400E-05	6.25965E-08
1400.00	4.68226E-05	4.67581E-05	6.45128E-08
1500.00	4.50979E-05	4.50316E-05	6.62543E-08
1600.00	3.58664E-05	3.57992E-05	6.71930E-08
1700.00	2.30135E-05	2.29466E-05	6.69399E-08
1800.00	1.08697E-05	1.08042E-05	6.55073E-08
1900.00	2.74348E-06	2.68014E-06	6.33390E-08
2000.00	6.11826E-08	1.17757E-32	6.11826E-08
2100.00	2.25379E-06	2.19395E-06	5.98376E-08
2200.00	7.29243E-06	7.23256E-06	5.98625E-08
2300.00	1.25974E-05	1.25360E-05	6.13467E-08
2400.00	1.59746E-05	1.59107E-05	6.38421E-08
2500.00	1.62779E-05	1.62114E-05	6.64915E-08
2600.00	1.36254E-05	1.35571E-05	6.83226E-08
2700.00	9.16539E-06	9.09679E-06	6.86075E-08
2800.00	4.53216E-06	4.46500E-06	6.71550E-08
2900.00	1.21487E-06	1.15045E-06	6.44190E-08

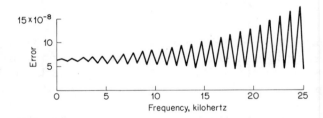

Fig. 5-67 Error in FFT approximation of triangular pulse of 2 msec duration ($N = 512$, $\Delta T = 19.531\ \mu$ sec).

Exercise 5-18.3

It is desired to use the FFT to compute the spectrum of a signal having a two-second duration. A resolution of 1/10 Hz is to be obtained up to a maximum frequency of 25 Hz. How closely should the waveform be sampled, how many samples will be required, and how many augmenting zeros will be required?

ANS: 500; 400; 0.02.

References

1. BRACEWELL, R. M., *The Fourier Transform and Its Applications*. New York: McGraw-Hill Book Company, Inc., 1965.
   The book is written at a senior or graduate level. Although accurate and precise in its discussions, this book is written more from an engineer's than a mathematician's point of view. Much detailed information on transform methods of analysis is included, and many interesting practical applications are discussed. This book is an excellent reference for those who plan to carry out practical analysis employing Fourier transform methods. A number of tables are included.

2. PAPOULIS, A., *The Fourier Integral and Its Applications*. New York: McGraw-Hill Book Company, Inc., 1962.
   This book is written at a senior-graduate level. It provides a comprehensive treatment of the Fourier transform. More mathematical in approach than Bracewell, this book is an excellent reference for those planning to carry out mathematical investigations using the Fourier transform.

3. CAMPBELL, G. A., and R. M. FOSTER, *Fourier Integrals for Practical Applications*. New York: D. Van Nostrand Company, 1948.
   This book contains an extensive table of Fourier transform pairs.

4. GUILLEMIN, E. A., *The Mathematics of Circuit Analysis*. New York: John Wiley & Sons, Inc., 1949.
   This book contains much material relevant to circuit and system analysis, presented in a manner that aids in developing an intuitive feeling for the various techniques considered.

5. ERDÉLYI, A., W. MAGNUS, F. OBERHETTINGER, and F. G. TRICOMI, *Table of Integral Transforms*, Vol. 1, New York: McGraw-Hill Book Company, Inc., 1954.
   This book contains tables of Fourier, Laplace, and Mellin transforms. It is recommended for persons interested in more advanced work in transform techniques.

6. COOLEY, J. W., P. A. W. LEWIS, and P. D. WELCH, "The Fast Fourier Transform and Its Applications," *IEEE Trans. on Education*, Vol 12, No. 1, March 1969.

7. GOLD, B., and C. M. RADER, *Digital Processing of Signals*. New York: McGraw-Hill Book Company, Inc., 1969.
   This book contains much information on discrete signal representation and processing, including the DFT and a variety of FFT algorithms. Many references to current literature are given.

Problems

5-1. Compute the exponential Fourier series expansion for the periodic waveforms shown and sketch the amplitude spectrum of each.

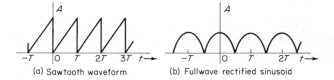

(a) Sawtooth waveform   (b) Fullwave rectified sinusoid

5-2. Compute the trigonometric Fourier series expansion for the periodic waveforms shown using the results given in Table 5-1.

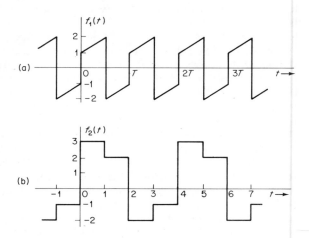

5-3. A rule-of-thumb that is often used in the design of amplifiers for square waves is that the amplifier should pass all harmonics up to the tenth in order to obtain good reproduction of the waveform. (a) Investigate the validity of this rule-of-thumb by computing the distortion in the output of this amplifier as a percentage of the total power in the output. (b) Plot the waveform resulting from retention of 5, 10, and 15 harmonics of the square wave. (*Hint*: Use a digital computer.)

5-4. A certain amplifier is nonlinear in the sense that the output is not proportional to the input. The relationship between input and output is given mathematically by $v_0(t) = [v_i(t)]^7$. (a) For a sine wave of unit amplitude as the input, compute the *percent harmonic* distortion, which is defined as 100 times the ratio of the rms value of the harmonic components to the rms value of the fundamental component. (b) An approximate way of measuring the percent harmonic distortion of a component is to measure the ratio of the rms amplitude of residual harmonics when the fundamental is eliminated (for example, by a filter) to the rms value of the fundamental and harmonics combined. Verify that the error of such an estimate is less than 5 percent if the harmonic distortion does not exceed 30 percent.

5-5. Complete the following table relating the symmetry properties of a waveform to the components present and the expressions for the Fourier coefficients.

Mathematical Expression	Symmetry Condition	Special Characteristics	a_n	b_n
$f(t) = f(-t)$		Cosine terms only	$\dfrac{4}{T}\displaystyle\int_0^{T/2} f(t)\cos\left(\dfrac{2\pi nT}{T}\right)dt$	
	$f(t)$ odd	•		
	$f(t)$ even half wave			
$f\left(t \pm \dfrac{T}{2}\right) = -f(t)$		Odd n only		
$f(t) = f(-t)$ and $f\left(t \pm \dfrac{T}{2}\right) = f(-t)$				

5-6. The first quarter cycle of a periodic waveform is shown in the accompanying figure. Sketch the waveform over one complete cycle for each of the following cases:

a. $f(t)$ is even and contains only even harmonics.

b. $f(t)$ is even and contains only odd harmonics.

c. $f(t)$ is odd and contains only even harmonics.

d. $f(t)$ is odd and contains only odd harmonics.

e. $f(t)$ is odd and contains both even and odd harmonics.

f. $f(t)$ is neither even nor odd and contains only odd harmonics.

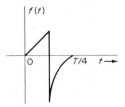

5-7. For the periodic waveform shown, compute the total power in the waveform and the power that would be passed by a unity gain amp-

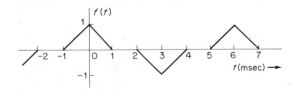

lifier that passes all frequency components up to 800 Hz and none above that frequency.

5-8. When a waveform can be accurately represented by straight line segments (called a *piecewise linear approximation*), the Fourier series coefficients can be calculated readily by differentiating twice to obtain a series of impulses and then making use of the relationship

$$f(t) = \sum_{n=-\infty}^{\infty} \frac{-\beta_n}{n^2 \omega_0^2} \epsilon^{jn\omega_0 t}$$

where β_n is the exponential Fourier series coefficient for $f''(t)$.
a. Derive the above relationship.
b. Use the procedure described above to derive the Fourier series coefficients for the following waveforms.

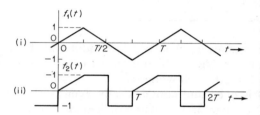

5-9. Show that the following relationships represent valid Fourier transformations.

$$\mathcal{F}\{\delta(t - t_o)\} = \epsilon^{-j\omega t_o}$$

$$\mathcal{F}\left\{\frac{d^n}{dt^n} f(t)\right\} = (j\omega)^n F(\omega)$$

5-10. Using the transform relationships given in Problem 5-9, it is possible to simplify computation of Fourier transforms of signals that can be represented by piecewise linear approximations. The procedure is to differentiate the piecewise linear approximation once or twice so that only impulses or doublets remain, to write down the transform of the derivative by inspection, and then to convert to the transform of the original time function by dividing by $(j\omega)^n$. Using this method, find the transforms of the signals shown.

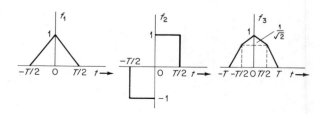

5-11. When a function has odd or even symmetry about the origin, it is possible to compute the transform of the complete function from only the portion of the function corresponding to positive time only, $f_+(t)$. The required relationships are as follows:

$$f_e(t) = f_e(-t) = f_+(-t)u(-t) + f_+(t)u(t)$$
$$\Leftrightarrow 2Re\mathcal{F}\{f_+(t)\}$$

$$f_0(t) = -f_0(-t) = -f_+(-t)u(-t) + f_+(t)u(t)$$
$$\Leftrightarrow 2j\ Im\ \mathcal{F}\{f_+(t)\}$$

a. Derive the above relationships.
b. Use the above expressions to compute the Fourier transforms of the waveforms shown starting from the transforms of the causal functions.

$$u(t) - u(t-T) \Leftrightarrow Te^{-j\omega T/2}\frac{\sin \omega T/2}{\omega T/2}$$

$$\epsilon^{-\alpha t}u(t) \Leftrightarrow \frac{1}{j\omega + \alpha}$$

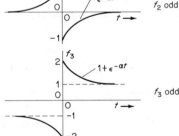

5-12. Find the system function corresponding to each of the following impulse responses:
a. $h(t) = [\epsilon^{-t} - \epsilon^{-2t}]u(t)$
b. $h(t) = [\delta(t) + \epsilon^{-t}]u(t)$
c. $h(t) = 2\epsilon^{-2t}\cos 2t\ u(t)$

5-13. Find the system function, $H(\omega) = v_0(\omega)/v_i(\omega)$, for the network shown and sketch the amplitude response and the phase response.

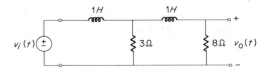

5-14. Compare the frequency responses of the two networks shown. The difference in the two responses is due to the "loading" effect of the two cascaded networks. It is a commonly used rule of thumb that changing the impedance level of one network relative to the other by a factor of 10 will eliminate this effect except for a loss in gain. Check the validity of this procedure for the network shown.

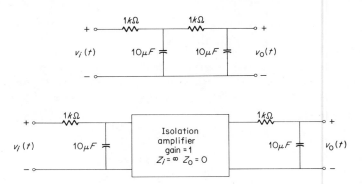

5-15. For many communications problems it is desirable to have a pulse signal with a large fraction of its energy concentrated within a specific bandwidth. (a) Compare the fraction of the energy contained in the frequency band $|f| < (1/T)$ for three pulses shown. (b) The energy contained in a particular pulse for a given amplitude is also important. Compute the relative energies of each of the pulses shown. (c) Which pulse would you select for signaling purposes and why?

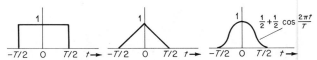

5-16. A time function of the form $te^{-2t}u(t)$ is applied to an ideal lowpass filter for which

$$H(\omega) = 1 \qquad -2\pi B \le \omega \le 2\pi B$$
$$= 0 \qquad \text{elsewhere}$$

For what value of B does this filter pass exactly one-half of the energy (on a one-ohm basis) of the input signal?

5-17. A time function, $f(t)$, has a Fourier transform of

$$F(\omega) = \frac{2 + j\omega}{4 - \omega^2 + j5\omega}$$

Using the results of Section 5-9 write the Fourier transform of each of the following time functions:

a. $2f(3t - 1)$ **d.** $f(-\frac{1}{2}t)$
b. $\epsilon^{-j2t}f(t - 2)$ **e.** $f(1 - t)$
c. df/dt **f.** $f(t) \cos t$

5-18. An ideal delay line is a device for which the output signal has the same size and shape as the input signal but occurs at a later time. Consider the following systems, each of which contains an ideal delay line with a delay of one second. Determine the system function relating the input and output voltages.

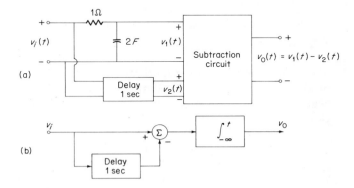

(a)

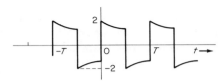

(b)

5-19. Find the Fourier transforms of the following time functions:
   a. $f(t) = 0 \quad -\infty < t \le 0$
   $\qquad = A \cos \omega_o t \quad 0 \le t < \infty$
   b. $f(t) = (1 + m \cos \omega_1 t) \cos \omega_o t \quad -\infty < t < \infty$
   c. $f(t) = \dfrac{a_0}{2} + \sum\limits_{n=1}^{\infty} a_n \cos n\omega_o t + b_n \sin n\omega_o t \quad -\infty < t < \infty$

5-20. Find the Fourier transform of the following periodic time function, which is composed of segments of exponentials with time constants of T:

5-21. Find the Fourier transforms of the following time functions by using frequency convolution:
   a. $f(t) = [1 + m(t)] \sin \omega_o t$
   b. $f(t) = \epsilon^{-\alpha t} \cos \omega_o t u(t)$
   c. $f(t) = \cos \omega_o t \sin \omega_o t$

5-22. A frequently used method of comparing two finite energy signals is by computing their correlation function, which is defined to be

$$r(\tau) = \int_{-\infty}^{\infty} f_1(t) f_2^*(t - \tau) dt$$

The correlation coefficient or normalized correlation function is defined to be

$$\rho(\tau) = \frac{r(\tau)}{\sqrt{\int_{-\infty}^{\infty} f_1^2(t) dt \int_{-\infty}^{\infty} f_2^2(t) dt}}$$

Signals are said to be correlated (that is, alike) to the extent that their correlation coefficient approaches unity.

a. Prove the validity of the following expression for the correlation function:

$$r(\tau) = \mathscr{F}^{-1}\{F_1(f) F_2(f)^*\}$$

b. Show that the area under a correlation function is equal to the product of the areas of the two correlated signals.

c. Show that the following relationship among derivatives of functions is valid for a correlation function:

$$r^{(m+n)}(\tau) = \int_{-\infty}^{\infty} f_1^{(m)}(t) f_2^{*(n)}(t - \tau) dt$$

d. Compute the autocorrelation function (that is, the correlation of a function with itself) of a rectangular pulse of unit height and duration T.

5-23. The triangular-shaped pulse shown in the accompanying figure is applied to an ideal lowpass filter having a cutoff frequency of 100 Hz. Compute the output waveform of the filter.

5-24. The high-frequency performance of an amplifier can be measured in terms of its response to a unit step input. One measure of performance that can be obtained from such a test is the rise time, which is defined as the time required for the output to go from 10 to 90 percent of its final value.

a. Compute the rise time of an ideal lowpass filter having a bandwidth of W Hz.

b. Determine the half-power bandwidth of a single-stage lowpass $R-C$ filter having the same rise time as the ideal filter.

5-25. It is required to design a third-order lowpass filter with a maximally flat voltage transfer function having a cutoff frequency of 1000 Hz.

a. Determine the expression for the filter transfer function.

b. Sketch the circuit configuration and specify the component values for the filter.

5-26. Transmission of a modulated carrier through a bandpass system can often be analyzed very simply in terms of an analogous lowpass system. The procedure is as follows: A time function identical to the modulation signal is applied to the lowpass analog of the bandpass system. The output of the lowpass system is identical to the modulation that would be present on the carrier at the output of the bandpass system. What restrictions on the signal spectrum and system function are required for this procedure to give valid results? How is the lowpass analog related to the bandpass system function?

5-27. An algorithm for generating a bandpass circuit from a lowpass circuit is as follows:

a. Leave all resistances unchanged.

b. Reduce all capacitances by a factor of 2 and place an inductor in parallel with each such that the two elements resonate at the center frequency, ω_0, of the bandpass system.

c. Reduce all inductances by a factor of 2 and place in series with each a capacitance of the proper value to resonate at ω_0.

The resulting bandpass system will have a response to a modulated carrier whose envelope (modulation) is approximately the same as the response of the lowpass system to the envelope alone. The approximation is very good for narrowband systems. There is no formal procedure for going from a bandpass circuit to an analogous lowpass circuit, since such analogs do not always exist. Using the foregoing procedure find the equivalent bandpass filter corresponding to the lowpass filters shown in (a) and (b). Determine and sketch the response of the circuit in (c).

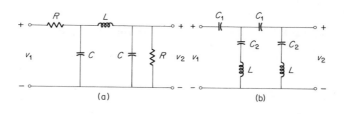

(a)   (b)

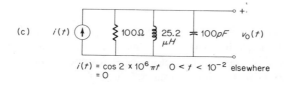

(c)

$i(t) = \cos 2 \times 10^6 \pi t \quad 0 < t < 10^{-2}$ elsewhere
$= 0$

5-28. A lowpass waveform, $m(t)$, has a maximum frequency content of 1000 hertz. After sampling, the signal is to be reconstructed by passing the samples through a single-section R–C lowpass filter. Specify an appropriate sampling rate and filter bandwidth and discuss the relationship of these parameters to the distortion of the reconstructed signal.

5-29. A bandpass waveform, $f(t)$, is to be sampled and then reconstructed by passing the samples through a bandpass filter. If the samples are to be taken every T seconds, and the sampling width is Δt, determine the minimum sampling rate if $\mathscr{F}\{f(t)\} = F(f)$ exists only for $100 \le |f| \le 120$ Hz. Does there exist a maximum sampling rate?

5-30. The time function $f(t) = 5 \cos(1000\pi t) \cos^2(2000\pi t)$ is sampled 4500 times per second. If reconstruction is to be accomplished by passing the sampled signal through an ideal lowpass filter of bandwidth 2600 Hz, determine the output time function, assuming the filter has zero phase shift and unity gain over its passband. Compute the mean square error of the output time function. What is the minimum sampling rate that permits the signal to be uniquely reconstructed?

5-31. Compute the DFT of the following sequences:
   a. $\{f(nT)\} = \{1, 0, 0\}$ $T = 1/2\,s$
   b. $\{f(nT)\} = \{1, 0, 0, 1, 0, 0\}$ $T = 2\,s$
   c. $f(nT) = \cos n$ $n = 0, 1, 2, \ldots$

5-32. Carry out, in the time domain, the periodic and aperiodic convolution of the following sequences and sketch the results. Using the DFT and properly augmented sequences, compute the periodic and aperiodic convolutions of the sequences

$$\{f_1(nT)\} = \{2, 2, 2, 2\}   T = 1s$$
$$\{f_2(nT)\} = \{1, 1, 2, 2\}$$

5-33. Using the FFT algorithm compute the amplitude and phase spectra of the following signals:

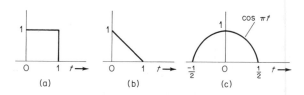

(a)   (b)   (c)

5-34. Two commonly used window functions are the Hamming window and the Hanning window. These functions possess the desirable property of being compact in the frequency domain and having low sidelobes in the time domain. These functions are as follows:

Hanning window $F_1(f) = \frac{1}{2} + \frac{1}{2}\cos(\pi f/f_0)$ $|f| < f_0$
   $= 0$ elsewhere
Hamming window $F_2(f) = 0.54 + 0.46\cos(\pi f/f_0)$ $|f| < f_0$
   $= 0$ elsewhere

Using the FFT, compute and plot the time function corresponding to these spectra.

5-35. Carry out the numerical convolution of the two functions shown using a sampling interval of 0.01 s. Repeat, using the FFT with exactly the same spacing of samples. Compare the results by the two methods and also with the exact solution.

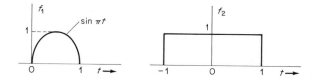

5-36. Find the error in computing the spectrum of a rectangular pulse, $u(t + \frac{1}{2}) - u(t - \frac{1}{2})$, using the FFT. Use a sampling interval of 0.125 s and durations of 4 s and 16 s and then repeat with a sampling interval of 0.0625 s. Compute the error at each point of the FFT spectrum using the known analytical solution and plot the result.

CHAPTER

6

Laplace Transforms

6-1 Introduction

The frequency domain methods based on the Fourier transform, as discussed in Chapter 5, provide powerful analytical tools for the study of signals and systems. However, in order to utilize these methods it is necessary that a signal be of such a nature as to have a Fourier transform. It was pointed out in Chapter 5 that one set of sufficient conditions for the Fourier transform to exist requires that the signal be absolutely integrable. In certain special cases where the power of the time function was concentrated at specific frequencies, it was found possible to circumvent this requirement by utilizing the concept of δ functions in the frequency domain. Unfortunately, even this extension leaves out many important signals. For example, the ramp function, the positive exponential, and the entire class of sample functions from stationary random processes do not meet the conditions required for representation by the ordinary Fourier transform. Two approaches are possible to permit the handling of such functions. One approach is to extend the applicability of the Fourier transform by interpreting it in terms of generalized functions based on the theory of distributions. This leads to very powerful methods of analysis but requires use of advanced mathematical concepts in order to be properly understood and applied. The other approach is to introduce a con-

200

vergence factor into the Fourier integral to ensure that the integral will converge and then reinterpret the results in the light of this convergence factor. It is this latter procedure that is considered here. When the convergence factor is taken as an exponential time function with a negative real exponent, the resulting transformation leads to the well-known Laplace transform under a simple change of variable. All of the results required for using the Laplace transform can be obtained by starting from a simple definition and proceding from that point on, without mention of the Fourier transform. However, much insight can be gained by starting with the Fourier transform and showing how it evolves into the Laplace transform. This insight is particularly useful when it is desired to interpret results of Laplace transform analyses in terms of the frequency domain.

Consider a time function $f(t)$ that is sufficiently well-behaved so that $\epsilon^{-\sigma t}f(t)$ is absolutely integrable if σ is sufficiently large. Now compute the Fourier transform of $\epsilon^{-\sigma t}f(t)$.

$$\mathscr{F}\{\epsilon^{-\sigma t}f(t)\} = \int_{-\infty}^{\infty} f(t)\epsilon^{-(\sigma+j\omega)t}dt$$

$$= F\left(\frac{\sigma+j\omega}{j}\right) = F_1(\sigma+j\omega) \tag{6-1}$$

where $F(\omega)$ is the Fourier transform of $f(t)$, and $F_1(x) = F(x/j)$. The corresponding inverse transformation is

$$f(t)\epsilon^{-\sigma t} = \mathscr{F}^{-1}\{F_1(\sigma+j\omega)\}$$

$$= \frac{1}{2\pi}\int_{-\infty}^{\infty} F_1(\sigma+j\omega)\epsilon^{j\omega t}d\omega \tag{6-2}$$

Taking the convergence factor to the right-hand side of (6-2) gives

$$f(t) = \frac{1}{2\pi}\int_{-\infty}^{\infty} F_1(\sigma+j\omega)\epsilon^{(\sigma+j\omega)t}d\omega \tag{6-3}$$

Since σ and $j\omega$ always appear together, it is convenient to define a new variable, $s = \sigma + j\omega$. This gives $d\omega = -jds$, and the transform pair can be written as

$$F_1(s) = \int_{-\infty}^{\infty} f(t)\epsilon^{-st}dt \tag{6-4}$$

$$f(t) = \frac{1}{2\pi j}\int_{\sigma-j\infty}^{\sigma+j\infty} F_1(s)\epsilon^{st}ds \tag{6-5}$$

Equation (6-4) is called the *two-sided*, or *bilateral*, *Laplace transform*, and it exists whenever the integral exists. Equation (6-5) is called the *inverse Laplace transform* and will be put in a more general form shortly.

The two-sided Laplace transform is so called because it includes both the positive and negative portions of the time axis and can, therefore, be employed in the analysis of signals having both positive and negative portions. Such functions and their transforms are considered later in this chapter. For the present it is convenient to limit consideration to functions existing only for positive time or to the positive time portion of functions existing for both positive and negative time. The Laplace transform of such functions is defined as

$$F(s) = \int_{-\infty}^{\infty} f(t)u(t)\epsilon^{-st}dt = \int_{0}^{\infty} f(t)\epsilon^{-st}dt \qquad (6\text{-}6)$$

This is known as the *one-sided*, or *unilateral, Laplace transform* or more frequently is referred to merely as the Laplace transform of the function $f(t)$.[1] The original time function is obtained by application of (6-5) and leads to the function $f(t) u(t)$, that is, a function defined only for positive time.

In (6-6) the lower limit is taken to include the origin. Any discontinuities in the function or impulses at the origin are included in the integral. Symbolically, this may be written as

$$F(s) = \int_{0}^{\infty} f(t)\epsilon^{-st}dt = \int_{0^-}^{\infty} f(t)\epsilon^{-st}dt = \lim_{a \to 0} \int_{-a}^{\infty} f(t)\epsilon^{-st}dt \qquad (6\text{-}7)$$

The use of this convention leads to a consistent interpretation of the integral and simplifies handling of initial conditions and singularities occurring at the origin. The importance and utility of this convention is seen in later sections.

The definition of the Laplace transform as given in (6-5) and (6-7) is the same as the classical mathematical definition except for the convention regarding the lower limit. In the classical definition the lower limit is taken as 0^+ and excludes the point at the origin. In most cases of interest there is no contribution to the integral resulting from inclusion of the origin, and in such cases the two definitions lead to identical expressions for the Laplace transform. This will be true whenever

$$\int_{0^-}^{0^+} |f(t)|dt = 0 \qquad (6\text{-}8)$$

In such cases all of the results of classical Laplace transform theory will be identical to those obtained when the lower limit 0^- is used. Whenever the integral of (6-8) is different from zero, there will be a difference in the transforms obtained by the two definitions.

Unless otherwise stated, all Laplace transforms considered in the following sections are assumed to be one-sided as defined by (6-6), and

---

[1] Note that the subscript was dropped from $F(s)$ in (6-6) for convenience. There remains, however, a distinct difference between the functional form of the Fourier transform, $F(\omega)$, and that of the Laplace transform, $F(s)$. This matter is discussed in considerable detail in Section 6-15, where the problem of converting between Laplace and Fourier transforms is discussed.

consideration of time functions is restricted to their positive time portions only. This restriction is not so severe as it might seem, since it permits handling almost all system analysis problems involving nonrandom signals. For example, the impulse response of any physical system satisfies this restriction. Likewise, all practical signals must start at some finite time that can usually be taken to be zero or later.

In using the Laplace transform, it will be found that the integral relationships (6-4) and (6-5) do not actually enter into the calculations very often. They are of great importance in transform theory and can be used to carry out fundamental operations. However, it will be found that much simpler computational methods are available that do not involve direct evaluation of these integrals.

The Laplace transform has certain advantages over the Fourier transform. Among these are the ease of computing transforms; the simplicity of the transforms themselves; the ease of including initial conditions in the solution of differential equations; the insight into system performance that is possible through use of the complex frequency concept; and the ability to deal with time functions that are not absolutely integrable. However, the Fourier transform is still required for certain kinds of signal and system analysis problems. After we examine the conditions under which a function may be Laplace-transformed, a number of elementary transforms are computed and a number of theorems useful in the application of transforms developed. Following this, a variety of techniques for finding the time function corresponding to a given transform are considered, and a number of applications of the Laplace transform to system analysis problems are discussed.

6-2 Existence of the One-Sided Laplace Transform

Requirements on $f(t)$ for the existence of the one-sided Laplace transform $F(s)$ are not at all severe. In order for $F(s)$ to exist, it is necessary that the Laplace integral (6-8) converge. The following set of conditions on $f(t)$ is sufficient to assure the convergence of the integral and covers virtually all waveforms and signals likely to be encountered in the application of Laplace transform theory to physical problems.

Theorem 1.[2] If $f(t)$ is integrable in every finite interval $a < t < b$ (where $0 \le a < b < \infty$), and for some value of c the limit

$$\lim_{t \to \infty} \epsilon^{-ct} |f(t)| \qquad (6\text{-}9)$$

[2]N. M. Nicholson, *Fundamentals and Techniques of Mathematics for Scientists.* New York: John Wiley & Sons, Inc., 1961, p. 330. W. M. Brown. *Analysis of Linear Time Invariant Systems.* New York: McGraw-Hill Book Company, Inc., 1963, pp. 26–39.

exists, then the Laplace integral converges absolutely and uniformly for $\text{Re}(s) > c$; that is,

$$F(s) = \int_0^\infty f(t)\epsilon^{-st}dt < \infty \qquad \text{Re}(s) > c \qquad (6\text{-}10)$$

There are two important properties of this theorem that should be noted: the presence of a finite number of infinite discontinuities is permissible so long as they have finite areas under them; and because of the uniform convergence, it is permissible to invert the order of integration in multiple integrals without altering the result. Most functions encountered in the analysis of engineering problems satisfy the requirements of Theorem 1. Even such functions as ϵ^{100t} or t^n are seen to be sufficiently well-behaved to have Laplace transforms. It is possible to specify functions that do not have a Laplace transform, such as ϵ^{t^2}, but they are of little practical importance in system analysis.

As in the case of Fourier transforms, some short-hand notation is very convenient. To this end, the Laplace transform of a function $f(t)$ will be indicated by

$$\mathcal{L}\{f(t)\} = F(s) = \int_0^\infty f(t)\epsilon^{-st}dt \qquad (6\text{-}11)$$

Similarly, for the inverse transform,[3]

$$\mathcal{L}^{-1}\{F(s)\} = f(t) = \frac{1}{2\pi j}\int_{c-j\infty}^{c+j\infty} F(s)\epsilon^{st}ds \qquad (6\text{-}12)$$

Another notation for the transform relationship is the following:

$$f(t) \Leftrightarrow F(s) \qquad (6\text{-}13)$$

Since the relationship indicated in (6-13) implies a transformation in either direction, it is necessary that the function $f(t) = 0$, for $t < 0$, when the one-sided Laplace transform is being considered. Unless the functional form of $f(t)$ contains this restriction, it must be incorporated by multiplying $f(t)$ by $u(t)$. This matter is discussed further in the next section.

When a Laplace transform is computed, there is a restriction on the value of σ for which the transform is valid. This value of σ determines the region in the complex s-plane in which the integral converges. The value of σ is of importance in determining whether or not a Fourier transform exists for the function and in determining a path of integration that could be used for evaluation of (6-12) by contour integration in the complex plane. However, in most practical calculations, the restrictions on σ are clear and are not tabulated as part of the computation. When the restrictions are pertinent, they must be included. The restrictions on σ

---

[3]The limits on the integral were changed from $\sigma - j\infty$ and $\sigma + j\infty$ to $c - j\infty$, where it is assumed that $c > \sigma$. This is a more general formulation in that it allows any path of integration to the right of the line $s = \sigma$ in the complex plane.

will be given in a number of the derivations of the transforms that follow. Whenever the two-sided Laplace transform is being considered, it is necessary to include the restrictions on σ.

6-3   Direct Computation of Transforms

The defining integral, (6-8), can be used for the direct computation of transforms. As an example, consider the exponential function $\epsilon^{-\alpha t}u(t)$. The Laplace transform is given by

$$\mathcal{L}\{\epsilon^{-\alpha t}u(t)\} = \int_0^\infty \epsilon^{-\alpha t}\epsilon^{-st}dt$$

$$= \frac{\epsilon^{-(\alpha+s)t}}{-(\alpha+s)}\Bigg|_0^\infty \tag{6-14}$$

In order for this integral to converge, it is necessary for $\mathrm{Re}(s) = \sigma > -\alpha$, in which case the transform becomes

$$\epsilon^{-\alpha t}u(t) \Leftrightarrow \frac{1}{s+\alpha} \qquad \sigma > -\alpha \tag{6-15}$$

The transform given in (6-15) is valid for complex α provided $\sigma > \mathrm{Re}(-\alpha)$.

There is one aspect of the above calculation that deserves special emphasis since it is frequently overlooked or lost along the way. This is the question of the inclusion of the unit step multiplying the time function in (6-15). In computing the Laplace transform of a time function, the step is redundant since the transform is defined to include only the positive-time portion of the function. Thus,

$$\mathcal{L}\{f(t)u(t)\} = \mathcal{L}\{f(t)\} \tag{6-16}$$

In any specific case it will be found that when the inverse transformation is carried out, the resulting time function will be zero for negative time. This property is inherent in the transformation itself and is intimately related to the algebraic structure of the transform and the specified region of convergence. This matter is discussed in detail when methods for computing the inverse Laplace transform are considered in Sections 6-6, 6-7, 6-17, and 6-18. The important point at the moment is that, although $\mathcal{L}\{\epsilon^{-\alpha t}\} = 1/(s+\alpha)$, it does *not* follow that $\mathcal{L}^{-1}\{1/(s+\alpha)\} = \epsilon^{-\alpha t}$, but rather that $\mathcal{L}^{-1}\{1/(s+\alpha)\} = \epsilon^{-\alpha t}u(t)$. In many discussions of the Laplace transform it is implicitly assumed that only positive time is being considered and the unit step is omitted in computations. When time functions are being considered that can be nonzero for negative time, it is necessary to be more careful with the notation. In order to avoid any confusion, the unit step will be included whenever it is required to properly delineate the time interval over which the function is being considered. In general, this will be necessary only when an inverse transform is stated or implied as, for example, by the symbol $\Leftrightarrow$.

The Laplace transforms of the unit impulse and unit step are readily found by direct computation to be

$$\mathcal{L}\{\delta(t)\} = \int_0^\infty \delta(t)\epsilon^{-st}dt = 1 \tag{6-17}$$

$$\boxed{\delta(t) \Leftrightarrow 1} \quad \sigma > -\infty \tag{6-18}$$

$$\mathcal{L}\{u(t)\} = \int_0^\infty \epsilon^{-st}dt = \left.\frac{\epsilon^{-st}}{-s}\right|_0^\infty \tag{6-19}$$

$$\boxed{u(t) \Leftrightarrow \frac{1}{s}} \quad \sigma > 0 \tag{6-20}$$

This last result can also be obtained directly from (6-15) by allowing $\alpha \to 0$.

Many other transforms could be computed in this same manner; however, the integrals become very involved and the calculations tedious for even moderately complicated functions. A more practical procedure is to utilize certain operational properties of the Laplace transform to simplify the computations. In addition, familiarity with these operational properties of transforms provides considerable insight into the physical significance, in the time domain, that should be attached to various observed characteristics of the transforms in the complex frequency, or s-domain.

Exercise 6-3.1
Find the Laplace transforms of t and $u(t - t_0)$.

ANS: $\dfrac{1}{s^2}$; $\dfrac{1}{s}\epsilon^{-t_0 s}$; $\sigma > 0$

6-4 Transform Theorems

The utility of transform methods in system analysis stems from the fact that certain operations in one domain are different and oftentimes simpler operations in the other domain. For example, ordinary differential equations in the time domain become algebraic equations in the s-domain after being Laplace-transformed. In order to make effective use of the simplifications possible with transform methods it is necessary to establish the relationships between operations in the two domains. Furthermore, it is necessary to be able to move easily between the two domains. As will be evident shortly, a good understanding of the relationships between operations in the two domains greatly aids carrying out the transformation between the domains. It is this operational relationship between the original (time) domain and the transform (complex frequency) domain that is considered next.

Linearity. By direct application of the defining integral it can be shown that the Laplace transformation is a linear operation; that is, if $f_1(t) \Leftrightarrow F_1(s)$ for $\sigma > \sigma_1$ and if $f_2(t) \Leftrightarrow F_2(s)$ for $\sigma > \sigma_2$, then

$$Af_1(t) + Bf_2(t) \Leftrightarrow AF_1(s) + BF_2(s) \qquad \sigma > \sigma_1, \sigma_2 \qquad \text{(6-21)}$$

where A and B are constants. The main use of this theorem is to allow the linear decomposition of a function to be transformed one element at a time and then summed after transformation to give the transform of the total function. As an example, consider the problem of computing the Laplace transform of $\cos \omega_0 t$. This can be done by breaking up the function into a sum of exponentials and transforming them one at a time by use of (6-15). This gives

$$\mathcal{L}\{\cos \omega_0 t\} = \mathcal{L}\left\{ \frac{1}{2}\epsilon^{j\omega_0 t} + \frac{1}{2}\epsilon^{-j\omega_0 t} \right\}$$

$$= \frac{1}{2}\mathcal{L}\{\epsilon^{j\omega_0 t}\} + \frac{1}{2}\mathcal{L}\{\epsilon^{-j\omega_0 t}\} \qquad \text{(6-22)}$$

$$= \frac{1}{2}\left[\frac{1}{s - j\omega_0} \right] + \frac{1}{2}\left[\frac{1}{s + j\omega_0} \right]$$

$$\cos \omega_0 t u(t) \Leftrightarrow \frac{s}{s^2 + \omega_0^2} \qquad \text{(6-23)}$$

Exercise 6-4.1
Derive the transform relationship

$$\sin \omega_0 t u(t) \Leftrightarrow \frac{\omega_0}{s^2 + \omega_0^2}$$

Scaling. Multiplication by a constant of the variable in one domain affects the image function in a simple manner. For example, if $f(t)$ is replaced by $f(at)$, where $a > 0$ in the defining integral (6-6), we obtain

$$\mathcal{L}\{f(at)\} = \int_0^\infty f(at)\epsilon^{-st}dt = \frac{1}{a}\int_0^\infty f(\lambda)\epsilon^{-s\lambda/a}d\lambda \qquad \text{(6-24)}$$

$$\mathcal{L}\{f(at)\} = \frac{1}{a}F\left(\frac{s}{a}\right)$$

The reason that the constant must be positive is that use of a negative constant amounts to reflecting the time function around the origin so as to include under the integral that portion of $f(t)$ that lies in the interval $-\infty \rightarrow 0$. Clearly, the integral involving this portion of $f(t)$ is not, in general, related in any predictable way to the integral involving only the positive time portion of $f(t)$. In cases where the negative portion of the time function is related to the positive time portion by the simple replacement of t by $-t$ in the functional expression for the positive time portion and if the transform exists, then the above scale-change procedure is valid for negative constants also. Two examples will clarify this. First, let $f(t) = \epsilon^{-t}$ and compute $\mathscr{L}\{f(-t)\}$.

$$\mathscr{L}\{f(-t)\} = \mathscr{L}\{\epsilon^{t}\} = \int_{0}^{\infty} \epsilon^{t}\epsilon^{-st}dt$$

$$= \frac{\epsilon^{(1-s)t}}{1-s}\bigg|_{0}^{\infty} = \frac{1}{s-1} \quad \sigma > 1$$

(6-25)

This same result can be obtained from the scaling theorem as follows:

$$\mathscr{L}\{f(-t)\} = \frac{1}{-1}F(-s) = (-1)\frac{1}{(-s)+1} = \frac{1}{s-1}$$

(6-26)

Consider now the case of $f(t) = \epsilon^{-|t|}$. Now we have

$$\mathscr{L}\{f(-t)\} = \mathscr{L}\{\epsilon^{-|t|}\} = \mathscr{L}\{f(t)\} = \frac{1}{s+1}$$

(6-27)

Since $f(-t)$ cannot be obtained by replacing the argument of $f(t)$ with negative t, it is not possible to use (6-24). In fact, such indiscriminant use of (6-24) leads to an incorrect answer. The change of scale is very useful in extending tables of transforms to include unnormalized variables.

Exercise 6-4.2
Find the Laplace transforms of $\delta(at)$, $u(at)$, and $u(-at)$.

ANS: $\dfrac{1}{a}, 0, \dfrac{1}{s}$

Delay. The transform of a delayed time function can often readily be computed from the transform of the undelayed function. For example, if the original function is a causal time function, $f(t)u(t)$, then the transform of the delayed function is

$$\mathscr{L}\{f(t-t_0)u(t-t_0)\} = F(s)\epsilon^{-t_0 s} \quad t_0 > 0, \sigma > 0$$

(6-28)

This result may be readily shown by direct application of the defining integral and a simple change of variable.

When the original function is not causal (that is, not zero for $t < 0$), the delay brings part of the negative-time portion of the function onto the positive-time axis and can lead to confusion and errors unless care is taken. Some examples illustrate the utility of the t-shift, or $time$-$delay$, theorem in computing transforms as well as some of the precautions required in its application.

Consider the various ways in which the function $f(t) = t$ may be expressed with delays.

$$f_1(t) = f(t)u(t) = tu(t)$$

(6-29)

$$f_2(t) = f(t-t_0)u(t) = (t-t_0)u(t)$$

(6-30)

$$f_3(t) = f(t)u(t - t_0) = tu(t - t_0) \tag{6-31}$$

$$f_4(t) = f(t - t_0)u(t - t_0) = (t - t_0)u(t - t_0) \tag{6-32}$$

These functions are shown in Fig. 6-1. It is seen that the functions are all different, and accordingly, all will have different Laplace transforms.

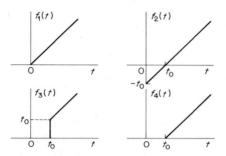

Fig. 6-1. Delayed functions.

These transforms can be computed as follows:

$$F_1(s) = \mathcal{L}\{tu(t)\} = \frac{1}{s^2} \tag{6-33}$$

$$F_2(s) = \mathcal{L}\{(t - t_0)u(t)\} = \mathcal{L}\{tu(t) - t_0 u(t)\}$$
$$= \frac{1}{s^2} - \frac{t_0}{s} \tag{6-34}$$

$$F_3(s) = \mathcal{L}\{tu(t - t_0)\} = \mathcal{L}\{(t - t_0)u(t - t_0) + t_0 u(t - t_0)\}$$
$$= \frac{1}{s^2}\epsilon^{-t_0 s} + \frac{t_0}{s}\epsilon^{-t_0 s} \tag{6-35}$$

$$F_4(s) = \mathcal{L}\{(t - t_0)u(t - t_0)\} = \frac{1}{s^2}\epsilon^{-t_0 s} \tag{6-36}$$

The various algebraic manipulations used in computing these transforms were made to change the form to one that had a transform simply related to that of the original function. This procedure can frequently be used to simplify calculations of transforms of complicated functions. When such simplifications are not possible, it is necessary to carry out the complete computation. For example, $F_3(s)$ would be computed as

$$F_3(s) = \mathcal{L}\{tu(t - t_0)\} = \int_{t_0}^{\infty} t\epsilon^{-st}dt = \frac{\epsilon^{-st}}{s^2}(-st - 1)\Big|_{t_0}^{\infty}$$
$$= \frac{st_0\epsilon^{-t_0 s}}{s^2} + \frac{\epsilon^{-t_0 s}}{s^2} \quad \sigma > 0 \tag{6-37}$$

The t-shift theorem is also particularly useful in calculations where a signal can be considered as being made up of the sum of delayed signals,

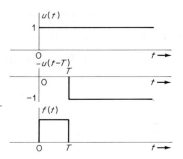

Fig. 6-2 Pulse signal as the sum of two unit steps.

each of which has a known transform. An example of this kind of calculation is the rectangular pulse shown in Fig. 6-2. This pulse can be considered as the sum of a unit step at the origin and a negative unit step delayed an amount T. The transform is readily found to be

$$\mathscr{L}\{u(t) - u(t - T)\} = \frac{1}{s} - \frac{1}{s}\epsilon^{-Ts}$$

$$= \frac{1}{s}(1 - \epsilon^{-Ts})$$

(6-38)

In a similar manner the Laplace transf   n of a single-loop sinusoidal pulse can be thought of as the sum of two sinusoids, as shown in Fig. 6-3. The Laplace transform is then

$$\mathscr{L}\{f(t)\} = \mathscr{L}\{\sin tu(t) + \sin(t - \pi)u(t - \pi)\}$$

$$= \frac{1}{s^2 + 1} + \frac{\epsilon^{-\pi s}}{s^2 + 1}$$

(6-39)

$$= \frac{1}{s^2 + 1}(1 + \epsilon^{-\pi s})$$

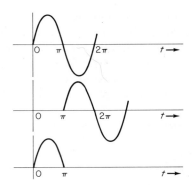

Fig. 6-3 Sinusoidal pulse as the sum of undelayed and delayed sine waves.

Exercise 6-4.3

Find the Laplace transforms of the signals shown.

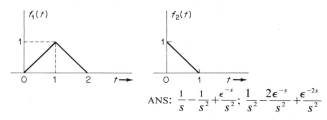

ANS: $\dfrac{1}{s} - \dfrac{1}{s^2} + \dfrac{\epsilon^{-s}}{s^2}$; $\dfrac{1}{s^2} - \dfrac{2\epsilon^{-s}}{s^2} + \dfrac{\epsilon^{-2s}}{s^2}$

Differentiation of time functions.   The Laplace transform of the derivative of a time function is very simply related to the Laplace transform of the original function. The relationship can be derived as follows:

$$\mathscr{L}\left\{\frac{df(t)}{dt}\right\} = \int_0^\infty \frac{df(t)}{dt}\epsilon^{-st}dt \tag{6-40}$$

Integrating by parts gives

$$\mathscr{L}\left\{\frac{df(t)}{dt}\right\} = f(t)\epsilon^{-st}\Big|_0^\infty + s\int_0^\infty f(t)\epsilon^{-st}dt \tag{6-41}$$

The quantity $f(t)\epsilon^{-st}$ must go to zero as $t \to \infty$ in order for the transform $F(s)$ to exist. Therefore, (6-41) becomes

$$\mathscr{L}\left\{\frac{df(t)}{dt}\right\} = sF(s) - f(0) \tag{6-42}$$

It is seen from (6-42) that the transform of the derivative of a function is obtained by multiplying the transform of the function by s and subtracting the value of the function at the origin. It is this simple relationship between the transforms of functions and their derivatives that makes handling of differential equations so simple by transform methods. This matter is discussed extensively in later sections of this chapter.

There is one word of caution that is required with regard to (6-42). In order for this expression to be consistent with the definition of the Laplace transform used here, it is necessary to interpret $f(0)$ as $f(0^-)$. This means that if the function $f(t)$ is a causal function, then $f(0) = f(0^-) = 0$. However, if $f(t)$ is noncausal, then the initial value corresponding to $f(0^-)$ must be included. In order to illustrate this point, consider the two functions $f_1(t) = \cos \omega_0 t$ and $f_2(t) = \cos \omega_0 t u(t)$, which have identical Laplace transforms. These functions along with their derivatives are shown in Fig. 6-4. The Laplace transform of the derivative of $f_1(t)$ is

$$\mathscr{L}\left\{\frac{d}{dt}f_1(t)\right\} = \mathscr{L}\left\{\frac{d}{dt}\cos \omega_0 t\right\} = s\left(\frac{s}{s^2 + \omega_0^2}\right) - \cos(0^-)$$

$$= \frac{s^2}{s^2 + \omega_0^2} - 1 = \frac{-\omega_0^2}{s^2 + \omega_0^2} \tag{6-43}$$

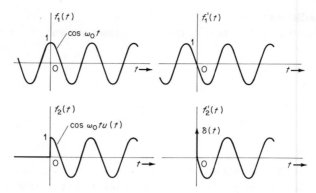

Fig. 6-4 The functions $\cos \omega_0 t$ and $\cos \omega_0 tu(t)$ and their derivatives.

This corresponds to the transform of $d/dt \cos \omega_0 t = -\omega_0 \sin \omega_0 t$.

The transform of the derivative of $f_2(t)$ is

$$\mathscr{L}\left\{\frac{d}{dt} \cos \omega_0 tu(t)\right\} = s\left(\frac{s}{s^2 + \omega_0^2}\right) - \cos(0^-)u(0^-)$$

$$= \frac{s^2}{s^2 + \omega_0^2} \qquad (6\text{-}44)$$

This same result can be obtained by differentiating first and then transforming. This gives

$$\mathscr{L}\left\{\frac{d}{dt} \cos \omega_0 tu(t)\right\} = \mathscr{L}\{-\omega_0 \sin \omega_0 tu(t) + \cos \omega_0 t\delta(t)\}$$

$$= \mathscr{L}\{-\omega_0 \sin \omega_0 tu(t)\} + \mathscr{L}\{\delta(t)\} \qquad (6\text{-}45)$$

$$= \frac{-\omega_0^2}{s^2 + \omega_0^2} + 1 = \frac{s^2}{s^2 + \omega_0^2}$$

Thus, the derivatives have considerably different transforms in the two instances but are consistent with the indicated mathematical operations. In the classical mathematical theory the transforms of the derivatives would be the same in these two cases because the initial condition would be evaluated at 0^+. This use of 0^+ avoids the problem of derivatives of discontinuous functions and, accordingly, avoids the use of impulses, which are often not considered to be legitimate mathematical functions. When one accepts the impulse as a useful concept to be employed in the analysis problem, a much more consistent methodology results from using the 0^- convention, as is done here.

In many instances the differentiation theorem is useful in the derivation of transforms or the interpretation of inverse transforms. However, its most important use is in the transformation of differential equations in the time domain into algebraic equations in the complex frequency domain.

Exercise 6-4.4

Using the differentiation theorem, compute the Laplace transforms of the unit impulse, $\delta(t)$, and unit doublet, $\delta'(t)$, from the transform of the unit step, $u(t)$.

ANS: $s, 1$

Integration.   The Laplace transform of the integral of a function can be derived from the defining equation for the Laplace transform and is given by

$$\mathscr{L}\left\{\int_0^t f(\lambda)d\lambda\right\} = \frac{1}{s}F(s) \tag{6-46}$$

This relationship is useful for transforming integro-differential equations to the s-domain and also in the recognition of certain properties of time functions from their transforms. For example, whenever a transform is such that $1/s$ can be factored out, it is evident from (6-46) that the corresponding time function is the integral of the inverse transform of $sF(s)$. As an illustration, consider the transform $F(s) = 1/s^2$. Since $1/s$ can be factored out, it follows that the inverse transform is

$$\mathscr{L}^{-1}\left\{\frac{1}{s^2}\right\} = \int_0^t \mathscr{L}^{-1}\left\{\frac{1}{s}\right\}d\lambda = \int_0^t u(\lambda)d\lambda = tu(t) \tag{6-47}$$

Similarly, it follows that

$$\mathscr{L}^{-1}\left\{\frac{1}{s^3}\right\} = \frac{1}{2}t^2 u(t) \tag{6-48}$$

s-Shift theorem.   Another theorem that follows readily from the defining equation is the so-called s-shift theorem. This theorem states that shifting $F(s)$ by an amount a in the s-domain corresponds to multiplying $f(t)$ by ϵ^{-at} in the time domain. This may be shown as follows:

$$F(s) = \int_0^\infty f(t)^{-st}dt \tag{6-49}$$

$$F(s+a) = \int_0^\infty f(t)\epsilon^{-(s+a)t}dt = \mathscr{L}\{\epsilon^{-at}f(t)\} \tag{6-50}$$

$$F(s+a) \Leftrightarrow \epsilon^{-at}f(t)u(t) \tag{6-51}$$

As an example, consider the problem of finding the Laplace transform of $\epsilon^{-at}\cos\omega_0 tu(t)$. Since

$$\cos\omega_0 tu(t) \Leftrightarrow \frac{s}{s^2+\omega_0^2} \tag{6-52}$$

it follows immediately that

$$\epsilon^{-at}\cos\omega_0 tu(t) \Leftrightarrow \frac{s+a}{(s+a)^2+\omega_0^2} \tag{6-53}$$

Exercise 6-4.5

Use the s-shift theorem to find the inverse transform of $F_1(s) = \dfrac{1}{s^2 + 2s + 1}$ and $F_2(s) = \dfrac{1}{s^2 + 2s + 2}$

ANS: $te^{-t}u(t)$; $\epsilon^{-t}\sin tu(t)$

Convolution. The Laplace transform of the convolution of two causal time functions is the product of the transforms of the functions taken separately. This may be shown in the following manner:

$$\mathscr{L}\{f_1(t) * f_2(t)\} = \int_0^\infty \left[\int_0^\infty f_1(\lambda)f_2(t-\lambda)d\lambda\right]\epsilon^{-st}dt \tag{6-54}$$

Reversing the order of integration gives

$$\mathscr{L}\{f_1(t) * f_2(t)\} = \int_0^\infty f_1(\lambda)\int_0^\infty f_2(t-\lambda)\epsilon^{-st}dt\,d\lambda$$

$$= \int_0^\infty f_1(\lambda)\epsilon^{-\lambda s}F_2(s)d\lambda \tag{6-55}$$

$$= F_1(s)F_2(s)$$

This theorem is useful in computing transforms and inverse transforms and in interpreting the physical significance of transforms. As an example of the use of this theorem, consider a simple lowpass filter with impulse response, $h(t) = \epsilon^{-t}u(t)$, and a damped exponential input, $v_i(t) = \epsilon^{-t}u(t)$, as shown in Fig. 6-5. The output is given by the convolution of the input and the impulse response. Thus,

$$v_0(t) = v_i(t) * h(t) \tag{6-56}$$

Taking the Laplace transform of both sides of (6-56) gives

$$V_0(s) = V_i(s)H(s)$$

$$= \mathscr{L}\{\epsilon^{-t}u(t)\} \cdot \mathscr{L}\{\epsilon^{-t}u(t)\} \tag{6-57}$$

$$= \frac{1}{s+1} \cdot \frac{1}{s+1} = \frac{1}{(s+1)^2}$$

This result is recognized as the transform of $tu(t)$ with $s + 1$ substituted

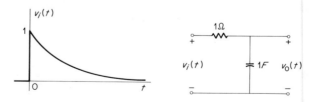

Fig. 6-5 Simple lowpass filter with an exponential input.

for s, and therefore by the s-shift theorem the inverse transform is

$$v_0(t) = \mathcal{L}^{-1}\left\{\frac{1}{(s+1)^2}\right\} = t\epsilon^{-t}u(t). \tag{6-58}$$

The solution of considerably more complicated circuit and systems problems is often no more involved than this example and is one of the reasons the Laplace transform has come to be so widely used.

Product theorem.   The Laplace transform of the product of two time functions is given by the convolution of the transforms of the functions. This result is derived as follows:

$$\mathcal{L}\{f_1(t)f_2(t)\} = \int_0^\infty f_1(t)\left[\frac{1}{2\pi j}\int_{c-j\infty}^{c+j\infty} F_2(\lambda)\epsilon^{\lambda t}d\lambda\right]\epsilon^{-st}dt$$

$$= \frac{1}{2\pi j}\int_{c-j\infty}^{c+j\infty} F_2(\lambda)\int_0^\infty f_1(t)\epsilon^{-(s-\lambda)t}dt\,d\lambda \tag{6-59}$$

$$= \frac{1}{2\pi j}\int_{c-j\infty}^{c+j\infty} F_1(s-\lambda)F_2(\lambda)d\lambda$$

The operation on the right-hand side of (6-59) can be thought of as complex convolution. Its use in evaluating transforms requires integration in the complex frequency plane, a subject that is discussed in Appendix B and in Section 6-18.

Transform of a periodic function.   The Laplace transform of a periodic function can be obtained readily by repeated application of the delay theorem. As an illustration, consider the waveform shown in Fig. 6-6. The complete time function can be expressed as the summation of delayed replicas of the waveform that occurs during the first period. Thus, if $f(t)$ is periodic

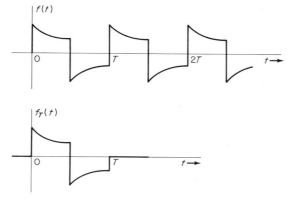

Fig. 6-6 Periodic function.

Table 6-1 Laplace Transforms corresponding to Mathematical Operations

	Time Domain		s-Domain	
Property	Time function	Laplace transform	Property	
Linearity	$a_1f_1(t) + a_2f_2(t)$	$a_1F_1(s) + a_2F_2(s)$	Linearity	
Time differentiation	$f'(t)$	$sF(s) - f(0^-)$	s Multiplication	
Time integration	$\int_0^t f(\xi)d\xi$	$\frac{1}{s}F(s)$	Division by s	
t Multiplication	$tf(t)$	$-\dfrac{dF(s)}{ds}$	Differentiation	
Division by t	$\frac{1}{t}f(t)$	$\int_s^\infty F(\xi)d\xi$	Integration	
Delay (t-shift)	$f(t - t_0)u(t - t_0)$	$\epsilon^{-st_0}F(s)$	Multiplication by exponential	
Multiplication by exponential	$\epsilon^{-at}f(t)$	$F(s + a)$	s-shift	
Scale change	$f(at),\ a > 0$	$\frac{1}{a}F\left(\frac{s}{a}\right)$	Scale change	
Time convolution	$f_1 * f_2 = \int_0^t f_1(\lambda)f_2(t - \lambda)d\lambda$	$F_1(s)F_2(s)$	Multiplication of transforms	
Initial value	$f(0^+)$	$\lim\limits_{s\to\infty} sF(s)$	Limit as $s \to \infty$	
Final value	$f(\infty)$	$\lim\limits_{s\to0} sF(s)$ [F(s) left half plane poles only]	Limit as $s \to 0$	
Second derivative	$f''(t)$	$s^2F(s) - sf(0^-) - f'(0^-)$	s Multiplication	
Multiplication of time functions	$f_1(t)f_2(t)$	$\dfrac{1}{2\pi j}\int_{c-j\infty}^{c+j\infty} F_1(s - \lambda)F_2(\lambda)d\lambda$	Complex convolution	

with period T and the first period is defined as

$$f_T(t) = \begin{cases} f(t) & 0 \leq t < T \\ 0 & \text{elsewhere} \end{cases} \tag{6-60}$$

then

$$f(t) = f_T(t) + f_T(t - T) + f_T(t - 2T) + \cdots \tag{6-61}$$

The Laplace transform of (6-60) is readily computed to be

$$
\begin{aligned}
F(s) &= \mathscr{L}\{f_T(t) + f_T(t - T) + \cdots\} \\
&= F_T(s) + \epsilon^{-Ts} F_T(s) + \epsilon^{-2Ts} F_T(s) + \cdots \\
&= F_T(s)[1 + \epsilon^{-Ts} + \epsilon^{-2Ts} + \cdots] \\
&= F_T(s)\left[\frac{1}{1 - \epsilon^{-Ts}}\right]
\end{aligned}
\tag{6-62}
$$

Thus, the Laplace transform of a periodic function is the transform of the function over one period divided by $1 - \epsilon^{-Ts}$.

6-5   Tables of Transforms

Extensive tables of Laplace transforms have been prepared by various authors, and a number of these are listed in the references at the end of the chapter. For many calculations only a relatively simple table of transforms is required since these can be extended by the principles just discussed to cover a wide variety of functions. In the latter portion of this chapter various methods for carrying out inverse transformation are considered, and it is found that relatively simple techniques handle a majority of the transforms encountered in practice. Because of this it is not necessary, in general, to resort to hunting through a table of Laplace transforms in order to carry out the analysis of a system using these methods. The accompanying, rather short table of transform operations (Table 6-1) and table of Laplace transforms (Table 6-2) will more than suffice for solving all of the problems in this text.

Not all of the transform operations listed in Table 6-1 are discussed in the text. The ones that are not considered are self-explanatory and can be established directly from the defining relationship for the Laplace transform. Their use will be illustrated in subsequent examples. Similarly, several of the transforms given in Table 6-2 have not been derived but can be obtained by elementary operations on transforms that have already been computed.

Table   6-2   Laplace   Transforms
of Elementary Functions

Time function	Transform
$\delta(t)$	1
$u(t)$	$\dfrac{1}{s}$
$tu(t)$	$\dfrac{1}{s^2}$
$t^n u(t)$	$\dfrac{n!}{s^{n+1}}$
$\epsilon^{-\alpha t} u(t)$	$\dfrac{1}{s+\alpha}$
$t\epsilon^{-\alpha t} u(t)$	$\dfrac{1}{(s+\alpha)^2}$
$\sin(\beta t)u(t)$	$\dfrac{\beta}{s^2+\beta^2}$
$\cos(\beta t)u(t)$	$\dfrac{s}{s^2+\beta^2}$
$\epsilon^{-\alpha t}\sin(\beta t)u(t)$	$\dfrac{\beta}{(s+\alpha)^2+\beta^2}$
$\epsilon^{-\alpha t}\cos(\beta t)u(t)$	$\dfrac{s+\alpha}{(s+\alpha)^2+\beta^2}$
$\sinh atu(t)$	$\dfrac{a}{s^2-a^2}$
$\cosh atu(t)$	$\dfrac{s}{s^2-a^2}$

6-6 Inverse Laplace Transform

Utility of the Laplace transform in the solution of the differential equations occurring in system analysis is directly related to the ease with which the transformations between the time domain and the complex frequency domain can be made. Some consideration has already been given to the computation of transforms, that is, going from the t-domain to the s-domain. It is now time to consider the reverse process in some detail, that is, going from the s-domain to the t-domain.

The Laplace transform has the important property of uniqueness that greatly simplifies the problem of obtaining transforms and inverse transforms. For any function, there is only one Laplace transform corresponding to that function. In the case of the inverse Laplace transform it is found that there can be more than one original function corresponding to a particular transform. However, these different original functions are found to differ only by a so-called null function, $n(t)$, that has the property

$$\int_0^t |n(\lambda)|\,d\lambda = 0 \qquad \text{for all } t > 0 \tag{6-63}$$

An example of two functions differing by a null function would be two functions that are defined to have different values at points of discontinuity but that are identical elsewhere. It is usual in mathematical analysis to consider functions to be the same when they differ by only a null function. With this assumption the inversion of the Laplace transform is also unique. This uniqueness of $\mathscr{L}$ and $\mathscr{L}^{-1}$ means that any valid mathematical operation may be used in obtaining the transform or inverse transform and that the answer so obtained will be the same as that obtained by any other method.

The most generally useful method of inverting Laplace transforms is to modify, expand, or otherwise alter the transforms so as to put them into forms that can be inverted by inspection. The principal tools for this procedure are the techniques of partial fraction expansion and the various theorems relating to transforms of mathematical operations.

A much more powerful method of obtaining inverse transforms is through evaluation of the defining integral (6-12), which is

$$f(t) = \frac{1}{2\pi j} \int_{c-j\infty}^{c+j\infty} F(s)\epsilon^{st}ds$$

The evaluation of (6-12) is generally accomplished by contour integration and requires some understanding of complex variable theory for its proper application. The use of this technique is discussed in Section 6-18. First, however, the simpler and more frequently used method of partial fraction expansion is considered.

6-7   Partial Fraction Expansion

The transform analysis of linear time-invariant systems with lumped parameters generally leads to transforms that are rational functions of s; that is, transforms that are ratios of polynomials in s. For example, the transfer function relating the current in the nth mesh of an electric circuit to a voltage source in the circuit is the ratio of two determinants, each of which is a polynomial in s. Rational functions of this form can be expressed as the sum of simple fractions whose denominators are the factors of the denominator of the original function. There will be one term in the expansion for each first-order pole in the original function. As an example, consider the following rational function:

$$F(s) = \frac{5s + 13}{s^2 + 6s + 5} = \frac{2}{s+1} + \frac{3}{s+5} \tag{6-64}$$

The inverse transform is easily obtained by inspection from the expanded form as $(2\epsilon^{-t} + 3\epsilon^{-5t})u(t)$.

In order to expand a rational function in partial fractions it is necessary that the numerator be of lower degree than the denominator. If this requirement is not satisfied by the given function, it can always be met by dividing the denominator into the numerator until the remainder is of lower degree than the divisor. The function can then be expressed as the sum of the quotient terms and the remainder terms divided by the original denominator. This procedure for reducing the degree of the numerator is illustrated by the following example:

$$F(s) = \frac{s^4 + 8s^3 + 23s^2 + 34s + 19}{s^3 + 7s^2 + 15s + 9} \tag{6-65}$$

$$
\begin{array}{r}
s + 1 \\
s^3 + 7s^2 + 15s + 9 \overline{\smash{\big)}\, s^4 + 8s^3 + 23s^2 + 34s + 19} \\
\underline{s^4 + 7s^3 + 15s^2 + 9s} \\
s^3 + 8s^2 + 25s + 19 \\
\underline{s^3 + 7s^2 + 15s + 9} \\
s^2 + 10s + 10
\end{array}
$$

Therefore,

$$F(s) = s + 1 + \frac{s^2 + 10s + 10}{s^3 + 7s^2 + 15s + 9} \tag{6-66}$$

The inverse transforms of the first two terms are obtained readily, and the primary job is to obtain the inverse transform of the remaining term, which is now a proper rational function. In future discussions of partial fraction expansions it is assumed that the function has first been reduced to a proper rational function.

Let the rational function to be expanded be of the form

$$F(s) = \frac{P(s)}{Q(s)} \tag{6-67}$$

where the denominator polynomial $Q(s)$ is now assumed to be of higher degree than the numerator polynomial $P(s)$. According to the fundamental theorem of algebra, if $Q(s)$ is an nth-degree polynomial, it will have n roots, not all of which need be different, and it can be factored into n factors, using these roots as follows:

$$
\begin{aligned}
Q(s) &= s^n + a_{n-1}s^{n-1} + a_{n-2}s^{n-2} + \cdots + a_1 s + a_0 \\
&= (s - \alpha_1)(s - \alpha_2) \cdots (s - \alpha_{n-1})(s - \alpha_n)
\end{aligned} \tag{6-68}
$$

where $\alpha_1, \alpha_2, \ldots, \alpha_n$ are the roots of $Q(s)$ and are solutions to the equation $Q(s) = 0$. Some of the α_k's may not be distinct, and some may be imaginary or complex. The variable s is the complex frequency, and the values of the complex frequency corresponding to the zeros of $Q(s)$ are called the *poles* of the function $F(s)$. Similarly, the complex frequencies corresponding to the roots of the numerator $P(s)$ are called the *zeros* of $F(s)$. The terminology is quite descriptive of the behavior of $F(s)$ at these critical frequencies, since when the denominator approaches zero the function increases without limit (a pole), and when the numerator approaches zero the function also approaches zero (a zero). In taking the inverse transform of a rational function, it is found that the locations of the poles in the complex s-plane completely determine the form of the solution, and the locations of the zeros affect only the amplitudes of the various components present in the solution. This will be apparent readily when the partial fraction expansion is considered in detail.

If it is desired to exclude imaginary and complex roots from the factorization of $Q(s)$, then the factored form can be expressed as the product of linear and quadratic factors, some of which may be repeated. The partial fraction expansion will then be given as a sum of simple fractions having these factors as denominators and each having a numerator of lower degree than its denominator. The most frequently occurring cases are as follows.

Nonrepeated linear factors. Assume that the denominator polynomial contains a nonrepeated linear factor $(s - \alpha_1)$. It is then possible to write the denominator as $Q(s) = (s - \alpha_1)Q_1(s)$, where $Q_1(s)$ is the remainder after $(s - \alpha_1)$ has been factored out of $Q(s)$. The rational function then becomes

$$F(s) = \frac{P(s)}{Q(s)} = \frac{P(s)}{(s - \alpha_1)Q_1(s)} = \frac{A}{s - \alpha_1} + \frac{P_1(s)}{Q_1(s)} \tag{6-69}$$

In (6-69), A is a constant [since this is the only polynomial of lower degree than $(s - \alpha_1)$ to be determined] and $P_1(s)$ is a polynomial of lower degree than $Q_1(s)$. The polynomial $P_1(s)$ will be ignored for the moment, since we are presently concerned with only the first term. Multiplying both sides of (6-69) by $(s - \alpha_1)$ gives

$$(s - \alpha_1)F(s) = \frac{P(s)}{Q_1(s)} = A + (s - \alpha_1)\frac{P_1(s)}{Q_1(s)} \tag{6-70}$$

If this equation is now evaluated at $s = \alpha_1$, the second term on the right-hand side goes to zero, and A is found to be

$$A = \frac{(s - \alpha_1)P(s)}{Q(s)}\bigg|_{s=\alpha_1} = \frac{P(\alpha_1)}{Q_1(\alpha_1)} \tag{6-71}$$

In words, (6-71) may be stated as follows: The constant A in (6-69) can be found by deleting the factor $(s - \alpha_1)$ from the denominator of $F(s)$ and then setting $s = \alpha_1$ in the remaining fraction. As an example, consider the following rational function:

$$F(s) = \frac{3s^2 + 2s + 1}{(s + 1)(s^2 + 12s + 36)} = \frac{A}{s + 1} + \text{other terms}$$

$$A = \frac{3s^2 + 2s + 1}{s^2 + 12s + 36}\bigg|_{s=-1} = \frac{2}{25}$$

Exactly the same procedure is used to evaluate the numerators of the partial fractions corresponding to other linear factors in the same expression. Consider the following example:

$$F(s) = \frac{s^2 + 3s}{(s+1)(s+2)(s+4)} = \frac{A}{s+1} + \frac{B}{s+2} + \frac{C}{s+4}$$

$$A = \left. \frac{s^2 + 3s}{(s+2)(s+4)} \right|_{s=-1} = -2/3$$

$$B = \left. \frac{s^2 + 3s}{(s+1)(s+4)} \right|_{s=-2} = 1$$

$$C = \left. \frac{s^2 + 3s}{(s+1)(s+2)} \right|_{s=-4} = 2/3$$

$$F(s) = \frac{-2/3}{s+1} + \frac{1}{s+2} + \frac{2/3}{s+4}$$

The expansion can always be checked by recombining the terms on the right. The foregoing expansion is valid for complex roots also, but this case will be discussed in connection with quadratic factors, since complex roots always occur as complex conjugate pairs and can be expressed as a single quadratic factor with real coefficients.

Repeated linear factors. If $(s - \alpha_1)^k$, with $k \geq 2$, is a factor of $Q(s)$, then the foregoing method is not sufficient, for there must be k terms in the partial fraction expansion corresponding to this repeated factor. These terms can be expressed in either of the following equivalent forms:

$$\frac{A_1}{(s - \alpha_1)} + \frac{A_2}{(s - \alpha_1)^2} + \cdots + \frac{A_k}{(s - \alpha_1)^k} \qquad \text{(6-72)}$$

or

$$\frac{B_0 + B_1 s + B_2 s^2 + \cdots + B_{k-1} s^{k-1}}{(s - \alpha_1)^k} \qquad \text{(6-73)}$$

where the A's or B's must be determined. Generally, it will be found that the form given in (6-72) leads to more convenient terms for actually carrying out the inverse transformation. Several methods are available for determining the A's or B's. Consider the following example:

$$\frac{s^2}{(s+1)^2(s+2)} = \frac{A_1}{s+1} + \frac{A_2}{(s+1)^2} + \frac{A_3}{s+2}$$

A_3 is determined as before and found to be $A_3 = 4$. The constant A_2 can be found in a similar manner; that is, by multiplying both sides by $(s+1)^2$ and then setting $s = -1$. This leads to a value for A_2 of 1. Unfortunately, the same procedure will not yeild the value of A_1, since multiplication by $(s+1)$ will not remove this factor from the denominator of the second term; so if s were set equal to -1, the expression would go to infinity. This problem can be handled directly by clearing fractions and then equating coefficients of like powers of s or by substituting convenient values of s into the expression to give numerical equations. In the present case this latter procedure may be carried out as follows:

$$\frac{s^2}{(s+1)^2(s+2)} = \frac{A_1}{(s+1)} + \frac{1}{(s+1)^2} + \frac{4}{s+2}$$

Setting $s = 0$ and solving the resulting equation for A_1 leads to the value $A_1 = -3$. Exactly the same result could be obtained by equating the coefficients of s^2, and so on. The final expansion is, therefore,

$$\frac{s^2}{(s+1)^2(s+2)} = \frac{-3}{(s+1)} + \frac{1}{(s+1)^2} + \frac{4}{s+2}$$

Another method that is very effective for the case of repeated roots involves differentiation. Assume that we have a rational function involving a linear factor in the denominator that is repeated k times; that is,

$$F(s) = \frac{P(s)}{Q(s)} = \frac{P(s)}{(s-\alpha)^k Q_1(s)} \tag{6-74}$$

$$= \frac{A_1}{s-\alpha} + \frac{A_2}{(s-\alpha)^2} + \cdots + \frac{A_k}{(s-\alpha)^k} + \frac{P_1(s)}{Q_1(s)} \tag{6-75}$$

If now we multiply both sides of (6-75) by the repeated factor $(s-\alpha)^k$, we obtain

$$(s-\alpha)^k F(s) = A_1(s-\alpha)^{k-1} + A_2(s-\alpha)^{k-2} + \cdots + A_k + \frac{P_1(s)(s-\alpha)^k}{Q_1(s)}$$

$$\tag{6-76}$$

Setting $s = \alpha$ gives the value of A_k as

$$A_k = (s-\alpha)^k F(s)\big|_{s=\alpha} \tag{6-77}$$

To simplify the notation, let $(s-\alpha)^k F(s) = F_1(s)$. We then have

$$A_k = F_1(s)\big|_{s=\alpha} \tag{6-78}$$

Differentiating (6-76) with respect to s and setting $s = \alpha$ gives

$$A_{k-1} = \frac{d}{ds} F_1(s)\big|_{s=\alpha} \tag{6-79}$$

Repeating once more, we obtain

$$A_{k-2} = \frac{1}{2} \frac{d^2}{ds^2} F_1(s)\big|_{s=\alpha} \tag{6-80}$$

The nth term becomes

$$A_{k-n} = \frac{1}{n!} \frac{d^n}{ds^n} F_1(s)\big|_{s=\alpha} \tag{6-81}$$

Applying this procedure to the previous problem, we have

$$F(s) = \frac{s^2}{(s+1)^2(s+2)} = \frac{A_1}{(s+1)} + \frac{A_2}{(s+1)^2} + \frac{A_3}{s+2}$$

As before, A_3 is found to be $A_3 = 4$. The function $F_1(s)$ is now formed, and A_1 and A_2 are evaluated:

$$F_1(s) = \frac{s^2}{s+2}$$

$$A_2 = F_1(s)|_{s=-1} = \frac{s^2}{s+2}\Big|_{s=-1} = 1$$

$$A_1 = \frac{d}{ds}\frac{s^2}{s+2}\Big|_{s=-1} = \frac{s^2+4s}{(s+2)^2}\Big|_{s=-1} = -3$$

In the case of repeated roots it is always possible to evaluate directly the constant in the numerator of the partial fraction corresponding to the highest power of the factor. This evaluation is made by multiplying through by the factor raised to the highest power and evaluating the resulting expression, with s set equal to the root corresponding to this factor. The remaining numerator terms will have to be evaluated by clearing fractions and equating powers of s, by choosing values of s to give simple equations in the unknowns, or by using the differentiation technique of (6-81).

Complex roots and quadratic factors. Some of the roots of the denominator polynomial may be complex. Such roots of polynomials with real coefficients always occur in complex conjugate pairs; that is, for each root $a_1 = -\alpha - j\omega_0$ there is another root $a_2 = a_1^* = -\alpha + j\omega_0$. These factors can be multiplied together to give a quadratic factor having real coefficients:

$$(s + \alpha + j\omega_0)(s + \alpha - j\omega_0) = (s + \alpha)^2 + \omega_0^2 \qquad \text{(6-82)}$$

Alternatively, this factor can be expressed as the quadratic factor:

$$(s + \alpha)^2 + \omega_0^2 = s^2 + 2\alpha s + \omega_0^2 + \alpha^2 = s^2 + as + b \qquad \text{(6-83)}$$

The corresponding partial fraction expansion can then take any of the following forms:

$$\frac{P(s)}{Q(s)} = \frac{P(s)}{Q_1(s)(s^2 + as + b)} \qquad \text{(6-84)}$$

$$= \frac{A_1 s + A_2}{s^2 + as + b} + \frac{P_1(s)}{Q_1(s)} \qquad \text{(6-85)}$$

$$= \frac{B_1 + jB_2}{s + (\alpha + j\omega_0)} + \frac{B_1 - jB_2}{s + (\alpha - j\omega_0)} + \frac{P_1(s)}{Q_1(s)} \qquad \text{(6-86)}$$

$$= \frac{C_1 s + C_2}{(s + \alpha)^2 + \omega_0^2} + \frac{P_1(s)}{Q_1(s)} \qquad \text{(6-87)}^4$$

---

[4]This expansion could also be written as

$$\frac{D_1(s + \alpha)}{(s + \alpha)^2 + \omega_0^2} + \frac{D_2}{(s + \alpha)^2 + \omega_0^2} + \frac{P_1(s)}{Q_1(s)}$$

which leads more directly to the transform in many cases. However, (6-87) is more consistent with previous procedures and easier to remember. Furthermore, it can be connected to this form by inspection.

In all cases the constants A, B, and C are real numbers. The last two forms are generally most convenient to use and are obtained from the first by factoring or completing the square. It should be noted that in (6-86) the numerators are complex conjugates of each other; thus, if one numerator is found, the other can be written down immediately. Also, the constants A_1 and A_2 are equal to C_1 and C_2, respectively, since (6-85) and (6-87) are identical expressions. The constants are determined by clearing fractions and equating powers of s, or in the case of (6-86), by multiplying them by one of the factors and then evaluating at the value of s corresponding to that factor. The following examples illustrate expansion of functions containing quadratic factors:

$$F(a) = \frac{2s^2 + 6s + 6}{(s+2)(s^2+2s+2)} = \frac{A}{s+2} + \frac{Bs+C}{s^2+2s+2}$$

The constant A is found by the usual method to be $A = 1$. The values of B and C are found by clearing fractions and equating the coefficients of powers of s.

$$2s^2 + 6s + 6 = s^2 + 2s + 2 + Bs^2 + 2Bs + Cs + 2C$$

$$B = 2 - 1 = 1$$

$$C = \frac{1}{2}(6-2) = 2$$

$$F(s) = \frac{1}{s+2} + \frac{s+2}{s^2+2s+2}$$

Another, and frequently simpler, method of determining the constants B and C is to set $s = 0$, obtaining $6/4 = 1/2 + C/2$, which readily gives $C = 2$. The value for B can be obtained by multiplying by s and letting $s \to \infty$. This leads to the equation $2 = 1 + B$, or $B = 1$. The inverse transform is found by completing the square in the denominator of the quadratic term and then adjusting the numerator by adding and subtracting quantities to make it complementary to the denominator. Thus,

$$F(s) = \frac{1}{s+2} + \frac{s+2}{(s+1)^2+1} = \frac{1}{s+2} + \frac{s+1}{(s+1)^2+1} + \frac{1}{(s+1)^2+1}$$

$$f(t) = (\epsilon^{-2t} + \epsilon^{-t}\cos t + \epsilon^{-t}\sin t)u(t)$$

In obtaining the final form, use was made of inverse transforms for exponentials and sinusoids and of the s-shift theorem.

This example can also be worked by using linear factors, as follows:

$$F(s) = \frac{2s^2 + 6s + 6}{(s+2)(s^2+2s+6)} + \frac{2s^2 + 6s + 6}{(s+2)[s+(1+j1)][(s+(1-j1)]}$$

$$= \frac{K_1}{s+2} + \frac{K_2}{s+1+j1} + \frac{K_3}{s+1-j1}$$

$K_1 = 1$, as before. K_2 and K_3 can be found by the same method because they are also associated with linear factors.[5] Thus,

$$K_2 = \frac{2s^2+6s+6}{(s+2)[s+1-j1]}\bigg|_{s=-1-j} = \frac{2(1+j2-1)-6-j6+6}{(-1-j+2)(-1-j+1-j)}$$

$$= \frac{1}{1-j} = \frac{1}{2}+j\frac{1}{2}$$

$$K_3 = K_2^* = \frac{1}{2}-j\frac{1}{2}$$

Note that the coefficients associated with complex conjugate roots must always be complex conjugates also in order for the sum of these factors to be real functions of s. Therefore,

$$F(s) = \frac{1}{s+2} + \frac{\frac{1}{2}+j\frac{1}{2}}{s+1+j} + \frac{\frac{1}{2}-j\frac{1}{2}}{s+1-j}$$

This expression could be inverted as it stands and would lead to complex exponentials, which could be converted to sinusoids. However, it is generally more convenient to combine the complex roots giving a term of the following form:

$$\frac{K}{s+(\alpha+j\omega_0)} + \frac{K^*}{s+(\alpha-j\omega_0)}$$

$$= \frac{Ks+K\alpha-jK\omega_0+K^*s+K^*\alpha+jK^*\omega_0}{(s+\alpha)^2+\omega_0^2} \tag{6-88}$$

$$= \frac{2\,\mathrm{Re}(K)(s+\alpha)}{(s+\alpha)^2+\omega_0^2} + \frac{2\mathrm{Im}(K)\omega_0}{(s+\alpha)^2+\omega_0^2} \tag{6-89}$$

---

[5]The evaluation of complicated polynomials, such as the numerator of $F(s)$, can often be most simply accomplished by use of the so-called remainder theorem, which may be stated as follows: If $D(s)$ is the quotient polynomial and R_0 the remainder after division of $F(s)$ by $(s-a)$, then the relationship $\quad F(s) = (s-a)D(s) + R_0$

is an identity. If s is set equal to a, then it follows that

$$R_0 = F(a)$$

Therefore, a polynomial can be evaluated at any point, a, by determining the remainder when the polynomial is divided by $(s-a)$. In the above example the numerator can be evaluated at $s=-1-j$ as follows:

```
                2s  +(4-2j)
        s+1+j | 2s²      +6s   +6
                2s² +(2+j2)s
                _____
                  (4-j2)s +6
                  (4-j2)s +6+j2
                  _____
                        -j2
```

Therefore, $R_0 = -j2 =$ value of numerator at $s=-1-j$. Although not so convenient in this instance, the remainder theorem is of great use in evaluating higher-order polynomials.

The inverse transform of these terms is

$$[2\text{Re}(K)\epsilon^{-\alpha t} \cos \omega_0 t + 2\,\text{Im}(K)\epsilon^{-\alpha t} \sin \omega_0 t]u(t) \tag{6-90}$$

Using this relationship, the inverse transform in the example is, thus, found to be

$$f(t) = (\epsilon^{-2t} + \epsilon^{-t} \cos t + \epsilon^{-t} \sin t)u(t) \tag{6-91}$$

Repeated quadratic factors. When a quadratic factor is repeated, it is necessary to use the same approach as for repeated linear factors. An example will illustrate the expansion in the case of a double quadratic factor.

$$F(s) = \frac{4s^2}{(s^2 + 1)^2(s + 1)}$$

$$= \frac{A}{s + 1} + \frac{Bs + C}{(s^2 + 1)^2} + \frac{Ds + E}{s^2 + 1}$$

A can be found using the normal procedure of multiplying through by $s + 1$ and setting $s = -1$. This gives $A = 1$. The remaining constants can be evaluated by clearing fractions and equating powers of s, as follows:

$$4s^2 = (s^2 + 1)^2 + (Bs + C)(s + 1) + (Ds + E)(s^2 + 1)(s + 1)$$

Equating powers of s gives the following equations:

$$
\begin{array}{ll}
D + 1 = 0; & \text{therefore, } D = -1 \\
D + E = 0; & \text{therefore, } E = -D = 1 \\
-2 + B + D + E = 0; & \text{therefore, } B = 2 \\
C + B + D + E = 0; & \text{therefore, } C = -B = -2 \\
1 + C + E = 0; & \text{therefore, } -1 - 2 + 1 = 0 \quad \text{(check)}
\end{array}
$$

$$F(s) = \frac{1}{s + 1} + \frac{2s - 2}{(s^2 + 1)^2} - \frac{s - 1}{s^2 + 1}$$

This expression can be inverted making use of elementary transform properties.

6-8 Solution of Differential Equations

The Laplace transform may be applied in a very straightforward manner to the solution of differential equations subject to specified initial conditions. As an example, consider the circuit shown in Fig. 6-7, in which it is desired to find the current in the circuit after the switch is closed at $t = 0$. The differential equation for this circuit (after the switch is closed) is readily obtained from Kirchhoff's voltage law as

$$2\frac{di}{dt} + 4i = 12$$

$$i(0^-) = 2A \tag{6-92}$$

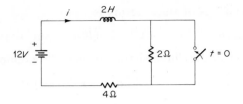

Fig. 6-7 Switched R-C circuit.

Taking the Laplace transform of each term in (6-92) gives

$$2[sI(s) - i(0^-)] + 4I(s) = \frac{12}{s} \qquad (6\text{-}93)$$

Substituting for $i(o^-)$ from (6-92) and solving for $I(s)$ gives

$$2[sI(s) - 2] + 4I(s) = \frac{12}{s}$$

$$I(s)[2s + 4] = \frac{12}{s} + 4 = \frac{4s + 12}{s} \qquad (6\text{-}94)$$

$$I(s) = \frac{4s + 12}{s(2s + 4)} = \frac{2s + 6}{s(s + 2)}$$

Expanding in partial fractions and taking the inverse transform gives the desired time function for the current.

$$I(s) = \frac{2s + 6}{s(s + 2)} = \frac{3}{s} - \frac{1}{s + 2} \qquad (6\text{-}95)$$

$$i(t) = (3 - \epsilon^{-2t})u(t) \qquad (6\text{-}96)$$

As another example, consider the two-node circuit shown in Fig. 6-8. In this circuit it is assumed that prior to opening the switch the capacitor voltages are zero. Assigning node voltages v_1 and v_2 as shown, the differential equations for the circuit can be written from Kirchhoff's current law as

$$2\frac{dv_1(t)}{dt} + v_1(t) - v_2(t) = 12 \qquad (6\text{-}97)$$

$$v_2(t) - v_1(t) + 5\frac{dv_2(t)}{dt} + 4v_2(t) = 0 \qquad (6\text{-}98)$$

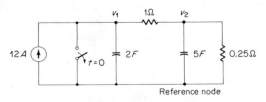

Fig. 6-8 Two-node R-C circuit.

Taking the Laplace transforms of the terms in these equations leads to

$$2sV_1(s) + V_1(s) - V_2(s) = \frac{12}{s} \qquad \text{(6-99)}$$

$$-V_1(s) + 5sV_2(s) + 5V_2(s) = 0 \qquad \text{(6-100)}$$

Collecting terms, (6-99) and (6-100) can be rewritten as

$$(2s + 1)V_1(s) - V_2(s) = \frac{12}{s} \qquad \text{(6-101)}$$

$$-V_1(s) + 5(s + 1)V_2(s) = 0 \qquad \text{(6-102)}$$

These equations can be solved readily for $V_1(s)$. The resulting expression is

$$V_1(s) = \frac{6(s + 1)}{s(s^2 + 1.5s + 0.4)}$$

Expanding in partial fractions and taking the inverse transform leads to the desired result:

$$V_1(s) = \frac{15.00}{s} - \frac{0.99}{s + 1.153} - \frac{14.01}{s + 0.347}$$

$$v_1(t) = (15.00 - 0.99\epsilon^{-1.153t} - 14.01\epsilon^{-0.347t})u(t) \qquad \text{(6-103)}$$

By checking the result against some known value of $v_1(t)$, it is possible to catch algebraic and other errors in a simple manner. In this instance it is possible to see by inspection from Fig. 6-8 that $v_i(0) = 0\,V$ and $v_1(\infty) = 15\,V$. These values are easily checked from (6-103) and show agreement.

Exercise 6-8.1
Find $v_2(t)$ for the circuit of Fig. 6-8.

ANS. $(3 + 1.29\epsilon^{-1.153t} - 4.29\epsilon^{-0.347t})u(t)$

6-9 Initial and Final Value Theorems

The technique of checking the computed solution of a differential equation against some easily determined value was mentioned in the previous section. This is a very valuable procedure and should be used whenever possible. As discussed so far, this technique is applicable only after the solution has been obtained as a time function. Unfortunately, if a mistake is found, there may be a great deal of effort that has been wasted in carrying out the inverse transformation of an erroneous transform. This difficulty may be avoided by checking the transform before carrying out the inversion operation. This check can be made with the aid of the following simple theorems.

Initial value theorem.

$$\lim_{t \to 0^+} f(t) = \lim_{s \to \infty} sF(s) \qquad \text{(6-104)}$$

Final value theorem.

$$\lim_{t \to \infty} f(t) = \lim_{s \to 0} sF(s) \qquad \text{(6-105)}$$

As an example of the application of these theorems, consider the transform of a damped cosine wave:

$$f(t) = \epsilon^{-\alpha t} \cos \omega_0 t \Leftrightarrow \frac{s + \alpha}{(s + \alpha)^2 + \omega_0^2}$$

From the initial value theorem it follows that

$$f(\infty) = \lim_{s \to 0} \frac{s(s + \alpha)}{(s + \alpha)^2 + \omega_0^2} = \lim_{s \to \infty} \frac{1 + \dfrac{\alpha}{s}}{1 + \dfrac{2\alpha}{s} + \dfrac{\alpha 2}{s^2} + \dfrac{\omega_0^2}{s^2}} = 1$$

Similarly, from the final value theorem,

$$f(\infty) = \lim_{s \to 0} \frac{s(s + \alpha)}{(s + \alpha)^2 + \omega_0^2} = 0$$

The use of these theorems as a partial check on the validity of a transform representation of a function is generally quite straightforward. As an example, consider the solution for the voltage, $v_1(t)$, in Fig. 6-8. The transform was found to be

$$V_1(s) = \frac{6(s + 1)}{s(s^2 + 1.5s + 0.4)}$$

From the initial and final value theorems the initial and final values of the time function are found to be

$$v_1(0^+) = \lim_{s \to \infty} \frac{6s(s + 1)}{s(s^2 + 1.5s + 0.4)} = 0 \text{ volts}$$

$$v_1(\infty) = \lim_{s \to 0} \frac{6(s + 1)}{s^2 + 1.5s + 0.4} = 15 \text{ volts}$$

These values check with the values that can be determined readily by inspection of the circuit diagram.

The proofs of these theorems follow directly from the expression for the Laplace transform of the derivative of a function. Assume that $f(t)$ and $f'(t)$ are Laplace transformable and that the function $f(t)$ may be discontinuous at the origin. Define a new function $f_1(t)$ such that

$$f(t) = f_1(t) + au(t) \qquad \text{(6-106)}$$

where $f_1(t)$ is continuous at the origin and a is the amount of discontinuity in $f(t)$ at the origin. These functions are shown in Fig. 6-9. Taking the derivative of (6-106) gives

$$f'(t) = f_1'(t) + a\delta(t) \qquad \text{(6-107)}$$

Using the derivative theorem of the Laplace transform, we obtain

$$sF(s) - f(0^-) = \int_{0^-}^{\infty} f'(t)\epsilon^{-st}dt$$

$$= \int_{0^-}^{\infty} [f_1'(t) + a\delta(t)]\epsilon^{-st}dt \qquad \text{(6-108)}$$

$$= \int_{0^-}^{\infty} f_1'(t)\epsilon^{-st}dt + a$$

Therefore, it follows that

$$sF(s) = \int_{0^-}^{\infty} f_1'(t)\epsilon^{-st}dt + a + f(0^-) \qquad \text{(6-109)}$$

The sum of the last two terms is readily seen from Fig. 6-9 to equal $f(0^+)$.

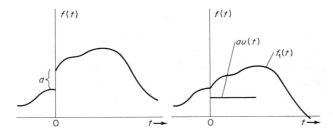

Fig. 6-9 Representation of a discontinuous function.

Taking the limit of (6-109) as $s \to \infty$ and assuming $f_1(t)$ to be of exponential order gives the desired result:

$$\lim_{s \to \infty} sF(s) = \lim_{s \to \infty} \int_{0^-}^{\infty} f_1'(t)\epsilon^{-st}dt + f(0^+)$$

$$= \int_{0^-}^{\infty} \lim_{s \to \infty} f_1'(t)\epsilon^{-st}dt + f(0^+) \qquad \text{(6-110)}$$

$$= f(0^+)$$

The proof of the final value theorem is much more straightforward and is left as an exercise for the student.

Exercise 6-9.1
Determine the initial and final values of the time functions corresponding to the transforms

$$\frac{3s^2 + 1}{4s^3 + 3s^2 + 2s} \quad \text{and} \quad \frac{3s^3 + 1}{4s^3 + 3s^2 + 2s + 1}$$

ANS. 0, 1/2, 3/4, ∞

Exercise 6-9.2
Prove the final value theorem.

6-10 Network Analysis

In dealing with circuit analysis problems it is frequently convenient to proceed directly with the transformed variables rather than starting with the time-domain specifications and subsequently transforming to the s-domain. This is generally very simple to do because of the direct relationships between the equations in the two domains. For example, in the case of an inductor the relationship between the voltage and current is

$$v(t) = L\frac{di(t)}{dt} \tag{6-111}$$

In this equation, L is the constant of proportionality between the voltage and the rate of change of current. The Laplace transform of (6-111) gives

$$V(s) = sLI(s) - Li(0) \tag{6-112}$$

The transformed equation contains an additional term resulting from the presence of an initial current in the inductor. In (6-112) this additional term adds directly to the transform of the voltage $V(s)$ and, therefore, can be represented as an ideal voltage generator having a transform of $-Li(0)$. The s-domain equivalent circuit of an inductance can, therefore, be thought of as an impedance sL in series with an ideal voltage source.

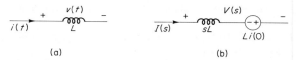

(a)   (b)

Fig. 6-10 The s-domain equivalent of an inductance. (a) Time domain. (b) s-Domain.

Such an equivalent circuit is shown in Fig. 6-10. Using Norton's theorem, the s-domain equivalent circuit of Fig. 6-10 can be converted to another form, as shown in Fig. 6.11. This second relationship can also be derived from the expression for the current through an inductance in terms of the voltage. Thus,

$$i(t) = \frac{1}{L}\int_0^t v(\lambda)d\lambda + i(0)$$

$$I(s) = \frac{1}{Ls}V(s) + \frac{1}{s}i(0)$$

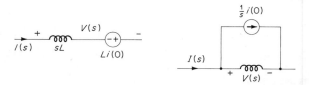

Fig. 6-11 Two s-domain equivalent circuits of an inductance.

Table 6-3 s-Domain Equivalents for Electric Circuit Elements (All Elements Are Represented in Terms of Their Impedance)

Time-domain representation

$$v(t) = Ri(t) \qquad v(t) = L\frac{di}{dt} \qquad v(t) = \frac{1}{C}\int_0^t i(\lambda)\,d\lambda + v(0)$$

$$i(t) = \frac{1}{R}v(t) \qquad i(t) = \frac{1}{L}\int_0^t v(\lambda)\,d\lambda + i(0) \qquad i(t) = C\frac{dv}{dt}$$

Complex frequency representation

$$V(s) = RI(s) \qquad V(s) = sLI(s) - Li(0) \qquad V(s) = \frac{1}{sC}I(s) + \frac{1}{s}v(0)$$

$$I(s) = \frac{1}{R}V(s) \qquad I(s) = \frac{1}{sL}V(s) + \frac{1}{s}i(0) \qquad I(s) = sCV(s) - Cv(0)$$

The second equation corresponds to the circuit on the right in Fig. 6-11. Similar s-domain circuits are readily found for resistances and capacitances and are tabulated in Table 6-3. By use of the equivalents of Table 6-3, it is very simple to sketch the s-domain or transform network, from which the circuit equations can be immediately written down, including all initial conditions. This method is particularly useful in the solution of transient problems. When the initial conditions are zero, the concept of the transfer function becomes very useful and usually leads to further simplifications in the solution of network response problems.

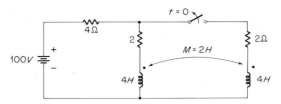

Fig. 6-12 Example of transient analysis.

Solution of transient problems involving initial conditions can be illustrated best by considering some specific examples. In Fig. 6-12 is shown a network that is assumed to have reached an equilibrium state prior to the time $t = 0$ when the switch is opened. It is required to find the current $i_1(t)$ for $t > 0$. The transform network is most easily obtained by replacing the coupled circuit by an equivalent T network, as shown in Fig. 6-13. The transform network is then as shown in Fig. 6-14.

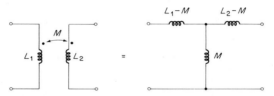

Fig. 6-13 Equivalent for an inductively coupled circuit.

From Fig. 6-14 the expression for $I_1(s)$ is found to be

$$I_1(s) = \frac{100/s + 20 + 40}{4 + 2 + 2s + 2s}$$

$$= \frac{15s + 25}{s(s + 3/2)} = \frac{50/3}{s} - \frac{5/3}{s + 3/2}$$

$$i(t) = \left(\frac{50}{3} - \frac{5}{3}\epsilon^{-3t/2}\right)u(t)$$

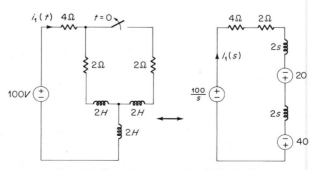

Fig. 6-14 Transform network of Fig. 6-13.

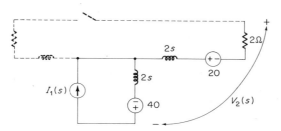

Fig. 6-15 Secondary circuit of Fig. 6-14.

Suppose it is desired to find the voltage across the secondary inductance. The equivalent transform circuit is shown in Fig. 6-15. Using the value for $I_1(s)$ found above, the transform of $V_2(s)$ is

$$V_2(s) = -20 - 40 + 2sI_1(s)$$

$$= -60 + \frac{30s + 50}{s + 3/2} - 30 + \frac{5}{s + 3/2}$$

Therefore,

$$v_2(t) = -30\delta(t) + 5\epsilon^{-3/2t}u(t)$$

6-11   The Transfer Function

The transfer function $H(s)$ of a network is defined as the ratio of the transforms of the output and input signals when the initial conditions are zero. If the excitation or input signal is $v_i(t)$ and the response or output signal is $v_o(t)$, then the transfer function $H(s)$ is given by

$$H(s) = \frac{\mathscr{L}\{v_o(t)\}}{\mathscr{L}\{v_i(t)\}} = \frac{V_o(s)}{V_i(s)} \tag{6-113}$$

This function will be independent of $V_i(s)$ since in a linear system the output is proportional to $V_i(s)$ and, therefore, $V_i(s)$ will cancel out in the ratio given in (6-113). The transfer function is a fundamental characteristic of a system and finds use in studies of system response, system stability, and system synthesis. For example, if one knows the desired output for a particular input signal, then it is possible to determine the transfer function of the system that will produce that output/input relationship by taking the ratio of the transforms of the signals. As an example of the computation of a transfer function, consider the network shown in Fig. 6-16, along with the corresponding transform network. For purposes of computing the transfer function it is not necessary to know the nature of $V_i(s)$. The input can be left arbitrary and the output computed in terms of this arbitrary input. Dividing this output by $V_i(s)$ then gives $H(s)$. The analysis proceeds essentially the same way as for steady-state ac analysis. In this example, the output can be found using the voltage division

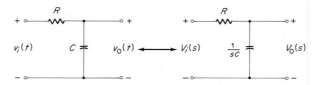

Fig. 6-16 Simple lowpass filter.

between the impedances R and $1/sC$ or by determining the current and multiplying by the impedance $1/sC$. In either case we find

$$V_0(s) = \frac{1/sC}{R + 1/sC} V_i(s) = \frac{1/RC}{s + 1/RC} V_i(s)$$

Therefore, the transfer function is

$$H(s) = \frac{V_0(s)}{V_i(s)} = \frac{1/RC}{s + 1/RC} \tag{6-114}$$

Once the transfer function has been determined, the output for any input can be found as the inverse transform of the product of the transfer function and the transform of the input signal. Thus,

$$V_0(s) = H(s)V_i(s) \tag{6-115}$$

$$V_0(t) = \mathscr{L}^{-1}\{V_0(s)\} = \mathscr{L}^{-1}\{H(s)V_i(s)\} \tag{6-116}$$

As an example of this use of the transfer function, we shall determine the output of the R–C circuit of Fig. 6-16 for an input of $u(t)\cos(100t)$ and a time constant $RC = 0.01$. Using the transfer function from (6-114) we have

$$V_0(s) = H(s)V_i(s) = \left[\frac{100}{s + 100}\right]\left[\frac{s}{s^2 + (100)^2}\right]$$

$$= \frac{A}{s + 100} + \frac{Bs + C}{s^2 + (100)^2} \tag{6-117}$$

$$A = \frac{100(-100)}{(100)^2 + (100)^2} = -\frac{1}{2}$$

Setting $s = 0$ in (6-117) gives $C = 50$, and multiplying through by s and letting $s \to \infty$ gives $B = \frac{1}{2}$. Therefore,

$$V_0(s) = \frac{1}{2}\left[\frac{-1}{s + 100} + \frac{s + 100}{s^2 + 10^4}\right]$$

$$v_0(t) = \frac{1}{2}[-\epsilon^{-100t} + \cos(100t) + \sin(100t)]u(t)$$

The transfer function can often be computed without carrying out a formal solution for the output in terms of the input. One of the simplest and quickest methods is to assume that the output is unity, and then compute what input must have been present to produce this output. The reciprocal of this input is the required transfer function. This method is especially useful for ladder networks because the computations are almost trivial. An example will clarify the method. In carrying out the actual computations of the input, the simplest procedure is to label the voltages and currents on the transform network. Figure 6-17(a) shows an $R-L-C$ circuit for which the voltage transfer function is required. The transform circuit is shown in Fig. 6-17(b) with an assumed output of $V_o(s) = 1$. The currents and voltages in the rest of the network are computed as shown in Fig. 6-17(c). Starting at the right-hand terminals, it is assumed that the voltage is 1 volt. Accordingly, the current in the resistance across the output terminals must be $1/R$, and this is also the current flowing through the inductance L_2. The voltage across L_2 is the product of the current $1/R$ and the impedance sL_2, or sL_2/R. The voltage across the capacitor is obtained as the sum of the voltage across the inductance and the output

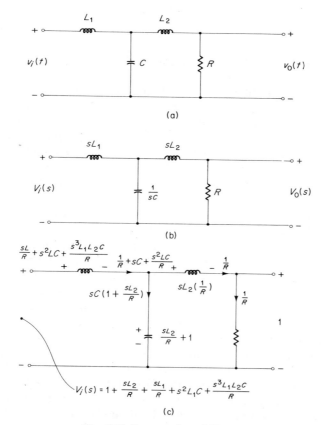

Fig. 6-17 Computation of $H(s)$.

voltage and is $1 + sL_2/R$. This computation process is continued until the input voltage is found to be

$$V_i(s) = 1 + \frac{sL_2}{R} + \frac{sL_1}{R} + s^2L_1C + \frac{s^3L_1L_2C}{R} \tag{6-118}$$

Since the transfer function is $V_o(s)/V_i(s)$, and $V_o(s)$ is unity, it follows that $H(s)$ is the reciprocal of (6-118), or

$$H(s) = \frac{V_o(s)}{V_i(s)} = \frac{1}{1 + \dfrac{sL_2}{R} + \dfrac{sL_1}{R} + s^2L_1C + \dfrac{s^3L_1L_2C}{R}} \tag{6-119}$$

$$= \frac{R/L_1L_2C}{s^3 + \dfrac{R}{L_2}s^2 + \left(\dfrac{1}{L_1C} + \dfrac{1}{L_2C}\right)s + \dfrac{R}{L_1L_2C}}$$

This method of computing transfer functions is also valid for ratios of currents or mixed variables, such as the ratio of the output voltage to input current. The method of computation is identical.

6-12   Steady-State Response to Periodic Waveforms

An important problem in network analysis is that of determining the response of a network to a periodic function other than a sinusoid. Often the desired result is the (periodic) waveform corresponding to the steady-state response. In theory, the solution to this problem is quite simple. However, in order to obtain useful results a considerable amount of computation can be required. The difficulty in arriving at the steady-state response comes about because of the one-sided nature of the Laplace transform. When the Laplace transform of the driving function is computed, it corresponds to that of a waveform that was zero for $t < 0$, and unless you know the precise initial conditions that would have existed in the network at $t = 0^-$, assuming the driving waveform had been applied since $t = -\infty$, there will be transient terms associated with the solution. The only way to know the correct initial conditions in order to eliminate the transient portion of the solution is to have already solved the problem. Fortunately, there is a simple way out of this dilemma. When a waveform is applied to a system, there are two parts to the response of the system. One part is the transient response that is produced by the shock excitation of the natural frequencies of the system when the waveform is applied. The other part is the steady-state response to the excitation. The sum of these two is the total response of the system. By use of the Laplace transform it is simple to calculate the total response and the transient response. The steady-state response can then be determined as the difference of these two responses.

It might appear to be more straightforward just to compute the total response and then evaluate it over a one-period interval for a very large value of t after all transients have died out. Such a procedure would give the correct result, but it turns out that in order to compute the response for a large value of t you must know it for all lesser values of t. This will become clearer when a specific example is considered in detail.

In order to successfully compute the steady-state response by subtracting the transient response from the total response, it is necessary to be able to identify the transient response. This response is actually the solution of the homogeneous differential equation describing the network, and therefore, its transform contains only the poles of the network response function. By separating out the portion of the response transform that contains only the poles of the system transfer function, a specific expression for the transient response can be obtained. The entire procedure can be explained most easily by carrying through an example.

Consider the circuit shown in Fig. 6-18. Assuming that the input is $v_1(t)u(t)$, the output can be calculated from the Laplace transform of the input and the system transfer function as

$$V_2(s) = V_1(s)H(s)$$

$$= \frac{1 - \epsilon^{-1/2s}}{s(1 - \epsilon^{-s})} \cdot \frac{1/2}{s + 1/2} \tag{6-120}$$

The transient portion of the response is that which results from the pole at $s = -\frac{1}{2}$. This portion of the response can be separated from the remainder by computing the partial fraction expansion terms corresponding to the poles of the transfer function. In the example we are considering, this leads to a transient term, $V_{\text{Tran}}(s)$, of

$$V_{\text{Tran}}(s) = \frac{(1 - \epsilon^{1/4})(1/2)}{(-1/2)(1 - \epsilon^{1/2})} \cdot \frac{1}{s + 1/2}$$

$$= -\frac{0.438}{s + 1/2}$$

$$v_{\text{Tran}}(t) = -0.438\epsilon^{-1/2t}u(t)$$

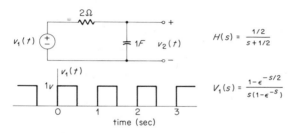

Fig. 6-18 Circuit with square-wave excitation.

The total response, $V_2(s)$, is the sum of the transient response and the steady-state response, $V_{SS}(s)$, and therefore,

$$V_{SS}(s) = V_2(s) - V_{\text{Tran}}(s)$$

$$= \frac{1/2(1 - \epsilon^{-1/2s})}{s(s + 1/2)(1 - \epsilon^{-s})} + \frac{0.438}{s + 1/2}$$

$$= \frac{0.5 + 0.438s - 0.5\epsilon^{-1/2s} - 0.438\epsilon^{-s}s}{s(s + 1/2)(1 - \epsilon^{-s})}$$

This can be evaluated to give the steady-state response over the first period by dropping those terms containing factors of ϵ^{-s} corresponding to delays greater than 1 second. The result is

$$V_{SST}(s) = \frac{0.5 + 0.438s - 0.5\epsilon^{-1/2s}}{s(s + 1/2)} \tag{6-121}$$

Considering the portion corresponding to the first one-half period, the inverse transform is found to be

$$v_{SST}(t) = \mathcal{L}^{-1}\left\{\frac{0.5 + 0.438s}{s(s + 1/2)}\right\}$$

$$= (1 - 0.562\epsilon^{-1/2t})u(t) \qquad 0 \le t < 1/2 \tag{6-122}$$

The time function for the second half of the period is

$$v_{SST}(t) = (1 - 0.562\epsilon^{-1/2t})u(t) - \mathcal{L}^{-1}\left\{\frac{0.5\epsilon^{-1/2s}}{s(s + 1/2)}\right\}$$

$$= (1 - 0.562\epsilon^{-1/2t})u(t) - (1 - \epsilon^{-1/2(t-1/2)})u(t - 1/2) \qquad 1/2 \le t < 1 \tag{6-123}$$

Cancelling the constant terms and multiplying and dividing the first term by $\epsilon^{1/4}$ allows the expression for the second half period to be written as

$$v_{SST}(t) = 0.562\epsilon^{-1/2(t-1/2)} \qquad 1/2 \le t < 1$$

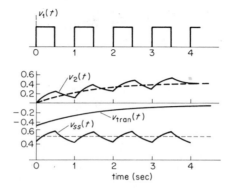

Fig. 6-19 Response of R-C network to a square-wave input.

The steady-state response over one period is, therefore,

$$v_{SST}(t) = \begin{cases} 1 - 0.562\epsilon^{-1/2t} & 0 \le t < 1/2 \\ 0.562\epsilon^{-1/2(t-1/2)} & 1/2 \le t < 1 \end{cases} \tag{6-124}$$

The total response at any time can be found by adding the transient response to the steady-state response. The waveforms, corresponding to these various responses are shown in Fig. 6-19.

6-13   Physical Realizability and Stability

It is possible to establish certain limitations on the form of the system transfer function if it is to correspond to a system that could actually occur in nature. The principal constraint for physical realizability is that the system be causal, or nonanticipatory. The requirement for causality in the time domain is that the impulse response, $h(t)$, be zero for negative time. The corresponding relationship in the frequency domain when the system function is expressed in terms of ω is discussed in Section 5-10. When the system function is expressed in terms of the complex frequency s as a one-sided Laplace transform, the causality requirement is automatically met since the inverse transform is identically zero for $t > 0$.

Another requirement of any practical physical system is that it be stable in the sense that the output of the system should remain bounded when the input is bounded (BIBO). It is shown in Section 4-6 that a sufficient condition for a system to be stable is that the area under the magnitude of the impulse response be finite. The significance of this requirement on $H(s)$ can be seen by examining the time functions that can result from various forms of transfer functions. In order to keep the discussion simple, as well as useful, the transfer functions will be taken as rational functions of s, corresponding to lumped parameter, time-invariant linear systems. Such transfer functions can be expanded in partial fractions, and the inverse transforms of the various terms in the expansion determine whether or not the system is stable.

The simplest kind of term that can occur is of the form $k_a/(s + a)$, which results from a pole in the transfer function at $s = -a$. The time function corresponding to this term is a decaying exponential, $k_a\epsilon^{-at}u(t)$. If a were a negative number ($a \neq 0$), the exponential would be increasing with time and would clearly correspond to an unstable system. From this discussion it is evident that in order for a system to be stable in the BIBO sense there can be no simple real poles in the right half of the s-plane. The same argument can be used to show that a stable transfer function can have no multiple-order real poles in the right half of the s-plane.

Another type of term that occurs in transfer functions is that in which the poles are complex. In physical systems described by real parameter values (as opposed to complex values) such poles always occur in pairs, with one element being the complex conjugate of the other. Terms of this type have the form

$$\frac{k_1}{s + \alpha + j\omega_0} + \frac{k_1^*}{s + \alpha - j\omega_0} = \frac{2(s + \alpha) \operatorname{Re} k_1 + 2\omega_0 \operatorname{Im} k_1}{(s + \alpha)^2 + \omega_0^2} \tag{6-125}$$

The inverse transform of (6-125) has the form of an exponentially damped sinusoid with the exponential factor being $\epsilon^{-\alpha t}$. Again, as long as α is positive, that is, as long as the poles are in the left half plane, the system will be stable. When the poles are in the right half plane (that is, $\alpha < 0$) the corresponding time function is an exponentially increasing sinusoid and corresponds to an unstable system. Repeated complex poles in the right half plane also give exponentially increasing time functions and, therefore, correspond to unstable systems.

A particularly interesting situation results when the poles are in neither the left half plane nor the right half plane but lie exactly on the $j\omega$-axis. A pole at the origin corresponds to a step function, while a conjugate pair of imaginary poles at $\pm j\omega_0$ corresponds to a sine or cosine wave. Terms of this type do not correspond to stable systems in the BIBO sense because an input at the natural frequency leads to an output whose magnitude is an increasing function of time. For example, consider a system function of the form $H(s) = 1/(s^2 + \omega_0^2)$ and let the input be a sinusoid of unit amplitude and having an angular frequency of ω_0. The system response would be

$$\mathcal{L}^{-1}\left\{\frac{1}{s^2 + \omega_0^2} \cdot \frac{\omega_0}{s^2 + \omega_0^2}\right\} = \frac{1}{2\omega_0^2}(\sin \omega_0 t - \omega_0 t \cos \omega_0 t)u(t)$$

which is seen to increase without bound. Only a signal at the resonant frequency of the system leads to this instability, and the responses to other bounded signals are well-behaved. Accordingly, a transfer function containing terms of this type is often called *marginally* stable. When repeated poles occur on the $j\omega$ axis, the corresponding terms in the impulse response increase as powers of t, and the system is always unstable.

In summary, the requirements for (marginal) stability in terms of the poles of the transfer function of a time-invariant, lumped parameter system are:

1. No poles in the right half of the s-plane.
2. No (repeated) poles on the $j\omega$-axis.

In formulating these criteria, it is implicitly assumed that the system function is a proper rational function. A general rational system function can always be expressed as the sum of a polynomial in s plus a proper rational function, as discussed in Section 6-7. If this polynomial has a term in s^k, where $k > 0$, with a nonzero coefficient, then a portion of the system is acting as a kth-order differentiator. Hence, the response to the bounded input, $\sin \omega t$, will include a term having amplitude proportional to ω^k, and therefore, it can be made arbitrarily large by increasing the

frequency of the unit amplitude sinusoid. Thus, we have a third requirement for stability:

3. The degree of the numerator of $H(s)$ cannot exceed that of the
denominator.

Taken together, conditions 1, 2, and 3 are both necessary and sufficient
for determining whether a rational system function corresponds to a stable
system.

The application of the above stability criteria requires knowledge of the
location of the poles of the transfer function. One way to obtain this
information is to factor the denominator polynomial of the transfer function. Such computations are very tedious if carried out by hand but can be
done quite readily with subroutines that are part of the standard software
package of most general-purpose digital computers.

If one is interested only in determining whether or not a system is
stable, then it is not necessary to determine the exact location of the poles,
but only if any of them are in the right half plane. There are several relatively simple techniques for obtaining this information, and some of these
will be considered briefly.

Let the denominator polynomial of the transfer function be designated
as

$$Q(s) = a_n s^n + a_{n-1} s^{n-1} + \cdots + a_1 s + a_0 \qquad \text{(6-126)}$$

A necessary, but not sufficient, condition that all the roots of $Q(s)$ lie in the
left half plane is that all of the coefficients of (6-126) be positive. Hence, if
one or more of the coefficients $(a_0, a_1, \ldots, a_n)$ are negative, then $Q(s)$ *must*
have right half plane roots, and the transfer function is definitely unstable.
No further computations are necessary. Another condition that must be
satisfied by $Q(s)$ for an unconditionally stable system is that all coefficients
(except possibly a_0) must be nonzero. Thus, if the polynomial has any
missing powers of s, then it must have roots in the right half plane or on the
$j\omega$-axis, and the transfer function will be either unstable or marginally
stable.

In the methods to be discussed below it is assumed that $Q(s)$ has
coefficients that are all positive and nonzero. Hence, it is not immediately
obvious whether or not the system is stable, and it is necessary to consider more powerful techniques. One such method is the Routh test.

Routh test.   This test forms a table (Routh table) from the coefficients of
$Q(s)$. The first two rows of the table are formed directly from the coefficients of the polynomial as

$$
\begin{array}{cccc}
a_n & a_{n-2} & a_{n-4} & \cdots \\
a_{n-1} & a_{n-3} & a_{n-5} &
\end{array}
$$

The third row of the table is formed by cross-multiplication as follows:

$$\frac{a_{n-1}a_{n-2} - a_n a_{n-3}}{a_{n-1}} \qquad \frac{a_{n-1}a_{n-4} - a_n a_{n-5}}{a_{n-1}}$$

The fourth row is found by using the same cross-multiplying rule on rows two and three. Additional rows are formed in this way until no term remains. In the course of this development the elements in any row may be multiplied or divided by any positive number without changing the result. This may simplify the numerical work of finding the elements of the succeeding row. After the table is completed, the first column is examined for *changes* in sign. If all the elements in the first column have the same sign, then $Q(s)$ has no roots in the right half plane. If there are changes of sign in the first column, the number of changes equals the number of roots in the right half plane.[6]

An example will serve to illustrate the Routh test. Let

$$Q(s) = s^3 + s^2 + 2s + 8$$

The Routh table becomes

$$
\begin{array}{ccc}
1 & 2 & 0 \\
1 & 8 & 0 \\
-6 & 0 & \\
8 & 0 &
\end{array}
$$

It is clear that there are two changes of sign in the first column, and hence, $Q(s)$ has two roots in the right half plane. The reader can verify easily that the roots of $Q(s)$ are in fact -2, $1/2 + j1/2\sqrt{15}$, and $1/2 - j1/2\sqrt{15}$.

Occasionally, the Routh test may fail to give any result because a zero is generated in the first column or because an entire row of zeros is generated. These cases may arise from roots on the $j\omega$-axis or from a root being the negative of another root. There are techniques for handling these special situations but they will not be discussed here.

Hurwitz test.[7]   Another test of system stability that is relatively easy to apply is the Hurwitz test in which a continued fraction expansion of the ratio of the even and odd parts of $Q(s)$ is formed. Let $Q(s)$ be written as

$$Q(s) = m(s) + n(s) \qquad \text{(6-127)}$$

where $m(s)$ contains all of the even powers of s, and $n(s)$ contains all the odd powers of s. When $m(s)$ is of higher degree than $n(s)$, form the ratio

[6]Since a_0 has been assumed positive, there must be an *even* number of roots in the right half plane if there are any. In some cases, one may be interested in applying the Routh test to polynomials having some negative coefficients (and, hence, some right half plane roots) simply to determine the number of right half plane roots. The requirement in this case is that, at least, a_n be positive.

[7]A polynomial, $Q(s)$, with all of its roots in the left half plane is frequently referred to as a *Hurwitz polynomial*.

$m(s)/n(s)$ and expand into a continued fraction of the form

$$\frac{m(s)}{n(s)} = b_1 s + \cfrac{1}{b_2 s + \cfrac{1}{b_3 s + \cfrac{1}{\begin{matrix}\cdot\\\cdot\\\cdot\\\cfrac{1}{b_k s}\end{matrix}}}} \tag{6-128}$$

The necessary and sufficient condition that $Q(s)$ have all of its roots in the left half plane is that all the coefficients $b_1, b_2, \ldots, b_k$ be positive. If $n(s)$ is of higher degree, the ratio $n(s)/m(s)$ is expanded. In order to illustrate this procedure, use the same example as before.

$$Q(s) = s^3 + s^2 + 2s + 8$$
$$\therefore m(s) = s^2 + 8$$
$$n(s) = s^3 + 2s$$

The continued fraction expansion of $n(s)/m(s)$ [since $n(s)$ is of higher degree here] can be carried out conveniently in the following form:

$$
\begin{array}{r}
s^2+8 \,\overline{\smash)\,s^3+2s} \,\big\lfloor s \\
\underline{s^3+8s} \\
-6s\,\overline{\smash)\,s^2+8}\,\big\lfloor -1/6s \\
\underline{s^2} \\
8\,\overline{\smash)\,-6s}\,\big\lfloor -6/8s \\
\underline{-6s} \\
0
\end{array}
$$

Since the coefficients are not all positive, $Q(s)$ must have roots in the right half plane.

It is often possible to use the Hurwitz test to investigate the range of some particular parameter that results in a stable system. Suppose, for example, that the denominator polynomial of a transfer function is

$$Q(s) = s^4 + 3s^3 + 3s^2 + s + k$$

and it is desired to find the range of k values for which the system is stable. In this case

$$m(s) = s^4 + 3s^2 + k$$
$$n(s) = 3s^3 + s$$

and the continued fraction expansion becomes

$$3s^3 + s \overline{\left)s^4 + 3s^2 + k\right.} \left|\frac{1}{3}s\right.$$

$$s^4 + \frac{1}{3}s^2$$

$$\frac{8}{3}s^2 + k \overline{\left)3s^3 + s\right.} \left|\frac{9}{8}s\right.$$

$$3s^3 + \frac{9}{8}ks$$

$$\left(1 - \frac{9}{8}k\right)s \overline{\left)\frac{8}{3}s^2 + k\right.} \left|\frac{64}{24 - 27k}s\right.$$

$$\frac{8}{3}s^2$$

$$k\overline{\left)\frac{8 - 9k}{8}s\right.} \left|\frac{8 - 9k}{8k}s\right.$$

In order to determine the range of k values for which all of the coefficients are positive, it is necessary to examine three cases:

$$27k < 24 \qquad \text{or} \qquad k < \frac{8}{9}$$

$$9k < 8 \qquad \text{or} \qquad k < \frac{8}{9}$$

$$8k > 0 \qquad \text{or} \qquad k > 0$$

Hence, the range of k values for stability is

$$0 < k < \frac{8}{9}$$

Exercise 6-13.1

Find the relationship among the coefficients of the polynomial

$$Q(s) = a_3 s^3 + a_2 s^2 + a_1 s + a_0$$

that will assure that there are no roots in the right half plane.

ANS. $a_1 a_2 > a_0 a_3$

6-14 Frequency Response from Pole-Zero Plots

The transfer function $H(s)$ evaluated along the $j\omega$-axis is the frequency response of the system. Certain general characteristics of the frequency response and an alternate method of evaluating the transfer function at a specific frequency can be obtained directly from a pole-zero plot of $H(s)$ in the s-plane. In order to estimate $H(s)$ along the $j\omega$-axis it is helpful to visualize $|H(s)|$ as a surface lying above the s-plane. The

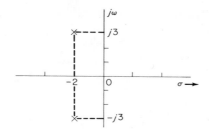

Fig. 6-20 Poles for plot of $H_1(s) = \dfrac{1}{(s+2)^2 + 9}$.

height of the surface above this plane at any point is equal to the magnitude of $H(s)$ at that point. As an example, consider the function

$$H_1(s) = \frac{1}{s^2 + 4s + 13} = \frac{1}{(s+2)^2 + 9}$$

The pole-zero plot of $H_1(s)$ is shown in Fig. 6-20 and consists of complex conjugate poles at $s = -2 \pm j3$. At each pole the magnitude of $H(s)$ becomes infinite. If we now visualize $|H(s)|$ as a surface above the s-plane, we obtain the surface shown in Fig. 6-21. In this figure only the portion to the left of the $j\omega$-axis is shown. The height along the $j\omega$-axis is the desired response. The surface in Fig. 6-21 can be thought of as an originally flat rubber sheet that has been stretched, as shown, by the poles pushing up at the critical frequencies. When there are zeros present, the rubber sheet is considered to be fastened down at each zero. As a further example, consider the transfer function

$$H_2(s) = \frac{s^2 + 2s + 2}{s^2 + 4s + 13} = \frac{(s+1)^2 + 1}{(s+2)^2 + 9}$$

The pole-zero plot of this transfer function is shown in Fig. 6-22. It is seen that this plot is very similar to the previous case considered except that now there are two zeros between the poles and the origin. Using the rubber-sheet analogy, it would be expected that since the zeros tack the

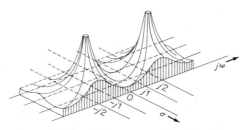

Fig. 6-21 Pictorial representation of $|H(s)|$.

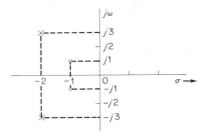

Fig. 6-22 Pole-zero plot of $H_2(s) = \dfrac{s^2 + 2s + 2}{s^2 + 4s + 13}$.

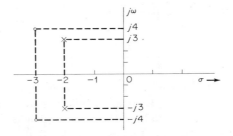

Fig. 6-23 Pole-zero of $H_3(s) = \dfrac{(s+3)^2 + 16}{(s+2)^2 + 9}$.

sheet down at $s = 1 \pm j1$, the height along the $j\omega$-axis from the origin on out for some distance would be reduced from what it is without the zeros. If the zeros were moved to the other side of the poles, as in Fig. 6-23, it is clear that these zeros would have little effect on the amplitude at the origin because of the "shielding" effect of the poles that lie between the zeros and the origin. The effect of the zeros would be more pronounced at frequencies slightly greater than $\omega = 3$. The magnitudes of the frequency responses for these cases are shown in Fig. 6-24 and clearly indicate the effects that have been discussed. Much additional insight into system behavior can be obtained from pole-zero plots, and they are widely used in

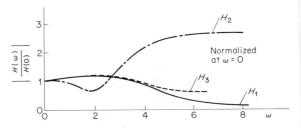

Fig. 6-24 Amplitude spectra of $H_1(j\omega)$, $H_2(j\omega)$, and $H_3(j\omega)$.

system analysis and design. Further discussion of this method is contained in the references at the end of the chapter.

Before leaving the subject of pole-zero plots, one additional use of this representation should be discussed. Consider a general transfer function of the form

$$H(s) = k_0 \frac{(s - \gamma_0)(s - \gamma_2) \cdots (s - \gamma_m)}{(s - \gamma_1)(s - \gamma_3) \cdots (s - \gamma_n)} \qquad \text{(6-129)}$$

where γ_k are the complex roots of the numerator and denominator polynomials, and k_0 is a real constant. Suppose now that it is required to evaluate $H(s)$ at some complex frequency $s = s_0 = \sigma_0 + j\omega_0$. This may be done algebraically by substituting $(\sigma_0 + j\omega_0)$ for s in (6-129) and evaluating the resulting expression. If this procedure is carried out, each factor in the numerator and denominator will be a complex number and can be expressed as a magnitude $M_k = |s_0 - \gamma_k|$ and an angle

$$\gamma_k = \tan^{-1}\left[\frac{\text{Im}(s_0 - \gamma_k)}{\text{Re}(s_0 - \gamma_k)}\right]$$

The complete expression for $H(s_0)$ can then be written as

$$H(s_0) = k_0 \frac{M_0 \cdot M_2 \cdots M_m \,\underline{/\psi_0 + \psi_2 + \cdots + \psi_m}}{M_1 \cdot M_3 \cdots M_n \,\underline{/\psi_1 + \psi_3 + \cdots + \psi_n}} \qquad \text{(6-130)}$$

The magnitude and phase angles in (6-130) can be obtained from the pole-zero diagram of $H(s)$ in a very simple manner. Consider a typical pole-zero plot as shown in Fig. 6-25 and suppose that it is desired to evaluate $H(s)$ at $s = -1 + j1$. By drawing directed line segments from each critical frequency to the point $s_0 = -1 + j1$, we will obtain a set of vectors that represent the factors in (6-130). The magnitudes are equal to the lengths of the lines, and the phase angles are the angles the vectors make with the positive σ-axis. For rapid computation the lengths and angles can be measured directly from an accurate plot, or if desired, the plot can be used to clarify the geometric and trigonometric relationships necessary to allow analytical computations. For the example shown, it is

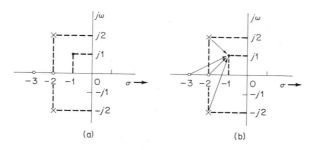

(a)　　　　　　　　　(b)

Fig. 6-25 Pole-zero diagram.

easily seen that the magnitude is given by

$$|H(-1+j1)| = k_0\frac{\sqrt{2^2+1^2}\sqrt{1^2+1^2}}{\sqrt{1^2+1^2}\sqrt{1^2+3^2}} = k_0\sqrt{1/2}$$

and the phase angle by

$$\underline{/H(-1+j1)} = \tan^{-1}\left(\frac{1}{1}\right) + \tan\left(\frac{1}{2}\right) - \tan^{-1}\left(\frac{-1}{1}\right) - \tan^{-1}\left(\frac{3}{1}\right)$$

$$= 45° + 26.6° + 45° - 71.6° = 45°$$

Rapid estimates of the magnitude of the frequency response can be made by moving the point of evaluation along the $j\omega$-axis. Another place where this method is particularly useful is in evaluating the residue at a pole in the partial fraction expansion of a rational function of s. As an example of this application, consider a transfer function of the form

$$H(s) = \frac{2s(s+2)}{(s+3)(s^2+2s+5)}$$

The pole-zero plot of $H(s)$ is shown in Fig. 6-26(a). In order to find the residue at a pole, the procedure is to multiply the gain constant (in this case, 2) by the vectors from each zero to the pole in question and to divide by the vectors from each pole to the pole in question. Using this proce-

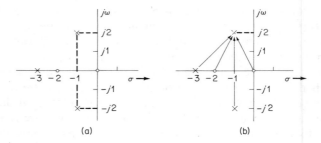

(a)   (b)

Fig. 6-26 Evaluation of the residue at a pole.

dure, the residue at the pole located at $-1+j2$ is found from Fig. 6-26(b) to be

$$k_{-1+j2} = \frac{2(\sqrt{5}\underline{/63.4°})(\sqrt{5}\underline{/116.6°})}{(2\sqrt{2}\underline{/45°})(4.00\underline{/90°})}$$

$$= 0.884\underline{/45°}$$

The residue at $s = -1 - j2$ is found as the complex conjugate of this quantity; that is, $k_{-1-j2} = 0.884\underline{/-45°}$.

6-15   Relationship between the Laplace and Fourier Transforms

The Laplace transform and the Fourier transform are very closely related, as is evident from the derivation of the Laplace transform from the
Fourier transform given at the beginning of this chapter. The one-sided, or
unilateral, Laplace transform, which has been considered so far, is concerned only with the behavior of time functions for $t > 0$. The Fourier
transform, on the other hand, includes both positive and negative time but
cannot handle functions having finite average power without using special
limiting operations which are accounted for by means of δ functions. In
carrying out computation, it is generally more convenient to use the variable s rather than the variable ω, since the algebra is much simpler. However, whenever it is required to consider frequency spectra, it is necessary to return to the variable ω. It is, therefore, useful to be able to move
easily between the variables s and ω in working with transforms. For
certain classes of functions, the change from $\mathscr{L}\{f(t)\}$ to $\mathscr{F}\{f(t)\}$ is quite
simple; for others, it is more involved. The important situations are given
in the discussion that follows. In all cases it is assumed that $F(s)$ is a
legitimate unilateral Laplace transform.

F(s) with poles in the left half plane only.   When $F(s)$ has poles only in the
left half of the s-plane (not including the $j\omega$-axis), the transformation between $F(s)$ and $F(\omega)$ is accomplished simply by making the substitution
$s = j\omega$. These transforms include all finite energy signals defined for positive time only. As an example, consider the following Laplace transform:

$$\mathscr{L}\{\epsilon^{-at}u(t)\}\,W = \frac{1}{s + \alpha} \qquad \mathrm{Re}(s) > -\alpha \qquad \text{(6-131)}$$

If α is a positive constant, the only pole in $F(s)$ is in the left half plane
(LHP) at $s = -\alpha$, and it follows that

$$\mathscr{F}\{\epsilon^{-at}u(t)\} = F(s)\big|_{s=j\omega} = \frac{1}{s+\alpha}\bigg|_{s=j\omega} = \frac{1}{j\omega + \alpha} \qquad \text{(6-132)}$$

The same simple relationship holds for all other transforms meeting the
conditions stated above.

There is one aspect of relating the Fourier transform and the Laplace
transform that requires some further discussion, and that is the matter of
proper interpretation of $F(s)$ and $F(\omega)$ when the original function $f(t)$
was identical in each case. The problem is easily seen from the results just
obtained, where we have shown that if $f(t) = \epsilon^{-at}u(t)$, then

$$F(s) = \frac{1}{s + \alpha} \qquad \text{(6-133)}$$

$$F(\omega) = \frac{1}{j\omega + \alpha} \qquad \text{(6-134)}$$

Clearly, the function $F(\cdot)$ is different in (6-133) and (6-134), and what is meant by $F(\cdot)$ is determined by whether we use the variable s or ω. Much confusion can be avoided by considering the notation to be a symbolic representation rather than a mathematical function. Thus, $F(\omega)$ is interpreted as $\mathscr{F}\{f(t)\}$, and $F(s)$ is interpreted as $\mathscr{L}\{f(t)\}$. The mathematical relationships vary with the particular time functions involved. In the case just discussed it is found that $F(\omega) = F(s)|_{s=j\omega}$ or $F(s) = F(\omega)|_{\omega=-js}$. In the following subsection it is found that a different relationship holds when there are poles on the $j\omega$-axis. Many books attempt to avoid this difficulty by writing the Fourier transform as $F(j\omega)$. However, this leads to notational problems with the Fourier transform itself. If any confusion arises, it can be eliminated by using the operational notation $\mathscr{L}\{f(t)\}$ or $\mathscr{F}\{f(t)\}$ instead of $F(s)$ or $F(\omega)$.

F(s) with poles in the left half plane and on the $j\omega$-axis. By partial fraction expansion, transforms of this class can be separated into terms having poles in the LHP and terms having poles on the $j\omega$-axis. The terms with LHP poles can be handled in the same manner described in the preceding paragraph. The terms containing poles on the $j\omega$-axis require modified handling and, in general, lead to Fourier transforms containing δ functions. For example, if

$$F(s) = \frac{\omega_0}{s^2 + \omega_0^2}$$

we immediately recognize this as the transform of $\sin \omega_0 t u(t)$. By the methods discussed in Chapter 5, it is readily shown that

$$\mathscr{F}\{\sin \omega_0 t u(t)\} = \frac{\omega_0}{\omega_0^2 - \omega^2} + j\frac{\pi}{2}[\delta(\omega + \omega_0) - \delta(\omega - \omega_0)]$$

To obtain this result directly from $F(s)$ we must recognize that *each simple pole* on the $j\omega$-axis leads to two terms in the Fourier representation: One is obtained by the substitution $s = j\omega$; the other is a δ function having a strength of π times the residue at the pole. This can be written in equation form as:

$$F(s) = \sum_n \frac{k_n}{s - j\omega_n} \text{ poles at } s = j\omega_n \tag{6-135}$$

$$F(\omega) = F(s)|_{s=j\omega} + \pi \sum_n k_n \delta(\omega - \omega_n) \tag{6-136}$$

The residue at a simple pole is just the numerator of the term in the partial fraction expansion containing the pole. Consider the Laplace transform given above:

$$F(s) = \frac{\omega_0}{s^2 + \omega_0^2} = \frac{j/2}{s + j\omega_0} - \frac{j/2}{s - j\omega_0}$$

Using (6-136), the Fourier transform is found to be

$$F(\omega) = \frac{\omega_0}{(j\omega)^2 + \omega_0^2} + j\frac{\pi}{2}[\delta(\omega + \omega_0) - \delta(\omega - \omega_0)]$$

as previously noted. As a further example, consider the following transform:

$$F(s) = \frac{2s^2 + \omega_0^2}{s(s^2 + \omega_0^2)}$$

Expanding in partial fractions gives

$$F(s) = \frac{1}{s} + \frac{1/2}{s + j\omega_0} + \frac{1/2}{s - j\omega_0}$$

The corresponding Fourier transform is

$$F(\omega) = \frac{-2\omega^2 + \omega_0^2}{j\omega(-\omega^2 + \omega_0^2)} + \pi[\delta(\omega) + \frac{1}{2}\delta(\omega + \omega_0) + \frac{1}{2}\delta(\omega - \omega_0)]$$

$$= j\frac{\omega_0^2 - 2\omega^2}{\omega(\omega^2 - \omega_0^2)} + \pi\delta(\omega) + \frac{\pi}{2}[\delta(\omega + \omega_0) + \delta(\omega - \omega_0)]$$

When $F(s)$ contains higher-order poles, the relationship becomes somewhat more complicated and leads to higher-order singularity functions. Such cases can always be handled by converting from $F(s)$ back to the time domain and then taking the Fourier transform.

F(s) with poles in the right half plane.   When $F(s)$ has poles in the right half plane (RHP) and is a legitimate unilateral Laplace transform, it means that the convergence region $\text{Re}\,s = \sigma > 0$. Therefore, the integral does not converge on the $j\omega$-axis, and no corresponding Fourier transform exists. An example of such a function is

$$F(s) = \frac{1}{s - 1} \qquad \sigma > 1$$

This is the Laplace transform of the function $\epsilon^t u(t)$, and we have not defined a Fourier transform for such a function.

There are legitimate Laplace transforms, having poles in the right half of the s-plane, that are defined so that their strip of convergence includes the $j\omega$-axis. Such transforms correspond to noncausal time functions and are based on the bilateral, or two-sided, Laplace transform. These transforms are closely related to the Fourier transform, but details of the relationship must be deferred until the two-sided Laplace transform is discussed.

6-16   Two-Sided Laplace Transform

There are certain instances in which it is convenient to consider functions that extend into negative time. Such cases arise in connection with signals from stationary random processes, with periodic signals, and with the study of noncausal systems. Such time functions exist outside the range of the time variable that can be handled by the conventional Laplace transform. They can, however, be studied and analyzed by means of the two-sided Laplace transform that was briefly mentioned at the beginning of this chapter. The two-sided Laplace transform is defined as

$$F_{II}(s) = \mathscr{L}_{II}\{f(t)\} = \int_{-\infty}^{\infty} f(t)\epsilon^{-st}dt \tag{6-137}$$

and exists for all values of s for which the integral converges. The subscript II is included to distinguish this transform from the normal one-sided Laplace transform. The need to distinguish the two transforms results from the fact that the inverse transform of $F_{II}(s)$ is not unique unless the region of convergence is specified, while that of $F(s)$ is unique. This matter is discussed in some detail in the next two sections.

The values of s for which (6-137) exists can be determined by considering the positive-time and negative-time portions of $f(t)$ separately. Thus,

$$F_{II}(s) = \int_{-\infty}^{0} f(t)\epsilon^{-st}dt + \int_{0}^{\infty} f(t)\epsilon^{-st}dt \tag{6-138}$$

The first integral will exist for all values of s for which the real part, σ (Re $s = \sigma$), is *less* than some value β, while the second integral exists for σ *greater* than some value α. These regions are shown in Fig. 6-27 for a case in which

$$\alpha < \sigma < \beta$$

Since the two regions overlap, there is a *strip of convergence* in which $F_{II}(s)$ exists. If, on the other hand, the time function were such that $\beta < \alpha$, there would be no region of overlap, and $F_{II}(s)$ would not exist.

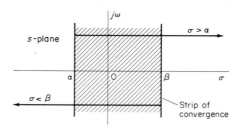

Fig. 6-27 Illustrating the strip of convergence for $F_{II}(s)$.

The foregoing discussion may be made somewhat clearer if some specific cases are considered. In the first case, let

$$f(t) = \epsilon^{at} \qquad a > 0,\ t < 0$$
$$= \epsilon^{-bt} \qquad b > 0,\ t \geq 0$$

Then

$$F_{II}(s) = \int_{-\infty}^{0} \epsilon^{at}\epsilon^{-st}dt + \int_{0}^{\infty} \epsilon^{-bt}\epsilon^{-st}dt$$

$$= \int_{-\infty}^{0} \epsilon^{-(s-a)t}dt + \int_{0}^{\infty} \epsilon^{-(s+b)t}dt$$

It is clear that the first integral will exist if Re $s = \sigma$ is *less* than a, while the second one exists for Re $s = \sigma$ greater than $-b$. Thus, the strip of convergence is defined by

$$-b < \text{Re } s < a$$

and includes the $j\omega$-axis. The corresponding two-sided Laplace transform becomes

$$F_{\text{II}}(s) = \frac{-1}{s-a} + \frac{1}{s+b} = \frac{-(a+b)}{(s-a)(s+b)} \qquad -b < \text{Re } s < a$$

As a second example, let

$$f(t) = 1 \qquad t < 0$$
$$= \epsilon^{-bt} \qquad b > 0, t \geq 0$$

Then

$$F_{\text{II}}(s) = \int_{-\infty}^{0} \epsilon^{-st} dt + \int_{0}^{\infty} \epsilon^{-bt} \epsilon^{-st} dt$$

$$= \int_{-\infty}^{0} \epsilon^{-st} dt + \int_{0}^{\infty} \epsilon^{-(s+b)t} dt$$

The first integral converges only when Re $s < 0$, whereas the second one converges for Re $s > -b$. Thus, the strip of convergence is defined by

$$-b < \text{Re } s < 0$$

and does not include the $j\omega$-axis. The resulting two-sided Laplace transform is

$$F_{\text{II}}(s) = \frac{-1}{s} + \frac{1}{s+b} = \frac{-b}{s(s+b)} \qquad -b < \text{Re } s < 0$$

As a final example, let

$$f(t) = \epsilon^{-bt} \qquad b > 0, -\infty < t < \infty$$

Then

$$F_{\text{II}}(s) = \int_{-\infty}^{0} \epsilon^{-bt} \epsilon^{-st} dt + \int_{0}^{\infty} \epsilon^{-bt} \epsilon^{-st} dt$$

$$= \int_{-\infty}^{0} \epsilon^{-(s+b)t} dt + \int_{0}^{\infty} \epsilon^{-(s+b)t} dt$$

The first integral converges only when Re $s < -b$ and the second integral converges only when Re $s > -b$. Thus, the two regions of convergence do not overlap, and the two-sided Laplace transform does not exist.

Many of the properties of two-sided Laplace transforms are the same as for one-sided transforms. These are summarized as follows:

Linearity. Equation (6-21) holds for two-sided Laplace transforms provided that the strip of convergence of the sum is taken as the *common* area of the strips of convergence of the individual transforms.

Time translation.   Equation (6-28) holds for two-sided Laplace transforms, except that the restriction $t_0 > 0$ is no longer necessary.

Scale change.   Equation (6-24), when employed with the two-sided Laplace transform, can be extended to include negative constants. The relationship is

$$\mathscr{L}_{\text{II}}\{f(at)\} = \frac{1}{|a|} F_{\text{II}}\left(\frac{s}{a}\right)$$

Differentiation.   For two-sided Laplace transforms, (6-42) must be modified to read

$$\mathscr{L}_{\text{II}}\left\{\frac{df}{dt}\right\} = sF_{\text{II}}(s) \tag{6-139}$$

Note that there are *no* terms corresponding to the difference between $f(0^-)$ and $f(0^+)$, even though such a discontinuity contributes a δ function to the derivative.

Integration.   In the case of the two-sided Laplace transform, the integration of the time function is assumed to start at $-\infty$. Hence, (6-46) is modified to read

$$\mathscr{L}_{\text{II}}\left\{\int_{-\infty}^{t} f(\lambda)d\lambda\right\} = \frac{1}{s} F_{\text{II}}(s) \tag{6-140}$$

The s-shift theorem.   Equation (6-51) holds for the two-sided Laplace transform except that the strip of convergence for $\epsilon^{-\alpha t}f(t)$ is that for $f(t)$ but shifted to the *left* by the amount Re α.

Convolution.   Equation (6-55) holds for two-sided Laplace transforms, except that the strip of convergence of the product is the common area of the strips of convergence of the individual transforms.

t Multiplication.   For the two-sided Laplace transform

$$\mathscr{L}_{\text{II}}\{tf(t)\} = -\frac{dF_{\text{II}}(s)}{ds} \tag{6-141}$$

The strip of convergence for this transform is the same as that for $F_{\text{II}}(s)$.

Initial- and final-value theorems.   These theorems do not hold in general for the two-sided Laplace transform. They can, however, be applied separately to the portions of $F_{\text{II}}(s)$ corresponding to negative and positive time. One result of such an analysis is the following expression for a discontinuity at the origin:

$$f(0^+) - f(0^-) = \lim_{s\to\infty} sF_{\text{II}}(s) \tag{6-142}$$

Exercise 6-16.1
Find the two-sided Laplace transforms of the following functions and indicate the strip of convergence in each case: $u(t + T/2) - u(t - T/2)$; $(t - a)u(-t)$.

ANS. $\left(\frac{2}{s}\right) \sinh\left(\frac{sT}{2}\right)$ $-\infty < \sigma < \infty, \frac{1}{s}\left[a - \frac{1}{s}\right]$ $\sigma < 0$

6-17   Inverse Two-Sided Laplace Transforms

Although the inversion integral for two-sided Laplace transforms is identical to that for the one-sided transform, the methods for carrying out the inversion are somewhat more involved—partly because the positive-time and negative-time portions must be handled separately and partly because of a fundamental ambiguity that exists in the interpretation of a two-sided Laplace transform when the region of convergence is not specified.

In order to examine this ambiguity in more detail, consider the two time functions defined by

$$f(t) = \epsilon^{-\alpha t} \qquad t > 0$$
$$ = 0 \qquad t < 0$$

and

$$g(t) = -\epsilon^{-\alpha t} \qquad t < 0$$
$$ = 0 \qquad t > 0$$

and sketched in Fig. 6-28. The two-sided Laplace transforms of these two time functions are

$$F_{\text{II}}(s) = \int_0^\infty \epsilon^{-\alpha t} \epsilon^{-st} dt = \frac{1}{s + \alpha} \qquad \text{Re}(s) > -\alpha$$

and

$$G_{\text{II}}(s) = \int_{-\infty}^0 -\epsilon^{-\alpha t} \epsilon^{-st} dt = \frac{1}{s + \alpha} \qquad \text{Re}(s) < -\alpha$$

and these are seen to be identical, except for the regions of convergence. Thus, the unique relationship that existed between a function and its Laplace transform in the one-sided case does *not* exist for the two-sided transform. Hence, if one is given a two-sided Laplace transform, *with no indication of the region of convergence*, it is impossible to determine the corresponding time function uniquely.

Fortunately, in the practical use of the two-sided Laplace transform the ambiguity problem usually can be resolved on the basis of physical

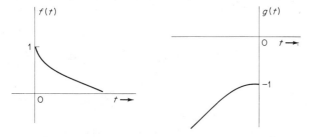

Fig. 6-28 Two time functions having the same Laplace transform.

considerations, since time functions that increase without limit as t approaches either $+\infty$ or $-\infty$ do not exist in reality. Thus, it is usually a simple matter to select the one practical time function even though others may be mathematically possible.

In the analysis of linear, time-invariant systems, the two-sided Laplace transforms that arise are almost always rational functions of s, just as in the one-sided case. These can be expanded into partial fractions in exactly the same way as has been discussed in the earlier part of this chapter. Each term in the partial fraction expansion will correspond to a pole of the transform, and some of these will be in the left half of the s-plane, whereas others will be in the right half. Furthermore, it has been shown that LHP poles correspond to decaying exponentials in positive time, whereas RHP poles yield exponentials that vanish at $-\infty$. Thus, it is possible to select realistic time functions for the inverse transform if one accepts the following conventions:

1. All terms in the partial fraction expansion that come from LHP poles will be assumed to yield time functions that exist only for $t \geq 0$.

2. All terms in the partial fraction expansion that come from RHP poles will be assumed to yield time functions that exist only for $t \leq 0$.

This procedure can be illustrated by means of the following example. Let the two-sided transform be

$$F_{\mathrm{II}}(s) = \frac{2s - 5}{(s - 1)(s + 2)} = \frac{-1}{s - 1} + \frac{3}{s + 2}$$

in which the partial fraction expansion has been obtained using the procedure discussed in Section 6-7. By analogy to the foregoing example and by employing the stated convention, we find that

$$\begin{aligned} f(t) &= \epsilon^t & t < 0 \\ &= 3\epsilon^{-2t} & t > 0 \end{aligned}$$

and this function is sketched in Fig. 6-29. Note that this convention leads to only one of *four* apparently possible time functions. The other three,

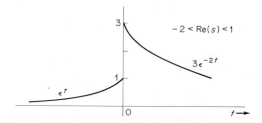

Fig. 6-29 Time function that vanishes at $\pm\infty$.

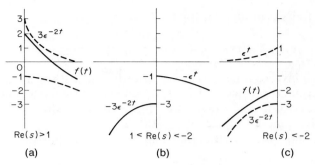

Fig. 6-30 Three other time functions from the same transform.

sketched in Fig. 6-30, are unbounded. The regions of convergence are also indicated for each sketch, and it is seen that there is *no* region of convergence for Fig. 6-30(b). Hence, this is not really a possible inverse transform. The other two sketches in Fig. 6-30 do have regions of convergence and are, therefore, mathematically acceptable inverse transforms even if they are not physically acceptable.

A useful trick that can be applied to RHP terms automatically enforces the convention stated above. It utilizes the techniques and tabulated transform pairs for the one-sided transform. This can be illustrated by writing the transform for a function that exists in negative time and then making a change in variable. Thus,

$$F(s) = \int_{-\infty}^{0} f(t)\epsilon^{-st}dt = \int_{\infty}^{0} f(-\lambda)\epsilon^{-(-s)\lambda}(-d\lambda)$$

in which $\lambda = -t$. Using the minus sign to interchange the limits gives

$$F(s) = \int_{0}^{\infty} f(-\lambda)\epsilon^{-(-s)\lambda}d\lambda \qquad \text{(6-143)}$$

Equation (6-143) may be interpreted as meaning that the Laplace transform of a function existing only in negative time can be obtained by (1) reflecting the negative-time function into positive time, (2) taking the ordinary one-sided Laplace transform for this positive-time function, and (3) replacing s by $-s$.

This procedure may be illustrated by considering the function

$$f(t) = \epsilon^{\alpha t} \qquad t < 0$$
$$= 0 \qquad t > 0$$

Its reflection, $f_+(t)$, is

$$f_+(t) = f(-t) = \epsilon^{-\alpha t} \qquad t > 0$$
$$= 0 \qquad t < 0$$

and this function has a one-sided transform of

$$F_+(s) = \frac{1}{s + \alpha}$$

Thus, the transform of $f(t)$, for $t < 0$, is

$$F(s) = F_+(-s) = \frac{1}{-s + \alpha} = \frac{-1}{s - \alpha}$$

In taking inverse transforms, the procedure is just reversed. Thus, for an RHP term, (1) replaces s by $-s$, (2) takes a normal one-sided inverse Laplace transform, and (3) reflects the resulting positive-time function into negative time by replacing t with $-t$.

The example discussed earlier may be used to illustrate this procedure.

$$F_{II}(s) = \frac{2s - 5}{(s - 1)(s + 2)} = \frac{-1}{s - 1} + \frac{s}{s + 2}$$

The RHP term is

$$F(s) = \frac{-1}{s - 1}$$

from which

$$F(-s) = F_+(s) = \frac{-1}{-s - 1} = \frac{1}{s + 1}$$

and

$$f_+(t) = \epsilon^{-t} \qquad t > 0$$

The resulting negative-time function is

$$f_-(t) = f_+(-t) = \epsilon^{t} \qquad t < 0$$

The LHP term is handled in the usual manner, so the total time function becomes

$$\begin{aligned} f(t) &= \epsilon^{t} & t < 0 \\ &= 3\epsilon^{-2t} & t > 0 \end{aligned}$$

which agrees with the preceding result.

So far, the discussion of the inverse transform has carefully avoided any mention of poles on the $j\omega$-axis. In this case, practical considerations of accepting only bounded functions give no clue as to how to proceed. For example, if

$$F(s) = \frac{1}{s}$$

the inverse transform may be taken as either

$$f(t) = -u(-t)$$

or

$$f(t) = u(t)$$

and these are both bounded. Likewise, a pair of conjugate imaginary poles on the $j\omega$-axis would lead to constant-amplitude sinusoids in either positive time or negative time. Thus, it might appear that there is no way of resolving the ambiguity that arises from $j\omega$-axis poles. This, in fact, would be the case if one were presented with a transform with no inkling as to how it arose. In practical system analysis problems, however, one

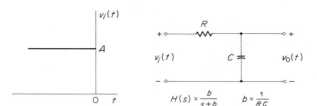

Fig. 6-31 Simple circuit responding to a negative-time input.

starts with time functions and derives the transforms. Hence, with a little care in notation, one can keep track of how a particular $j\omega$-axis pole was created and how it must be interpreted in the final result.

As a very simple illustration of this, consider the $R-C$ circuit and input waveform shown in Fig. 6-31. Although there are some very simple ways of handling this problem without using the two-sided Laplace transform, it is instructive to do so. The input signal is

$$v_i(t) = Au(-t)$$

and its transform is

$$V_i(s) = \frac{-A}{s_-} \qquad \mathrm{Re}(s) < 0$$

Note the minus subscript on the s to indicate that this arose from a negative-time function. The transform of the output signal is

$$V_0(s) = V_i(s)H(s) = \frac{-Ab}{(s+b)s_-} \qquad -b < \mathrm{Re}(s) < 0$$

and the partial fraction expansion of this is

$$V_0(s) = \frac{-A}{s_-} + \frac{A}{s+b}$$

Note that this is a valid transform because there is a strip of convergence with nonzero width.

In obtaining the inverse transform of $V_0(s)$, the minus subscript on s_- makes it clear that this is to be interpreted as a negative-time function. Hence, the resulting time function is

$$v_0(t) = Au(-t) + A\epsilon^{-bt}u(t)$$

which is sketched in Fig. 6-32(a). Without this indication, one might be tempted to interpret this term as a positive-time function, and this would lead to

$$v_0(t) = -Au(t) + A\epsilon^{-bt}u(t)$$

which is shown in Fig. 6-32(b). Since both responses are bounded, the only way one would know which is correct is by utilizing one's knowledge of how the pole at the origin arose.

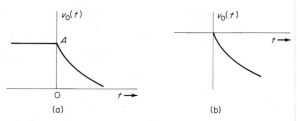

Fig. 6-32 Response of circuit in Fig. 6-31 for two different interpretations. (a) Correct response. (b) Incorrect response.

One way of easily keeping track of $j\omega$-axis poles is to use the notation s_+ when they arise from positive-time functions and s_- when they arise from a negative-time function. In this way the proper inverse transform can be obtained in even very complicated cases. It should be emphasized, however, that this mechanical procedure does not accomplish anything that could not be done by keeping track of the regions of convergence. The usefulness of this procedure lies solely in the fact that it requires less effort than the more fundamental procedure.

6-18   Complex Inversion Integral

The application of contour integration to evaluation of the integral representation of the inverse Laplace transform is discussed in Appendix B. The following important result is presented there. If $\lim_{|s|\to\infty} F(s) = 0$ uniformly along any line in a half plane to the left or right of the strip of convergence of the Laplace transform, then the complex inversion integral can be evaluated by closing the contour with an infinite semicircular arc. The contribution to the integral over the semicircular arc will be zero when the closure is to the left and $t \geq 0$ and, similarly, will be zero when closure is to the right and $t < 0$. In equation form this can be stated as

$$f(t) = \frac{1}{2\pi j} \int_{c-j\infty}^{c+j\infty} F(s)\epsilon^{st}ds = \frac{1}{2\pi j} \oint F(s)\epsilon^{st}ds \qquad t > 0 \qquad \text{(6-144)}$$

$$f(t) = \frac{1}{2\pi j} \int_{c-j\infty}^{c+j\infty} F(s)\epsilon^{st}ds = \frac{1}{2\pi j} \oint F(s)\epsilon^{st}ds \qquad t < 0 \qquad \text{(6-145)}$$

(Note the circular arrows on the integral signs and their directions.) The paths of integration corresponding to (6-144) and (6-145) are shown in Fig. 6-33.

When the transform is single-valued and analytic except at isolated poles, the integral can be evaluated by the theory of residues to give

$$\mathcal{L}^{-1}\{F(s)\} = f(t) = \begin{cases} 0 & t < 0 \\ \Sigma \text{ residues to left of } c & t > 0 \end{cases} \qquad \text{(6-146)}$$

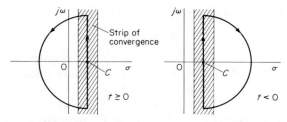

Fig. 6-33 Paths of integration for complex inversion integral.

$$\mathcal{L}^{-1}\{F_{II}(s)\} = f(t) = \begin{cases} -\Sigma \text{ residues to right of } c & t < 0 \\ \Sigma \text{ residues to left of } c & t > 0 \end{cases} \quad \text{(6-147)}$$

where c is the abscissa of a vertical line lying in the strip of convergence.

The significance of these results is best illustrated by some examples. Consider the Laplace transform $F_{II}(s) = 2s/(s^2 - 1)$ with a strip of convergence $-1 < \sigma < 1$. This is shown in Fig. 6-34. From (6-146) and (6-147) the inverse transform is seen to be

$$f(t) = \frac{1}{2\pi j} \int_{-j\infty}^{j\infty} \frac{2se^{st}}{s^2 - 1} ds = \begin{cases} \text{residue in pole at } s = -1 \text{ for } t > 0 \\ -\text{residue in pole at } s = 1 \text{ for } t < 0 \end{cases}$$

The method of evaluating residues is discussed in Appendix B, where the following formula is given. If $G(s) = F(s)e^{st}$ has an nth order pole at $s = s_0$, then form the function $\phi(s) = (s - s_0)^n G(s)$ and compute the residue as

$$K_{s_0} = \frac{\phi^{(n-1)}(s_0)}{(n - 1)!} \quad \text{(6-148)}$$

where $\phi^{(n-1)}$ is the $(n - 1)$th derivative of $\phi(s)$. Since $n = 1$ in the present instance, no derivative is required and the residues are found to be

$$K_{-1} = \frac{2se^{st}}{s - 1}\Big|_{s=-1} = \epsilon^{-t}$$

$$K_{+1} = \frac{2se^{st}}{s + 1}\Big|_{s=1} = \epsilon^{t}$$

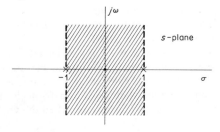

Fig. 6-34 Pole-zero plot of $\dfrac{2s}{s^2 - 1}$.

The corresponding time function is, therefore,

$$f(t) = \epsilon^{-t}u(t) - \epsilon^{t}u(-t)$$

Note the change of sign of the residue for $t < 0$ as required by (6-147).

If this same transform had been given as a one-sided Laplace transform with no strip of convergence specified, a different answer would be obtained. In the case of the one-sided transform it is known that $f(t)$ is zero for $t < 0$, and therefore, there can be no singularities to the right of the abscissa of integration. The strip of convergence is the half plane to the right of all singularities. In this case the residues at both poles would correspond to positive-time functions, and the correct inverse transform would be $f(t) = (\epsilon^{t} + \epsilon^{-t})u(t)$. This is not a very practical function but is valid mathematically.

From the above it is clear that for one-sided Laplace transforms it is not necessary to specify the region of convergence since it must be to the right of all poles. However, for the two-sided Laplace transform the region of convergence must be specified, or otherwise, confusion can result. In most instances the constraints of physical possibility will allow selection of a suitable strip of convergence. This procedure is discussed in the previous section and is not pursued further here.

Exercise 6-18.1

Using contour integration, find the inverse transform of $F_{11}(s) = -2/s(s+1)^2$, $-1 < \sigma < 0$.

ANS: $2u(-t) + 2\epsilon^{-t}(t+1)u(t)$

References

There are many references covering the material on Laplace transforms discussed in this chapter. The following list is representative of the references available and contains much information beyond that presented in this chapter.

1. CHURCHILL, R. W., *Operational Mathematics*, 2d ed. New York: McGraw-Hill Book Company, Inc., 1958.
   This book presents the mathematical theory of the Laplace transform at the senior or graduate level. The applications discussed are from the general field of physical science, and only a few relate directly to electrical engineering. An extensive table of transforms is included.
2. DOETSCH, G., *Guide to the Applications of Laplace Transforms.* New York: D. Van Nostrand Company, 1961.
   Written by a leading authority on the Laplace transformation, this book contains in summary fashion most of the results of classical Laplace transform theory required in the study of engineering problems. A very wide range of topics is covered, and an extensive table of transforms is appended to the text.
3. KUO, E. F., *Network Analysis and Synthesis*, 2d ed. New York: John Wiley & Sons, Inc., 1966.
   This book is written as a junior- or senior-level text. The Laplace transform is considered in some detail, along with a variety of other methods useful in

network analysis. Many applications of the Laplace transform to circuit analysis problems are considered. The last third of the book provides an excellent introduction of synthesis procedures carried out completely in the complex frequency domain (s-plane).

4. LATHI, B. P., *Signals, Systems, and Communication.* John Wiley & Sons, Inc., 1965.
Written as a junior- or senior-level text, this book covers many aspects of signal and system analysis, particularly as they apply to problems arising in communication engineering.

5. LE PAGE, W., *Complex Variables and the Laplace Transform for Engineers.* New York: McGraw-Hill Book Company, Inc., 1961.
This book presents a detailed but very readable development of the mathematical basis for the Laplace transform. Many of the fine points of the theory such as contour integration are discussed in such a manner as to be readily handled by junior or senior engineering students.

6. MC COLLUM, P. A., and B. F. BROWN, *Laplace Transform Tables and Theorems.* New York: Holt, Rinehart and Winston, Inc., 1965.
This book contains conveniently arranged tables of Laplace transform pairs, along with the most important theorems relating to the Laplace transform.

7. TRUXAL, J., *Control System Synthesis.* New York: McGraw-Hill Book Company, Inc., 1955.
This classic in the field of automatic control contains an excellent discussion of the Laplace transform and includes many examples of the use of transform methods in system design. In addition to treatment of a wide range of topics relating to automatic control, there is much information lucidly presented on networks and network synthesis.

8. VAN DER POL, B., and H. BREMMER, *Operational Calculus Based on the Two-Sided Laplace Integral*, 2d ed. New York: Cambridge University Press, 1955.
This book is highly recommended to anyone wishing to pursue study of applications of the Laplace transform to a wider range of problems. Although written carefully from a mathematical point of view, the book contains a vast amount of material that is of vital concern to engineers. It contains (for the average engineer) the most readable discussions of advanced topics in Laplace transform theory that are available today. Much of the material in the book relates to the one-sided Laplace transform, which is considered to be a special case of the two-sided transform. One word of caution is in order in that all transforms are multiplied by an extra p (that is, s in our notation). This was done to make the transform pairs identical with the older Heaviside operational calculus transform pairs and causes little confusion.

The two references listed below present an entirely different approach to the solution of system analysis problems by means of an operational calculus. The methods discussed do not make use of the Laplace transform in any way. However, the formal manipulations of derivatives, functions, and integrals is virtually identical to that which results from employing the Laplace transform. The developments presented in these books represent a different and probably more powerful justification of Heaviside's operational calculus than does the Laplace transform. The book by Mikusinski is somewhat easier to read, but the interested student will find either quite fascinating.

9. MIKUSINSKI, J., *Operational Calculus*. New York: Pergamon Press, The Macmillan Co. Inc., 1959.

10. KRABBE, G., *Operational Calculus*. Berlin: Springer-Verlag, 1970.

Problems

6-1. Find the Laplace transform of each of the following time functions, which are assumed to exist for $t \geq 0$ only:

a. e^{-2t} **d.** $2e^{-2t} - e^{-t}$

b. $e^{-t} - e^{-2t}$ **e.** $1 - e^{-2t}$

c. t^2

6-2. Find the one-sided Laplace transform of each of the following time functions, which are assumed to exist for all time:

a. $e^{-2|t|}$ **d.** $u(t - 1)u(1 - t)$

b. t^2 **e.** $u(t - 1) + u(1 - t)$

c. $te^{-2|t|}$

6-3. For each of the following time functions state whether or not the one-sided Laplace transform exists:

a. e^{-t^2} **d.** $\dfrac{1}{\sqrt{t}}$

b. e^{t^2} **e.** $\dfrac{1}{t}$

c. $\dfrac{1}{\sin t}$ **f.** $e^{-1/t}$

6-4. For each of the time functions in Problem 6-1 indicate the range of σ values for which the transform exists.

6-5. Find the one-sided Laplace transform of each of the following time functions:

a. $e^{-2t}u(1 - t)$ **c.** $\cosh 3t$

b. $\sinh 3t$ **d.** $10u(t) - 10u(t - 2)$

6-6. Using linearity, find the one-sided Laplace transforms of the following time functions:

a. $t^3 + t^2 + t + 1$ **c.** $\sum\limits_{n=1}^{\infty} \dfrac{1}{n} e^{-nt}$

b. $10e^{-3t} - 4e^{-2t} - 6e^{-t}$ **d.** $t + e^{-2t}$

6-7. Using Table 6-2 and the scaling theorem, find the one-sided Laplace transforms of the following functions:

a. $2te^{-4t}$ **c.** $e^{-2\alpha t} \cos(2\beta)$

b. $\cos(2\beta t)$ **d.** $(2t)^n$

6-8. Using the t-shift theorem, find the Laplace transforms of the following time functions:

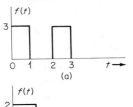
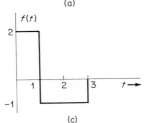

(a)

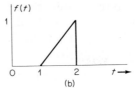

(b)

(c)

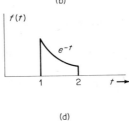

(d)

6-9. Using two different methods, find the Laplace transform of

$$f(t) = \frac{d}{dt}[e^{-\alpha t}u(t)]$$

6-10. If the time function

$$f(t) = e^{-\alpha t}$$

is noncausal, find the one-sided Laplace transform of $df(t)/dt$.

6-11. Consider the two time functions

$$f_1(t) = te^{-\alpha t}u(t), \qquad f_2(t) = te^{-\alpha t}$$

where $f_2(t)$ is noncausal. Find the Laplace transforms of the derivatives of both of these functions.

6-12. Derive equation (6-46) for the Laplace transform of the integral of a causal time function.

6-13. For a noncausal time function $f(t)$ and for $t \geq 0$, show that

$$\mathscr{L}\left\{\int_{-\infty}^{t} f(t)dt\right\} = \frac{1}{s}F(s) + \frac{1}{s}\int_{-\infty}^{0^-} f(t)dt$$

6-14. Consider the two functions

$$f_1(t) = e^{-\alpha t}u(t), \qquad f_2(t) = e^{-\alpha |t|}$$

where $f_2(t)$ is noncausal. Find the Laplace transforms of the integrals of both of these time functions when the integration is:
a. Over the range $0^- \leq t$ **b.** Over the range $-\infty < t$.

6.15. Using the s-shift theorem, find the Laplace transforms of the following time functions:
a. $e^{-(t-1)}u(t)$ **c.** $e^{-\alpha t}[u(t) - u(t-1)]$
b. $e^{-\alpha t}\sin\omega_0 t u(t)$ **d.** $\sinh at[u(t) - u(t-1)]$

6-16. Using the s-shift theorem, find the inverse Laplace transform of
   a. $1/(s^2 + 4s + 4)$ **b.** $2/(s^2 + 6s + 13)$

6-17. Find the Laplace transforms of the nth derivatives of the functions $f(t)$ and $f(t)u(t)$.

6-18. A set of functions that is orthonormal on the interval $0 < t < \infty$ can be used as basis functions for the orthogonal expansion of Laplace transformable signals. Such a set is the following:

$$\phi_n(t) = \frac{\epsilon^{-t/2}}{n!} L_n(t)$$

where $L_n(t)$ is the nth order Laguerre polynomial defined by

$$L_n(t) = \epsilon^t \frac{d^n}{dt^n}(t^n \epsilon^{-t})$$

Find the Laplace transform of $\phi_n(t)$.

6-19. Use the convolution theorem to find the Laplace transforms of the following functions of time:

   a. $f(t) = \displaystyle\int_0^t f_1(\lambda)d\lambda$

   b. $f(t) = \displaystyle\int_t^{t+a} f_1(\lambda)d\lambda$

   c. $f(t) = \displaystyle\int_0^\infty u(\lambda)u(t-\lambda)d\lambda$

6-20. Using the convolution theorem, write the Laplace transform of the output of the lowpass filter shown below.

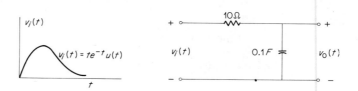

6-21. Find the Laplace transforms of the periodic waveforms shown.

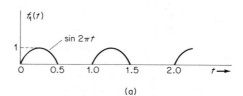

(a)

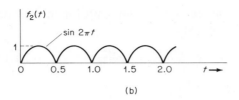

(b)

6-22. Find the inverse Laplace transform of each of the following functions:

a. $\dfrac{s}{(s+1)(s+4)}$

b. $\dfrac{s^3+6s^2+6s}{s^2+6s+8}$

c. $\dfrac{s+1}{s^2+2s}$

d. $\dfrac{e^{-2s}}{(s+1)(s+2)^2}$

e. $\dfrac{5s-12}{s^2+4s+13}$

f. $\dfrac{1-e^{-4s}}{3s^2+2s^2}$

6-23. Find the inverse Laplace transform of each of the following functions:

a. $\dfrac{s^2+3s+5}{s^2+3s+2}$

b. $\dfrac{s^2+s+1}{s^2+1}e^{-sT}$

c. $\dfrac{se^{-s}+2s^2+9}{s(s^2+9)}$

d. $\dfrac{s}{(s+1)(s^2+1)}$

e. $\dfrac{s^5+2s^3+s^2+s+4}{(s^2+4)(s^2+1)^2}$

f. $\dfrac{s^3+4s^2+6s-6}{(s^2+2s+10)(s^2+4s+4)}$

6-24. When $F(s)$ is the ratio of two polynomials, the residues at simple poles can be evaluated by the following expression:

$$k_n = \dfrac{P(s)}{Q'(s)}\bigg|_{s=\alpha_n}$$

where α_n is the root of $Q(s)$ at $s=\alpha_n$, and $Q'(s)$ is the derivative of $Q(s)$.

a. Use this expression to find the inverse transform of the functions of Problem 6-22(a) and (b).

b. How is this formula related to the procedure of multiplying through by a factor and then evaluating the resulting expression at the root of the factor as discussed in the text?

c. Write a general expression for the inverse transform of a rational function of s having only first-order poles in the LHP. The resulting expression is known as Heaviside's expansion theorem.

6-25. Solve the following differential equation using Laplace transform methods:

$$\dfrac{d^2x(t)}{dt^2}+4x(t)=0$$

where $x(0)=1$ and $x'(0)=0$.

6-26. Find the values of $y(0)$ and $y'(0)$ for the system described by the differential equation

$$\frac{d^2y(t)}{dt^2} + 2\frac{dy(t)}{dt} + y(t) = 2u(t)$$

where $y(1) = e^{-1}$ and $y'(1) = 0$.

6-27. Solve the following differential equations using Laplace transform methods:

a. $x''(t) + x'(t) = t^2 + 2t$ **b.** $2x'(t) + 4x(t) + y'(t) - y(t) = 0$
where $x(0) = 4x'(0) = -2$ $x'(t) + 2x(t) + y'(t) + y(t) = 0$
                                  where $x(0) = 0$ $x'(0) = 2$
                                        $y(0) = 1$ $y'(0) = -3$

6-28. Prove the final-value theorem, which states that

$$\lim_{t\to\infty} f(t) = \lim_{s\to 0} sF(s)$$

6-29. Using the initial- and final-value theorems, find the initial and final values associated with the transforms of Problem 6-22 [except part (b)].

6-30. Using the initial- and final-value theorems, find the initial and final values associated with the transforms of Problem 6-23, parts (c) through (f).

6-31. a. Sketch and label the transform network for the circuit shown, in which the switch is closed at $t = 0$.
b. Write the Laplace transform for the current in the switch, $i(t)$.
c. Find the current in the switch as a function of time.

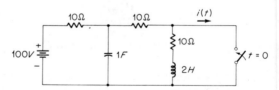

6-32. The switch in the circuit shown is moved from position 1 to position 2 at $t = 0$. The capacitor C_2 is initially uncharged.
a. Find the voltage $v_0(t)$ that appears across capacitor C_2.
b. At $t = 100$ the switch is returned to position 1, and at $t = 200$ it is moved again to position 2. Write an approximate expression for the voltage $v_0(t)$ for $t \leq 200$.

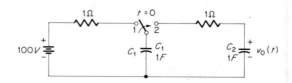

6-33. Determine the current $i_1(t)$ in the circuit shown when the input is a unit voltage pulse. Assume an ideal transformer.

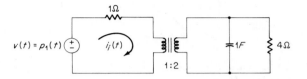

6-34. Compute the voltage $v_2(t)$ in the circuit shown. Assume that the switch has been closed a long time prior to $t = 0$ and that after $t = 0$ it remains open.

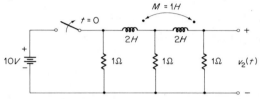

6-35. For the circuit shown, find each of the transfer functions defined as follows:

a. $H_a(s) = \dfrac{V_2(s)}{V_1(s)}$ **c.** $H_c(s) = \dfrac{I_2(s)}{V_1(s)}$

b. $H_b(s) = \dfrac{I_2(s)}{I_1(s)}$ **d.** $H_d(s) = \dfrac{V_2(s)}{I_1(s)}$

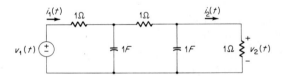

6-36. a. Determine the transfer function of the system shown.
b. Find the response of this system to a unit step and a unit ramp for $K = 0$ and $K = 1/8$. [*Note:* $H_2(s) = K$.]

$$X(s) \longrightarrow \xrightarrow{+} \boxed{H_1(s) = \dfrac{10}{s(s+1)}} \longrightarrow Y(s)$$

$$\boxed{H_2(s) = K}$$

6-37. For the circuit and periodic input signal shown below, find the steady-state response of the output in a single period.

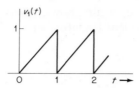

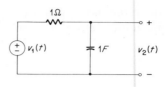

6-38. For each of the polynomials below, use the Routh test to determine the number of roots in the right half plane.

a. $s^3 + 6s^2 + 11s + 6$ d. $s^4 + 5s^3 + 5s^2 - 5s - 6$
b. $s^3 + 2s^2 - s - 2$ e. $s^4 + 7s^3 + 17s^2 + 17s + 6$
c. $s^3 + s^2 - 10s + 8$ f. $s^4 + 4$

6-39. Apply the Hurwitz test to each of the polynomials in Problem 6-38 and verify the results obtained there.

6-40. For the transfer function

$$H(s) = \frac{s^2 + 2s + 1}{s^4 + s^3 + 2s^2 + s + K}$$

find the range of K values for which it is stable.

6-41. Consider a transfer function of the form

$$H(s) = \frac{6s}{s^2 + 6s + 25}$$

a. Sketch the pole-zero plot for this transfer function.
b. Sketch $|H(j\omega)|$.
c. At what value of ω does $|H(j\omega)|$ have its maximum value, and what is this maximum value?

6-42. For the transfer function of Problem 6-41, use the pole-zero plot to evaluate the residues at all the poles.

6-43. Write the Fourier transforms for the time functions having the following one-sided Laplace transforms:

a. $\dfrac{s}{(s+8)^2}$ c. $\dfrac{s+3}{(s^2+9)(s+4)}$

b. $\dfrac{s^2+9}{s(s^2+4s+3)}$ d. $\dfrac{1}{s(s^2+9)}$

6-44. Find the two-sided Laplace transform of each of the following functions and indicate the strip of convergence:

a. $p_T(t + T/2)$ c. $e^{-\alpha|t|} \cos \omega_0 t$
b. $(t - a)u(-t)$ d. $e^{-\alpha t^2}$

6-45. A time function is defined as

$$f(t) = te^{-\alpha|t|} \qquad -\infty < t < \infty$$

a. Find the two-sided Laplace transform.
b. Find the strip of convergence.
c. Repeat parts (a) and (b) for the function $e^{\beta t} f(t)$, $\beta < \alpha$.

6-46. Find the two-sided Laplace transforms of the following functions. (*Hint*: Treat them as limiting forms of functions having known transforms.)

a. $f(t) = K$ $-\infty < t < \infty$

b. $f(t) = \cos \omega_0 t$ $-\infty < t < \infty$

6-47. Find the bounded time functions corresponding to the following two-sided Laplace transforms:

a. $F_{\mathrm{II}}(s) = \dfrac{-10}{s^2 - 4}$

b. $F_{\mathrm{II}}(s) = \dfrac{s - 2}{(s - 2)^2 + 9}$

c. $F_{\mathrm{II}}(s) = \dfrac{-16s}{(s^2 - 16)^2}$

6-48. Find $v_o(t)$ at the output of the circuit shown when the input is $v_i(t)$. Use the two-sided Laplace transform.

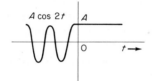 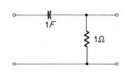

6-49. A given system has a transfer function of

$$H(s) = \frac{V_o(s)}{V_i(s)} = \frac{10s(s+2)}{s^3 + 8s^2 + 19s + 12}$$

and the input voltage has the form

$$v_i(t) = 5\epsilon^{-2|t|} \qquad -\infty < t < \infty$$

Find the output voltage $v_o(t)$.

6-50. Consider an n-stage R–C-coupled amplifier with identical time constants in each stage and a gain of K in each stage.

a. Compute the step response of such an amplifier, neglecting all frequency-dependent quantities except the coupling circuits.

b. Sketch the response for one- and two-stage amplifiers, assuming a unity gain and unity time constant for each stage.

c. How is the initial slope of the step response of an n-stage amplifier related to the initial slope of a single stage?

d. Is there any relationship between the low-frequency, half-power bandwidth and the initial slope of the step response?

6-51. When pulse signals are passed through a resistive attenuator as shown in (a), a loss in rise time results because of stray capacitance across R_2 to ground. By adding a small capacitor shunting R_0, the circuit response can be speeded up. Investigate this effect and

determine a suitable value for the shunting capacitor C_1, for the
circuit shown in (b). Assume the input to be a unit step to simplify
calculations.

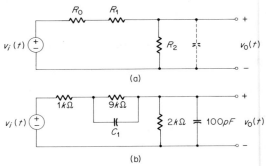

(a)

(b)

6-52. A frequently used definition of the "rise time" of a system is the
time required for the system response to go from 10 to 90 percent of
its final value when the excitation is a unit step.

a. For the circuit shown in (a), calculate the rise time and determine
the relationship between the rise time and half-power bandwidth
for circuits of this type.

b. Find the rise time for the "shunt peaked" circuit shown in (b) for
values of $L = 0.5$ mH (millihenry) and $L = 2$ mH. What limits the
amount of peaking that can be employed?

c. A general rule for the relationship between rise time T_r (seconds)
and half-power bandwidth B (Hz) of systems without exces-
sive overshoot is $T_rB = 0.35$ to 0.45. Using this relationship, compute the bandwidths of the two circuits shown.

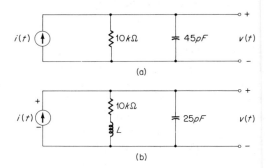

(a)

(b)

The z-Transform and Digital Filters

7-1 Introduction

The majority of signals occurring in electrical engineering are best approximated or modeled by a function of the continuous time variable, t. Typical of such signals are am and fm waveforms, radar and sonar returns, thermal noise, and a variety of others. In many cases these signals can be handled by analog processing techniques such as filters, envelope detectors, clippers, and so on. When this is the case special handling of the signals is not required. There are increasingly more situations arising, however, in which it is desired to process signals of this type in a digital computer. Such processing requires that the signal be represented by a set of values corresponding to a discrete-time variable. The most frequently used representation is the set of numbers corresponding to the signal amplitudes at equally spaced sampling instants. For convenience in mathematical analysis it is frequently useful to represent such signals as a sequence of impulses occurring at the sampling times, each having a strength equal to the signal amplitude at the particular sampling instant. These two representations are illustrated in Fig. 7-1.

The representation as a function of the continuous variable, t, is readily understood. The representation as a sequence of impulses can be visualized by considering it to result from passing the signal through an

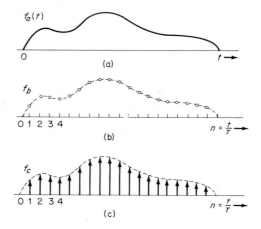

Fig. 7-1 Signal representation.

"impulse sampler" that generates impulses at a steady rate, with the strength of each impulse being given by the signal amplitude at the time of the impulse. Mathematically, this can be expressed as follows:

$$f_c(t) = f(t) \cdot \sum_{n=0}^{\infty} \delta(t - nT) = \sum_{n=0}^{\infty} f(nT)\delta(t - nT) \qquad (7\text{-}1)$$

The representation of Fig. 7-1(b) is not handled so readily mathematically, as it is not a function. It is, instead, a sequence.

$$f_b = [f(nT)] \qquad (7\text{-}2)$$

Note that in f_b and f_c nothing is stated or implied about the value of the original function for values of t other than integral multiples of the sampling period, T. It is obvious that f_b and f_c are intimately related, since either is readily obtained from the other. The nature of this relationship can be made more evident by considering the Laplace transform of f_c. Thus,

$$F_c(s) = \mathcal{L}\{f_c(t)\} = \mathcal{L}\left\{\sum_{n=0}^{\infty} f(nT)\delta(t - nT)\right\}$$

$$F_c(s) = \sum_{n=0}^{\infty} f(nT)\epsilon^{-nTs} \qquad (7\text{-}3)$$

An important property of this transform is its periodicity. This may be seen by considering $F_c[s + j(2\pi/T)]$:

$$F_c\left(s + j\frac{2\pi}{T}\right) = \sum_{n=0}^{\infty} f(nT)\epsilon^{-nT[s + j(2\pi/T)]}$$

$$= \sum_{n=0}^{\infty} f(nT)\epsilon^{-nTs} = F_c(s)$$

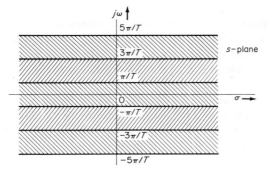

Fig. 7-2 Region of periodicity of $F_c(s)$.

Thus, $F_c(s)$ is periodic along any line parallel to the $j\omega$-axis in the s-plane. This is illustrated in Fig. 7-2. Many of the manipulative advantages of the Laplace transform arise because of the rational form of many of the most commonly used transforms. These advantages do not occur when the transform is periodic, since in computing inverse transforms, there are an infinite number of poles and zeros in the s-plane to contend with. This situation can be handled, but it is different from the operations normally carried out with the Laplace transform.

By making a further transformation, a considerable simplification of this situation can be brought about. This further transformation is to replace ϵ^{sT} in (7-3) by a new variable, z, giving

$$z = \epsilon^{sT} \tag{7-4}$$

$$F(z) = \sum_{n=0}^{\infty} f(nT)z^{-n} \tag{7-5}$$

The transformation, $F(z)$, is called the z-transformation (ZT) of the function $f(nT)$ and corresponds to mapping the left half of the s-plane into the interior of the unit circle and the right half of the s-plane into the exterior of the unit circle. This is illustrated in Fig. 7-3 and may be seen mathematically by noting that

$$|z| = |\epsilon^{sT}| = |\epsilon^{(\sigma + j\omega)T}| = \epsilon^{\sigma T}$$

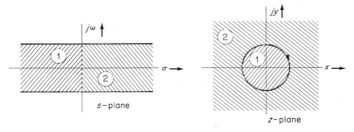

Fig. 7-3 Mapping of s-plane into a z-plane.

When σ is positive, $\epsilon^{\sigma T} > 1$, and when σ is negative, $\epsilon^{\sigma T} < 1$. For $\sigma = 0$, $z = 1$, and traveling from $-j\pi/T$ to $+j\pi/T$ along the $j\omega$-axis in the s-plane corresponds to traveling from $-j\pi$ to $+j\pi$ counterclockwise around the unit circle in the z-plane. Further excursions along the $j\omega$-axis merely retrace this same path.

The inverse z-transform is such that we recover the values of the original time function at the discrete-time instants. That is, we recover $\{f(nT)\}$. This is different from taking the inverse LT of $F_c(s)$, which is defined for all values of t. The values of $\mathscr{L}^{-1}\{F_c(s)\}$ for times different from nT can only be valid under very special circumstances (for example, band-limited signals) and so are not, in general, useful. It is, therefore, more convenient in most cases to work directly with the ZT, which involves only the values of the function at the sampling instants. The expression for the inverse z-transform ($\mathscr{Z}^{-1}$) is not readily obtained from the corresponding expression for the inverse Laplace transform. Instead, the expression will be stated and then shown to be valid. This is done in the next section.

7-2 Fundamentals of the z-Transform

Consider the sequence of numbers $\{x(nT)\}$, $n = 0, 1, 2, \ldots$. Such a sequence can be thought of as arising from a process whereby a continuous waveform (that is, one having a continuous-time variable) is sampled at times nT, $n = 0, 1, 2, \ldots$. The z-transform of this sequence is defined as

$$X(z) = \mathscr{Z}\{x(nT)\} = \sum_{n=0}^{\infty} x(nT)z^{-n} \tag{7-6}$$

Although this representation does not appear at first sight to have any striking utility, it will be seen shortly that it does indeed provide a representation allowing considerable insight and manipulative simplification to be obtained for discrete operations. For example, many useful discrete signals can be written in closed form using the z-transform. Four such signals are as follows:

1. Unit pulse.

$$x(nT) = 1 \quad n = 0$$
$$= 0 \quad n \neq 0 \tag{7-7}$$
$$X(z) = 1$$

2. Unit step.

$$x(nT) = 1 \qquad n \geq 0$$
$$= 0 \qquad n < 0$$

$$X(z) = \frac{1}{1 - z^{-1}}$$

(7-8)

3. Exponential.

$$x(nT) = \epsilon^{-\alpha n T} \qquad n \geq 0$$
$$= 0 \qquad n < 0$$

$$X(z) = \frac{1}{1 - \epsilon^{-\alpha T} z^{-1}}$$

(7-9)

4. Complex exponential.

$$x(nT) = \epsilon^{jn\omega_0 T} \qquad n \geq 0$$
$$= 0 \qquad n < 0$$

$$X(z) = \frac{1}{1 - \epsilon^{j\omega_0 T} z^{-1}}$$

(7-10)

Definition (7-6) is designated as the one-sided z-transform and is concerned only with the values of the sequence for positive n. When it is necessary to consider values of the sequence corresponding to negative n (for example, to include samples of functions extending over negative time) several different approaches may be utilized. For finite-duration signals redefining the origin of the sequence (for example, $m = n + k$) will eliminate all negative integers. A second, equivalent method is to use the translation theorem of the z-transform that will be discussed shortly. A third approach is to define a more general z-transform that includes both positive and negative integers. The z-transform defined in this manner is the two-sided z-transform and is given by

$$\mathcal{Z}_{\mathrm{II}}\{x(nT)\} = \sum_{n=-\infty}^{\infty} x(nT)z^{-n}$$

(7-11)

This definition is analogous to that of the two-sided Laplace transform.

As in the case of the Laplace transform, the one-sided representation is generally sufficient to handle problems arising from physical phenomena. However, for certain theoretical purposes, the two-sided transformation provides considerable simplification. When the functions involved are causal (that is, are zero for $n < 0$), the two transformations are identical. In most of the following discussions the one-sided z-transformation will be used; however, the extension to the two-sided transform is quite straightforward. If no statement is made to the contrary, it should be assumed that the one-sided z-transform is intended.

The mathematical process by which the original sequence is obtained from its z-transform is called *inversion*, and the inversion theorem can be established as follows: Both sides of (7-6) are multiplied by z^{k-1} and then integrated around a closed contour in the z-plane. These operations lead, formally, to the equality

$$X(z) = \sum_{n=0}^{\infty} x(nT)z^{-n}$$

$$\oint X(z)z^{k-1}dz = \oint \sum_{n=0}^{\infty} x(nT)z^{k-n-1}dz$$

For the equality to be valid it is necessary that the path of integration be in the region of convergence of $X(z)$. In general, this region of convergence will be everywhere outside some circle centered at the origin. In particular, if $\sum_{n=0}^{\infty}|x(nT)| < \infty$ [that is, if $x(nT)$ is absolutely summable], the region of convergence will include the unit circle. Under these conditions, interchanging the summation and integration is valid and gives

$$\oint X(z)z^{k-1}dz = \sum_{n=0}^{\infty} x(nT) \oint z^{k-n-1}dz$$

The integral on the right-hand side is zero except for $k = n$, in which case it has a value of $2\pi j$. Accordingly,

$$\oint X(z)z^{n-1}dz = 2\pi j x(nT)$$

from which
$$x(nT) = \frac{1}{2\pi j} \oint X(z)z^{n-1}dz \qquad (7\text{-}12)$$

As an illustration of the inversion theorem, consider the following example:
$$x(nT) = K^n \qquad (7\text{-}13)$$

$$X(z) = \sum_{n=0}^{\infty} K^n z^{-n} = 1 + Kz^{-1} + K^2 z^{-2} + \cdots$$
$$\qquad (7\text{-}14)$$

$$= \frac{1}{1 - Kz^{-1}}$$

The region of convergence of $X(z)$ is found readily by noting that, for this series to converge, $|Kz^{-1}| < 1$ or $|z| > |K|$. Applying the inversion theorem to (7-14),

$$x(nT) = \frac{1}{2\pi j} \oint \frac{z^{n-1}}{1 - Kz^{-1}} \, dz = \frac{1}{2\pi j} \oint \frac{z^n}{z - K} \, dz$$

There is a pole at $z = K$, with a residue of K^n. Choosing the path of integration as a circle of radius $K + \epsilon$, we obtain

$$x(nT) = 2\pi j \left(\frac{1}{2\pi j} K^n\right) = K^n \tag{7-15}$$

The inversion theorem, (7-12), is valid for the two-sided z-transform also, provided that the path of integration is in the region of convergence of $X_{II}(z)$.

Delay theorem. Let $X(z)$ be the transform of $\{x(nT)\}$. Consider now the transform of $\{x(nT - iT)\}$ for $i \geq 0$. Two sequences, $\{x(nT)\}$ and $\{x(nT - 3T)\}$, are shown in Fig. 7-4. Writing out the ZT gives

$$\mathcal{L}\{x(nT - iT)\} = \sum_{n=0}^{\infty} x(nT - iT)z^{-n} \tag{7-16}$$

Making the change of variable, $m = n - i$, gives

$$\mathcal{L}\{x(nT - iT)\} = \sum_{m=-i}^{\infty} x(mT)z^{-m-i}$$

$$= z^{-i} \sum_{m=0}^{\infty} x(mT)z^{-m} + z^{-i} \sum_{m=-i}^{-1} x(mT)z^{-m} \tag{7-17}$$

$$= z^{-i}X(z) + z^{-i} \sum_{m=-i}^{-1} x(mT)z^{-m}$$

Thus, a delay of i samples gives rise to a transform that is equal to z^{-i} times the undelayed transform plus some terms due to the portion of the sequence occurring for n negative. For $x(nT) = 0$, $n < 0$ this simplifies to

$$\mathcal{L}\{x(nT - iT)\} = z^{-i}X(z) \tag{7-18}$$

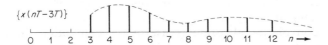

Fig. 7-4 Translated sequences.

Advance theorem. When the sequence is advanced instead of delayed, a similar result is obtained. In this case, for $i \geq 0$,

$$\mathscr{L}\{x(nT + iT)\} = z^i X(z) - z^i \sum_{k=0}^{i-1} x(kT)z^{-k} \tag{7-19}$$

Causality does not remove the right-hand terms in (7-19).

For the two-sided z-transform the extra terms do not appear in either case, and the translation theorem is

$$\mathscr{L}_{\mathrm{II}}\{x(nT + iT)\} = z^i X(z) \tag{7-20}$$

and is valid for i positive or negative.

Convolution theorem. The convolution theorem for the one-sided z-transform is most useful for causal sequences, and only this case is considered here; the more general case will be considered subsequently by means of the two-sided z-transform.

Consider the equality

$$Y(z) = H(z)X(z)$$

We wish to show that

$$y(nT) = \sum_{m=0}^{n} x(mT)h(nT - mT) = \sum_{m=0}^{n} x(nT - mT)h(mT) \tag{7-21}$$

These relationships are entirely analogous to the convolution theorem for causal, continuous-time signals. To show the validity of (7-21), consider the product $H(z)\,X(z)$ by using the infinite series representation. Thus,

$$
\begin{aligned}
Y(z) = H(z)X(z) &= [h(0) + h(T)z^{-1} + h(2T)z^{-2} + \cdots] \\
&\quad \cdot [x(0) + x(T)z^{-1} + x(2T)z^{-2} + \cdots] \\
&= x(0)h(0) + z^{-1}[x(0)h(T) + x(T)x(0)] \\
&\quad + z^{-2}[x(0)h(2T) + x(T)h(T) + x(2T)h(0)] + \cdots
\end{aligned}
\tag{7-22}
$$

But $Y(z)$ is also given by

$$Y(z) = y(0) + y(T)z^{-1} + y(2T)z^{-2} + \cdots \tag{7-23}$$

By equating coefficients of equal powers of z on the right-hand sides of (7-22) and (7-23), it is seen that the representation of (7-21) is correct. This operation is called *discrete* convolution and is denoted by $\{y(nT)\} = \{x(nT)\} * \{h(nT)\}$.

As an example of discrete convolution, consider the two finite sequences shown in Fig. 7-5.

The convolution theorem is useful for computing the response of discrete systems to discrete inputs. If $\{h(nT)\}$ is the response of a discrete system to a unit pulse input, then the response, $\{y(nT)\}$, for an arbitrary

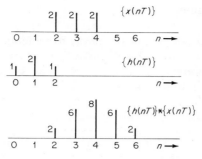

Fig. 7-5 Discrete convolution.

input, $\{x(nT)\}$, is given by

$$y(nT) = \sum_{m=0}^{n} h(mT)x(nT - mT)$$

Using the convolution theorem this can be interpreted in the z-domain in terms of a discrete transfer function, $H(z)$. Thus,

$$Y(z) = X(z)H(z) \qquad\text{(7-24)}$$

The transfer function can then be thought of as the ratio of the transforms of the output and input; that is,

$$H(z) = \frac{Y(z)}{X(z)} \qquad\text{(7-25)}$$

This interpretation is useful in analyzing systems, synthesizing filters, and obtaining physical interpretations of transform relationships.

Exercise 7-2.1
Find the z-transform of the following sequences:

$$\{x(nT)\} = 1, 1, 1$$

$$y(nT) = a^n u(n)$$

ANS. $\dfrac{1}{1 - az^{-1}}$; $1 + \dfrac{1}{z} + \dfrac{1}{z^2}$

Exercise 7-2.2
Find the inverse z-transform of the following functions:

$$X(z) = \frac{z^2}{(z - 1)^2}$$

$$Y(z) = \frac{1}{z + b}$$

ANS. $(-b)^{n-1}u(n - 1)$; $(n - 1)$

7-3   z-Transform Tables

Table 7-1 summarizes the most frequently encountered mathematical operations involving z-transforms. Most of these relationships were derived in the preceding section, and those that were not can be derived from the defining relationships.

Table 7-1   z-Transform Operations

Operation	Time function	z-Transform
Definition	$f(n)$	$\sum\limits_{n=0}^{\infty} f(n)z^{-n}$
Inversion	$\dfrac{1}{2\pi j} \oint F(z)z^{n-1}dz$	$F(z)$
Linearity	$af_1(n) + bf_2(n)$	$aF_1(z) + bF_2(z)$
Delay (right shift)	$f(n-k)u(n-k)$	$z^{-k}F(z)$
Advance (left shift)	$f(n+k)u(n)$	$z^k F(z) - z^k \sum\limits_{i=0}^{k-1} f(i)z^{-i}$
Multiply by a^n	$a^n f(n)$	$F(za^{-1})$
Multiplication of functions	$f_1(n)f_2(n)$	$\dfrac{1}{2\pi j} \oint \dfrac{F_1(\lambda)F_2(z\lambda^{-1})}{\lambda}d\lambda$
Convolution of functions	$\sum\limits_{k=0}^{n} f_1(n-k)f_2(k)$	$F_1(z)F_2(z)$
Initial value	$\lim\limits_{n\to 0} f(n)$	$\lim\limits_{z\to\infty} F(z)$
Final value	$\lim\limits_{n\to\infty} f(n)$	$\lim\limits_{z\to 1}(1-z^{-1})F(z)$

Table 7-2 provides a tabulation of the z-transforms of a number of discrete time functions. Many of the more complicated z-transforms can be calculated from the Laplace transform of the impulse-sampled, continuous function. The procedure is derived by means of the complex convolution theorem as follows:

$$\mathcal{L}\{f(t) \sum_{n=0}^{\infty} \delta(t-nT)\} = F(s) * \Sigma \epsilon^{-nTs}$$

$$= F(s) * \frac{1}{1-\epsilon^{-Ts}} \tag{7-26}$$

$$= \frac{1}{2\pi j} \int_{c-j\infty}^{c+j\infty} \frac{F(\xi)}{1-\epsilon^{-T(s-\xi)}} d\xi$$

The line integral of (7-26) can be evaluated by contour integration. The poles of the integrand, corresponding to the zeros of $1-\epsilon^{-T(s-\xi)}$, lie on a line to the right of the origin, $s - \xi = \pm j2\pi n/T$, or $\xi = s \pm j2\pi n/T$. The poles of $F(\xi)$ are in the LHP. Representative pole patterns are shown in Fig. 7-6.

Table 7-2 z-Transforms

$f(nT)$	$F(z) = \mathscr{Z}\{f(nT)\}$
$1, 0, 0, \cdots$	1
$1, 1, 1, \cdots$	$\dfrac{1}{1 - z^{-1}}$
nT	$\dfrac{Tz^{-1}}{(1 - z^{-1})^2}$
$(nT)^2$	$\dfrac{T^2 z^{-1}(1 + z^{-1})}{(1 - z^{-1})^3}$
ϵ^{-anT}	$\dfrac{1}{1 - \epsilon^{-aT} z^{-1}}$
$(K)^n$	$\dfrac{1}{1 - K z^{-1}}$
nTK^{nT}	$\dfrac{TK^T z^{-1}}{(1 - K^T z^{-1})^2}$
$\sin \omega nT$	$\dfrac{z^{-1} \sin \omega T}{1 - 2z^{-1} \cos \omega T + z^{-2}}$
$\cos \omega nT$	$\dfrac{(1 - z^{-1} \cos \omega T)}{1 - 2z^{-1} \cos \omega T + z^{-2}}$

By selecting c to the left of the poles of $1/(1 - \epsilon^{-T(s-\xi)})$ [corresponding to $\mathrm{Re}(s) > 0$] and closing the contour to the left, the only contribution to the integral will be that from $F(s)$. This procedure is valid if $F(s)$ falls off at least as rapidly as $1/s^2$ for large s. This corresponds also to $f(t)$ having an initial value of 0. Under these conditions (7-26) can be evaluated by the residue theorem to give

$$\mathscr{L}\left\{ f(nT) \sum_{n=0}^{\infty} \delta(t - nT) \right\} = \sum_{k} \text{ residues of } \left\{ \frac{F(\xi)}{1 - \epsilon^{-T(s-\xi)}} \right\} \text{ at poles of } F(s)$$

$$\text{inside contour.} \quad \textbf{(7-27)}$$

When $F(s)$ has only simple poles, (7-27) can be written as

$$\mathscr{L}\{f(nT) \sum_{n=0}^{\infty} \delta(t - nT)\} = \sum_{k} \frac{1}{1 - \epsilon^{-T(s-\xi)}} \text{ residues } F(\xi) \Big|_{\xi = s_k}$$

where s_k, $k = 0, 1, 2, \ldots$ are the poles of $F(s)$. Substituting $z = \epsilon^{sT}$, as

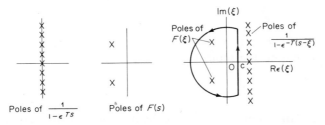

Fig. 7-6 Contour for evaluating equation (7-26).

discussed in Section 7-1, yields the z-transform as

$$\mathscr{Z}\{f(nT)\} = \sum_k \frac{1}{1 - \epsilon^{T\xi}z^{-1}} \text{ residues } F(\xi)\Big|_{\xi=s_k} \tag{7-28}$$

As an example of (7-28), consider the sampled sinusoid, $\sin \omega_0 nT$.

$$\mathscr{L}\{\sin \omega_0 t\} = \frac{\omega_0}{s^2 + \omega_0^2} = \frac{j\frac{1}{2}}{s + j\omega_0} - \frac{j\frac{1}{2}}{s - j\omega_0}$$

The poles of $F(s)$ are $s = \pm j\omega_0$, and the residues are $\pm j\frac{1}{2}$. The z-transform, therefore, is

$$\mathscr{L}\{\sin \omega_0 nT\} = \frac{j\frac{1}{2}}{1 - \epsilon^{-j\omega_0 T}z^{-1}} + \frac{-j\frac{1}{2}}{1 - \epsilon^{+j\omega_0 T}z^{-1}}$$

$$= \frac{z^{-1}\sin \omega_0 T}{1 - 2z^{-1}\cos \omega_0 T + z^{-2}}$$

Exercise 7-3.1
Find the z-transform of $f(nT) = nT$ by using the Laplace transform of the sampled continuous-time function.

ANS. See Table 7-2.

7-4 Solution of Difference Equations using the z-Transform

The general first-order difference equation can be written as

$$y(nT) = Ky(nT - T) + x(nT) \tag{7-29}$$

The present output, $y(nT)$, is expressed in terms of the present input, $x(nT)$, and the last previous value of the output, $y(nT - T)$. Multiplying both sides of (7-29) by z^{-n}, summing from $n = -\infty$ to $n = \infty$, and taking the two-sided z-transform gives

$$\sum_{n=-\infty}^{\infty} y(nT)z^{-n} = K \sum_{n=-\infty}^{\infty} y(nT - T)z^{-n} + \sum_{n=-\infty}^{\infty} x(nT)z^{-n}$$

$$Y(z) = z^{-1}KY(z) + X(z) \tag{7-30}$$

$$Y(z) = \frac{X(z)}{1 - Kz^{-1}}$$

The factor $1/(1 - Kz^{-1})$ can be thought of as a discrete transfer function relating the output and input sequences; that is,

$$H(z) = \frac{Y(z)}{X(z)} = \frac{1}{1 - Kz^{-1}}$$

This transfer function has a simple relationship to the response of the system [that is, the system characterized by (7-29)] to a sampled sinusoid.

This is demonstrated most easily by considering the response to the complex exponential ($\epsilon^{j\omega Tn}$). Using the expression for the z-transform of a complex exponential starting at $n = 0$, (7-30) can be written as

$$Y(z) = \frac{X(z)}{1 - Kz^{-1}} = \frac{1}{1 - \epsilon^{j\omega T}z^{-1}} \cdot \frac{1}{1 - Kz^{-1}}$$

$$y(nT) = \mathscr{L}^{-1}\left\{\frac{z^2}{(z - \epsilon^{j\omega T})(z - K)}\right\}$$

$$y(nT) = \frac{1}{2\pi j} \oint \frac{z^{n+1}}{(z - \epsilon^{j\omega T})(z - K)} dz \tag{7-31}$$

$$= \frac{(\epsilon^{j\omega T})^{n+1}}{\epsilon^{j\omega T} - K} + \frac{K^{n+1}}{K - \epsilon^{j\omega T}}$$

$$= \frac{\epsilon^{j\omega Tn}}{1 - K\epsilon^{-j\omega T}} - \frac{K^{n+1}\epsilon^{-j\omega T}}{1 - K\epsilon^{-j\omega T}}$$

For the system to be stable it is necessary that $|K| < 1$. In view of this, it is seen that for large n the second term goes to zero, leaving the first term as the steady-state solution. This term is of the form

$$y_{ss}(nT) = \epsilon^{j\omega Tn}H(z)|_{z=\epsilon^{j\omega T}} \tag{7-32}$$

Thus, $H(z)$ with $z = \epsilon^{j\omega T}$ can be interpreted as the frequency response of the discrete system. For the first-order system defined by (7-29), this becomes

$$H(z)|_{z=\epsilon^{j\omega T}} = \frac{1}{1 - K\epsilon^{-j\omega T}}$$

$$|H(\epsilon^{j\omega T})|^2 = \frac{1}{1 - 2K\cos \omega T + K^2} \tag{7-33}$$

The magnitude of the frequency response is sketched in Fig. 7-7 as a

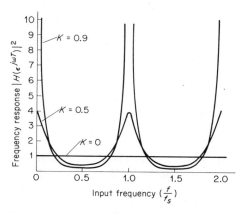

Fig. 7-7 Frequency response of first-order system.

function of input frequency normalized to the sampling frequency, $f_s = 1/T$.

The periodic nature of the frequency response of the discrete system is clearly evident from Fig. 7-7.

A geometrical interpretation of the discrete transfer function can be made by reference to Fig. 7-8. The frequency is measured in units of ωT along the unit circle. By the law of cosines the distance d from the frequency value on the unit circle to the location of the pole at $z = K$ is found to be $d = (1 + K^2 - 2K \cos \omega T)^{1/2}$, which is equal to the reciprocal of the magnitude of $H(\epsilon^{j\omega T})$.

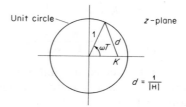

Fig. 7-8 Geometrical interpretation of $|H|$.

The second-order difference equation,

$$y(nT) = K_1 y(nT - T) + K_2 y(nT - 2T) + x(nT) \tag{7-34}$$

has the system function

$$H(z) = \frac{1}{1 - K_1 z^{-1} - K_2 z^{-2}} \tag{7-35}$$

Substituting $z = \epsilon^{j\omega T}$, multiplying by complex conjugate, and taking the square root gives the following expression for the magnitude of $|H(\epsilon^{j\omega T})|$:

$$|H(\epsilon^{j\omega T})| = \frac{1}{[(1 - K_1 \cos \omega T - K_2 \cos 2\omega T)^2 + (K_1 \sin \omega T + K_2 \sin 2\omega T)^2]^{1/2}} \tag{7-36}$$

With K_2 negative and greater in magnitude than $K_1^2/4$, the poles become complex conjugates, and as they approach the unit circle a resonance phenomenon is observed as a function of frequency. A typical response is sketched in Fig. 7-9.

In order for systems characterized by a discrete transfer function to be stable, it is necessary that all poles of $H(z)$ be inside the unit circle in the z-plane. An mth-order difference equation will have a z-transform system function containing m poles and r zeros, m being the number of unit delays of the input employed.

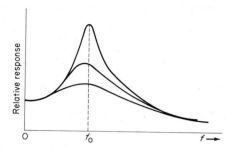

Fig. 7-9 Resonant response of second-order system.

Exercise 7-4.1

Solve the following difference equation:

$$x(n+2) - x(n+1) + \tfrac{1}{4}x(n) = 10\,u(n)$$

$$x(0) = 0$$

$$x(1) = 0$$

ANS $\qquad x(nT) = 40\left[1 - \dfrac{n+1}{2^n}\right], \qquad n \geq 0$

7-5 Digital Filters

By a digital filter is meant a discrete-time system that accepts an input sequence and delivers an output sequence that is some modification of the input. The operations performed by the filter can be represented mathematically by a suitable difference equation, or perhaps a set of difference equations. An ordinary difference equation can be written in general form as

$$y(nT) = \sum_{i=0}^{M} a_i x(nT - iT) - \sum_{k=1}^{N} b_k y(nT - kT) \qquad \text{(7-37)}$$

This equation relates the nth sample of the output to the N previous values of the output and the $M + 1$ most recent values of the input. If all of the b_k coefficients are zero, the filter corresponding to (7-37) is said to be of the *nonrecursive*, or *transversal*, type. In this type of filter the output is just a simple weighting of present and previous values of the input. The nonrecursive difference equation involving three delays (usually called a third-order system) may be written as

$$y(nT) = a_0 x(nT) + a_1 x(nT - T) + a_2 x(nT - 2T) + a_3 x(nT - 3T)$$

$$\text{(7-38)}$$

The realization of (7-38) as a filter is shown schematically in Fig. 7-10.

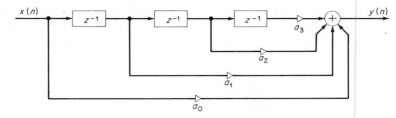

Fig. 7-10 Third-order nonrecursive filter. (Many different symbols are employed in the literature to represent digital processors and care must be taken to understand the representation being used. The digital processor diagrams in this chapter follow the recommendations in Rabiner et al., "Terminology in Digital Signal Processing," IEEE *Trans. Audio and Electroacoustics,* December 1972).

When one or more coefficients of each set $\{a_i\}$ and $\{b_k\}$ is nonzero, the system is said to be of the *recursive* type. In a recursive filter the present value of the output is determined from the $M + 1$ most recent values of the input plus the N previous values of the output. The general second-order recursive difference equation may be written as

$$y(nT) = a_0 x(nT) + a_1 x(nT - T) + a_2 x(nT - 2T)$$
$$- b_1 y(nT - T) - b_2 y(nT - 2T)$$

(7-39)

A filter corresponding to (7-39) is shown in Fig. 7-11. Several alternate and more useful realizations of recursive filters will be considered in a later section.

Several distinctly different properties are possessed by the recursive and nonrecursive filters. The nonrecursive filter has a finite memory because of the finite number of delays that can be realized in a practical implementation. This type of filter also generally has excellent phase charac-

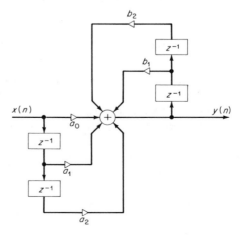

Fig. 7-11 A second-order recursive filter.

teristics. It is found that a large number of elements is required in order to obtain sharp cutoff characteristics in nonrecursive filters. On the other hand, filters of the recursive type have an infinite memory (because of the utilization of previous values of the output) and generally require significantly fewer elements to obtain a given cutoff characteristic. The phase characteristic of recursive filters is generally much poorer than that of nonrecursive filters.

Recursive filters are the discrete counterparts of linear, lumped-parameter, continuous-time systems. To see this analogy more clearly, it is convenient to consider the relationship between the output and input in the z-transform domain. Rearranging (7-37) by combining all the output samples under the summation sign and then taking the z-transform leads to the following representation:

$$\sum_{k=0}^{N} b_k y(nT - kT) = \sum_{i=0}^{M} a_i x(nT - iT)$$

$$\sum_{k=0}^{N} b_k \sum_{n=0}^{\infty} y(nT - kT) = \sum_{i=0}^{M} a_i \sum_{n=0}^{\infty} x(nT - iT)$$

$$\sum_{k=0}^{N} b_k z^{-k} Y(z) = \sum_{i=0}^{M} a_i z^{-i} X(z) \tag{7-40}$$

$$\frac{Y(z)}{X(z)} = \frac{\sum_{i=0}^{M} a_i z^{-i}}{\sum_{k=0}^{N} b_k z^{-k}}$$

This is recognized as the discrete transfer function, $H(z)$. From (7-37) it is known that $b_0 = 1$, and therefore, it is permissible to rewrite (7-40) as

$$\frac{Y(z)}{X(z)} = H(z) = \frac{\sum_{i=0}^{M} a_i z^{-i}}{1 + \sum_{k=1}^{N} b_k z^{-k}} \tag{7-41}$$

The relationship between the input and the output can be written from (7-41) as

$$Y(z) = X(z)H(z) \tag{7-42}$$

Because of the unique relationship between the z-transform and the sample values of the original sequence, it is seen that $y(nT)$ can be determined from $\{x(nT)\}$ and the discrete transfer function. From its derivation, it is seen that $H(z)$ is the ratio of two polynomials in z^{-1} with the coefficients in the polynomials corresponding to the coefficients in the original difference equation.

If the input sequence $\{x(nT)\}$ is chosen as the unit pulse

$$x(nT) = 1 \qquad n = 0$$
$$\qquad\quad = 0 \qquad n \neq 0$$

then $X(z) = 1$, and the response of the system is the inverse transform of $H(z)$. This is frequently called the impulse response of the digital filter but is more correctly the *unit pulse response*.

In order to proceed with the digital-filter design problem it is necessary to specify the characteristics that are desired. There are essentially two ways that this is done. The most fundamental specification would be the difference equation that is to be satisfied. Such a specification might arise directly from requirements in a discrete data-processing problem. In such a case the discrete transfer function could be derived and the filter synthesized directly by the methods to be described shortly. This is not the most common type of specification encountered, however. A much more common specification arises when a continuous-time signal is to be processed digitally, and it is desired to have the digital filter perform an operation comparable to that of an analog filter. The common requirement is to have the sample values of the digital filter output be equal to the values of the sampled output of the analog filter. This requirement may be satisfied by making the unit pulse response of the digital filter equal to the impulse response of the continuous-time filter at the sampling instants. The principal problem in digital-filter design is to obtain a suitable continuous-time-filter transfer function and then to convert it to a suitable digital transfer function. Once the conversion to $H(z)$ has been made, the filter realization can be obtained directly with no additional effort. In conventional design the realization problem is usually as difficult as obtaining a suitable functional approximation to the desired filter characteristic. For digital filters the realization poses no computational problem but does require judgment as to which of the various methods available is most appropriate. The design of recursive filters is considered in detail in the next section.

7-6 Recursive-Filter Design

It will be assumed that the desired digital filter is one that has a unit pulse response whose sample values equal the sampled-impulse response of a specified continuous filter. The most common specification for the continuous filter is in terms of its transfer function, $H(s)$. If some other specification is employed, [for example, $|H(s)|^2$] it may be converted to an appropriate $H(s)$. The problem is how to go from $H(s)$ to the required $H(z)$. Since digital filters are always bandlimited (as a result of the sampling process), and $H(s)$ is never strictly bandlimited, it is evident that some approximation in the digitization process must occur. Let the transfer function $H(s)$ be expanded in partial fractions to give

$$H(s) = \sum_{i=1}^{M} \frac{K_i}{s + s_i}$$

The impulse response of the continuous filter is

$$h(t) = \mathcal{L}^{-1}\left\{\sum_{i=1}^{M} \frac{K_i}{s + s_i}\right\} = \sum_{i=1}^{M} K_i \epsilon^{-s_i t}$$

It is desired that the unit pulse response of the discrete filter equal the sampled-impulse response. Therefore,

$$h(nT) = \sum_{i=1}^{M} K_i \epsilon^{-s_i nT} \tag{7-43}$$

Taking the z-transform of (7-43) gives

$$H(z) = \sum_{n=0}^{\infty} h(nT)z^{-n} = \sum_{n=0}^{\infty} z^{-n} \sum_{i=1}^{M} K_i \epsilon^{-s_i nT}$$

$$= \sum_{i=1}^{M} K_i \sum_{n=0}^{\infty} z^{-n} \epsilon^{-s_i nT} \tag{7-44}$$

$$= \sum_{i=1}^{M} \frac{K_i}{1 - \epsilon^{-s_i T}z^{-1}}$$

Thus, the transformation from $H(s)$ to $H(z)$ can be accomplished by the correspondence

$$\frac{1}{s + a} \rightarrow \frac{1}{1 - \epsilon^{-aT}z^{-1}} \tag{7-45}$$

As an example of the use of (7-45), consider the digitalization of a simple R–C lowpass filter. The expressions for $H(s)$ and $H(z)$ are

$$H(s) = \frac{a}{s + a} \tag{7-46}$$

$$H(z) = \frac{a}{1 - \epsilon^{-aT}z^{-1}} \tag{7-47}$$

The digital filter providing this transfer function is shown in Fig. 7-12.

As discussed in connection with the z-transform, the frequency response of a digital filter can be determined from the digital transfer function by making the substitution $z = \epsilon^{j\omega T}$. Making this substitution in (7-47) gives

$$H(\epsilon^{j\omega T}) = \frac{a}{1 - \epsilon^{-aT}\epsilon^{-j\omega T}} \tag{7-48}$$

$$|H(\epsilon^{j\omega T})| = \frac{a}{(1 - 2\epsilon^{-aT}\cos \omega T + \epsilon^{-2aT})^{1/2}} \tag{7-49}$$

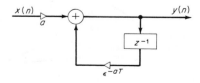

Fig. 7-12 Digital form of a simple lowpass filter.

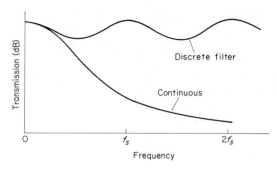

Fig-13 Frequency response of a digital filter.

The frequency response given by (7-49) is shown in Fig. 7-13 along with the corresponding frequency response of the continuous filter obtained by setting $s = j\omega$ in (7-46). The nature of the approximation involved in the digitalization process is clearly evident in Fig. 7-13 and results from the "folding," or "aliasing," effect that occurs as a result of sampling. The sampling operation can be thought of in the frequency domain as the superposition of an infinite number of replicas of the continuous spectrum, each one being displaced by an increment equal to the sampling frequency. Thus, the digital filter response shown in Fig. 7-13 can be thought of as arising from summation of the translated responses shown in Fig. 7-14. It is obvious from Fig. 7-14 that if the out-of-band transmission of the continuous filter is small, and f_s is large compared to the frequency band of interest, then the effects of folding will be small. Often this combination of characteristics can be achieved and excellent performance obtained with relatively simple filters. In cases where folding would be a serious problem, other design procedures must be considered. One such procedure employing the bilinear transformation is discussed in a later section.

Another characteristic to be observed from Fig. 7-13 is the continuous nature of the frequency response of a digital filter. This response is not discrete in any sense but is a continuous function of the frequency of the sampled continuous waveforms giving rise to the input sequence.

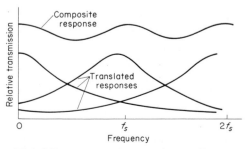

Fig. 7-14 Digital-filter response from continuous-filter responses.

Consider the design of a lowpass filter having an approximately flat frequency response over a bandwidth of zero to f_c. In some applications the sampling frequency is determined by the system, whereas in others this choice is up to the designer. For example, if a digital filter is to be used as part of a real-time system, the sampling rate must be high enough to permit the processing to occur as rapidly as the data become available. For the present example let the sampling frequency be chosen as $f_s = 1/T$. In order to limit the complexity of the computations a second-order Butterworth filter will be considered. By the methods discussed in Section 5-16 the appropriate transfer function is found to be

$$H(s) = \frac{\omega_c^2}{(s + 0.707\omega_c + j0.707\omega_c)(s + 0.707\omega_c - j0.707\omega_c)}$$

$$= \frac{\omega_c^2}{\left(s + \frac{\omega_c}{\sqrt{2}}\right)^2 + \left(\frac{\omega_c}{\sqrt{2}}\right)^2} \tag{7-50}$$

The next step would be to convert from $H(s)$ to $H(z)$ by the correspondence previously given. However, because of the frequent occurrence of complex conjugate poles, it is convenient to obtain an expression directly for such a transformation. The correspondences most often required are

$$\frac{s + a}{(s + a)^2 + b^2} \rightarrow \frac{1 - \epsilon^{-aT}\cos(bT)z^{-1}}{1 - 2\epsilon^{-aT}(\cos bT)z^{-1} + \epsilon^{-2aT}z^{-2}} \tag{7-51}$$

$$\frac{b}{(s + a)^2 + b^2} \rightarrow \frac{\epsilon^{-aT}(\sin bT)z^{-1}}{1 - 2\epsilon^{-aT}(\cos bT)z^{-1} + \epsilon^{-2aT}z^{-2}} \tag{7-52}$$

Applying (7-52) to (7-50) gives

$$H(z) = \frac{\sqrt{2}\omega_c\epsilon^{-0.707\omega_c T}\sin(0.707\omega_c T)z^{-1}}{1 - 2\epsilon^{-0.707\omega_c T}\cos(0.707\omega_c T)z^{-1} + \epsilon^{-1.414\omega_c T}z^{-2}} \tag{7-53}$$

The performance of this filter for $f_c = 100$ Hz and $f_s = 1$ kilo Hz and 10 kilo Hz is shown in Fig. 7-15.

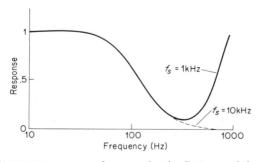

Fig. 7-15 Frequency response of a second-order Butterworth lowpass filter.

The above procedure can be extended readily to other configurations. However, in using this technique it is necessary to determine the effects of folding on the final response. Before describing a method of avoiding the folding problem, an examination of how one goes from $H(z)$ to the filter itself will be made.

Exercise 7-6.1

Derive the correspondence given in (7-51).

7-7   Digital-Filter Structures

Ordinary difference equations with constant coefficients lead to discrete transfer functions that are rational functions of z^{-1}. Similarly, digitization of rational approximations to continuous-filter functions leads to rational functions in z^{-1}. The general form of $H(z)$ is, from earlier results,

$$H(z) = \frac{\sum\limits_{i=0}^{M} a_i z^{-i}}{1 + \sum\limits_{k=1}^{N} b_k z^{-k}} \tag{7-54}$$

For a nonrecursive filter the b_i are zero, and $H(z)$ becomes

$$H(z) = \sum_{i=0}^{M} a_i z^{-i} = a_0 + a_1 z^{-1} + a_2 z^{-2} + \cdots + a_M z^{-M} \tag{7-55}$$

From (7-55) the transform of the output can be written as

$$Y(z) = [a_0 + a_1 z^{-1} + \cdots + a_M z^{-M}] X(z) \tag{7-56}$$

Inverting term by term gives

$$y(nT) = a_0 x(nT) + a_1 x(nT - T) + \cdots + a_M x(nT - MT) \tag{7-57}$$

This is seen to correspond to a summation of weighted values of the present and past inputs and can be implemented as shown in Fig. 7-16. One thing that must be considered in making filters of this kind is the gain through the filter. For example, consider putting a unit step sequence into

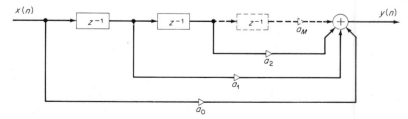

Fig. 7-16 Realization of an Mth-order nonrecursive filter.

the Mth-order filter corresponding to (7-55). The output will be

$$y(nT) = \mathscr{L}^{-1}\left\{(a_0 + a_1 z^{-1} + \cdots + a_M z^{-M})\frac{1}{1 - z^{-1}}\right\}$$

For large n (that is, $n > M$) this reduces to

$$y(nT) = \sum_{i=0}^{M} a_i \qquad\qquad \text{(7-58)}$$

The gain is therefore Σa_i and can be a substantial quantity for a high-order filter. Care must be taken to prevent overflow in the registers.

Four methods of realizing recursive filters are shown in Fig. 7-17. The first direct method [Fig. 7-17(a)] is easily derived from the general discrete transfer function (7-54). Writing the output in terms of the input and the transfer function we have

$$Y(z) = H(z)X(z) = \frac{\displaystyle\sum_{i=0}^{M} a_i z^{-i}}{1 + \displaystyle\sum_{k=1}^{N} b_k z^{-k}} X(z)$$

$$Y(z) + \sum_{k=1}^{N} Y(z)b_k z^{-k} = \sum_{i=0}^{M} a_i z^{-i} X(z) \qquad\qquad \text{(7-59)}$$

$$y(nT) = \sum_{i=0}^{M} a_i x(nT - iT) - \sum_{k=1}^{N} b_k y(nT - kT)$$

It is seen that this corresponds exactly to the system in Fig. 7-17(a). For any powers of z^{-1} not present in either the numerator or the denominator it merely is necessary to set that coefficient equal to zero to obtain the correct filter realization.

To see the basis for the second direct realization [Fig. 7-17(b)] it is helpful to define an auxiliary function, $W(z)$, corresponding to the trans-

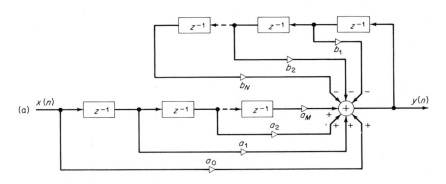

Fig. 7-17 Realization of the Nth-order recursive filter. (a) Direct.

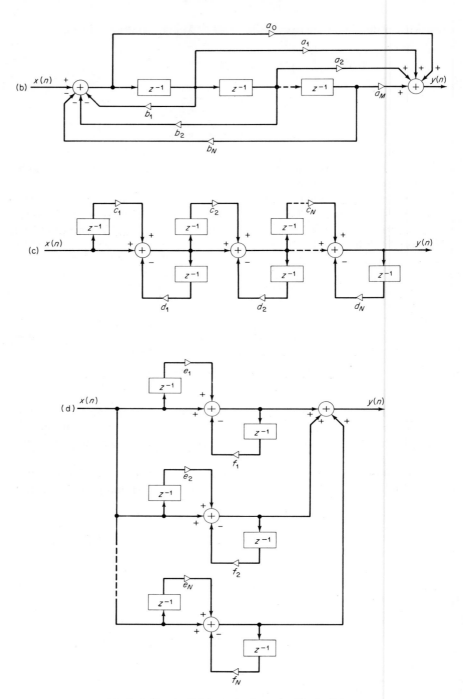

Fig. 7-17(Cont.) (b) Direct. (c) Cascade. (d) Parallel.

form of an intermediate sequence, $w(nT)$, such that

$$W(z) = X(z) \frac{1}{\sum\limits_{k=0}^{N} b_k z^{-k}} \tag{7-60}$$

$$w(nT) = x(nT) - \sum_{k=1}^{N} b_k w(nT - kT) \tag{7-61}$$

Using this relationship, it is then possible to write $Y(z)$ as

$$Y(z) = W(z) \sum_{i=0}^{M} a_i z^{-i} \tag{7-62}$$

$$y(nT) = \sum_{i=0}^{M} a_i w(nT - iT) \tag{7-63}$$

The processing operation is now described by the two simultaneous equations (7-61) and (7-63). Figure 7-18 shows one way to implement this system, and Fig. 7-17(b) is merely a straightforward modification in which common delay elements are used. In the realization of Fig. 7-18 it is seen that the numerator terms are realized by delays in the forward path, whereas the denominator terms are realized by delays in the feedback path.

The last two configurations, Fig. 7-17(c) and Fig. 7-17(d), are obtained by expressing $H(z)$ in factored form or in a partial fraction expansion. In Fig. 7-17(c), the cascade realization, each subsection realizes a single factor of the form given below and shown in Fig. 7-19.

$$H_1(z) = \frac{1 + cz^{-1}}{1 + dz^{-1}}$$

$$y_1(nT) = x_1(nT) + cx_1(nT - T) - dy_1(nT - T)$$

Cascading subsections of this type give an overall transfer function that is the product of the individual transfer functions.

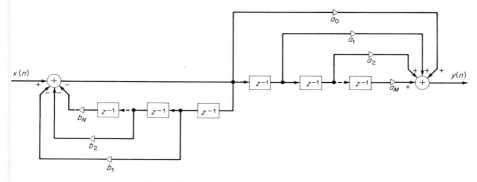

Fig. 7-18 Alternate realization of $H(z)$.

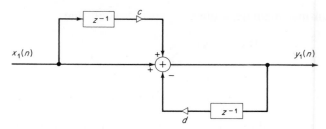

Fig. 7-19 Subsection of cascade realization of $H(z)$.

The parallel configuration of Fig. 7-17(d) is obtained directly by realizing each term in the partial fraction expansion of $H(z)$ by a separate subsection and then connecting the subsections in parallel to give an overall transfer function that is the sum of the individual transfer functions.

In the selection of the particular form most suitable in a given case several factors must be taken into account. The stability requirements of digital filters are such as to cause severe accuracy problems in connection with the coefficients in high-order systems realized in the direct form. For a filter with any significant complexity and having sharp transitions from passbands to stopbands, the use of the direct form should be avoided.

The choice between the cascade and parallel realizations is not clear-cut and is strongly influenced by the initial form of the continuous-filter transfer function. The parallel form is probably the most widely used realization.

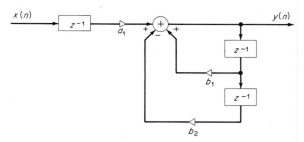

Fig. 7-20 Second-order Butterworth lowpass filter.

A realization of the second-order Butterworth lowpass filter synthesized in the previous section is shown in Fig. 7-20. The realization here is in the direct form. Attempting to go to the parallel or cascade realization leads to a requirement for complex coefficients.

Exercise 7-7.1

Sketch the digital filter sections for a sampling period of $T = 1$ corresponding to the continuous transfer functions

$$H_1(s) = \frac{s + 1}{(s + 1)^2 + 4}$$

$$H_2(s) = \frac{2}{(s + 1)^2 + 4}$$

ANS.

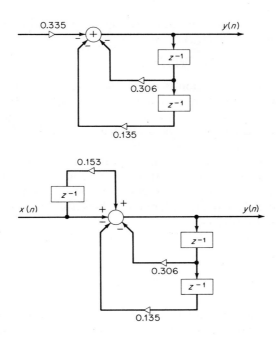

7-8 Bilinear Transformation

It is possible to avoid the "folding" problem that occurs when a direct transformation from $H(s)$ to $H(z)$ is used that only equates the sampled-impulse response of the continuous system to the samples of the discrete system. The technique is to first transform the continuous-transfer function $H(s)$ in the s-plane into a new transfer function $H(s_1)$ in the s_1-plane that is periodic in ω with period $\omega_s = 2\pi/T$. This amounts to mapping the entire s-plane into a strip $\pm j\omega_s/2$ around the origin and then repeating this mapping periodically. A transform that gives this result is

$$s = \frac{2}{T} \tanh \left(\frac{s_1 T}{2} \right) \tag{7-64}$$

Upon substituting $z = \epsilon^{s_1 T}$, equation (7-64) becomes

$$s = \frac{2}{T}\frac{z-1}{z+1} = \frac{2}{T}\frac{(1-z^{-1})}{(1+z^{-1})}$$ (7-65)

In terms of the z-plane, this transformation maps the entire left half of the s-plane into the interior of the unit circle and the entire right half of the s-plane into the exterior of the unit circle. There are no folding errors because there is no folding. This transformation is sometimes called the z-form.

The bilinear transformation is applied by making the substitution of (7-65) directly into $H(s)$; thus,

$$H(z) = H(s)|_{s=(2/T)(z-1)/(z-2)}$$ (7-66)

When this transformation is made, the $j\omega$-axis in the s-plane is mapped into the unit circle of the z-plane, and the resulting $H(z)$ evaluated along the unit circle will take on exactly the same values as $H(s)$ evaluated along the imaginary axis. The price that must be paid for this very desirable result is a serious warping of the frequency scale. Uniform increments along the $j\omega$-axis in the s-plane are far from uniform increments around the unit circle in the z-plane. If the continuous-system frequency variable is taken as ω_A and the digital frequency variable as ω_D, then the values of these variables that give identical values for the transfer function are established from (7-64):

$$\omega_A = \frac{2}{T}\tan\frac{\omega_D T}{2}$$ (7-67)

or

$$\frac{\omega_A}{\omega_s} = \frac{1}{\pi}\tan\frac{\omega_D \pi}{\omega_s}$$ (7-68)

This relationship is sketched in Fig. 7-21 and shows the departure from linearity that occurs. This nonlinear warping of the frequency scale can be serious if there is a requirement to control the response as a function of frequency in some precise manner. In such cases some type of prewarping of the transfer function is required. However, for filters having passbands and stopbands of essentially constant amplitude, compensation can

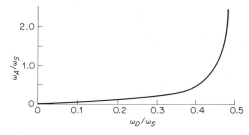

Fig. 7-21 Frequency warping of bilinear transformation.

be easily provided. All that is necessary is to change the critical frequency locations in the continuous filter so that after the transformation they will occur at the proper frequency. This is done readily from (7-67). Since the transformation is a rational function, the resulting transfer function will be rational, and the synthesis procedures previously discussed can be used.

As an example of the bilinear-transform method, consider modifying the second-order Butterworth lowpass filter to eliminate folding. Assume that $\omega_c/\omega_s = 1/10$. Then from (7-68),

$$\frac{\omega_A}{\omega_s} = \frac{1}{\pi}\tan\frac{\pi\omega_n}{\omega_s} = \frac{1}{\pi}\tan\frac{\pi}{10} = 0.103$$

The poles will be at $c = 0.103 \times (-0.707 \pm j0.707)$ and there are no zeros. The continuous-filter transfer function having unity gain is, with s being measured in units of ω_s,

$$H(s) = \frac{a_1 a_2}{(s-a_1)(s-a_2)} = \frac{0.0107}{s^2 + 0.146s + 0.0107}$$

Making the substitution $s = (1/\pi)(z-1)/(z+1)$ gives

$$H(z) = \frac{0.0107}{\dfrac{1}{\pi^2}\dfrac{(z-1)^2}{(z+1)^2} + \dfrac{0.146}{\pi}\dfrac{(z-1)}{(z+1)} + 0.0107}$$

$$= 0.0675\frac{1 + 2z^{-1} + z^{-2}}{1 - 1.144z^{-1} + 0.414z^{-2}}$$

The realization of this filter is shown in Fig. 7-22 and the frequency response in Fig. 7-23. It is evident that no folding occurs.

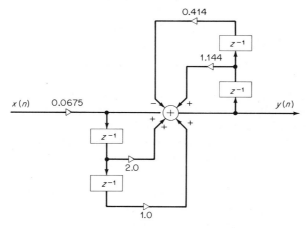

Fig. 7-22 Modified Butterworth filter.

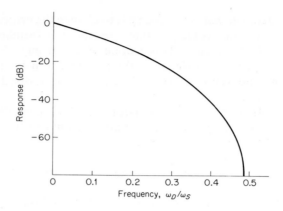

Fig. 7-23 Frequency response of filter based on bilinear transformation.

Exercise 7-8.1

Find the frequency for which the response of the filter shown in Fig. 7-22 is 1/10 of the response at zero frequency.

ANS. $\omega = 0.249\omega_s$

7-9 Nonrecursive Digital Filters

The design of nonrecursive filters can be broken down into two basic approaches: classical interpolation, differentiation, and integration methods; and Fourier series approximation to the desired transfer function. Only the latter method is considered here.

The Fourier series method consists of expanding the magnitude-frequency characteristic of the desired continuous filter in a Fourier series over the frequency interval $|\omega| < \omega_s/2$. Such expansions are of the form

$$H(\omega) = \sum_{n=0}^{\infty} a_n \cos n\omega T \tag{7-69}$$

$$H(\omega) = \sum_{n=0}^{\infty} b_n \sin n\psi T \tag{7-70}$$

The choice of which expansion to use is made on the basis of the behavior of $H(s)$ for small s. If $H(s) \simeq s^m$ for small s, then the cosine series is chosen for m even and the sine series for m odd. This choice assures the most rapid convergence of the series. Making the substitution $z = \epsilon^{j\omega T}$ leads directly to the digital-transfer function as

$$H(z) = a_0 + \frac{1}{2} \sum_{n=1}^{\infty} a_n(z^n + z^{-n}) \qquad m \text{ even} \tag{7-71}$$

$$H(z) = \frac{1}{2} \sum_{n=0}^{\infty} b_n(z^n - z^{-n}) \qquad m \text{ odd} \tag{7-72}$$

These expressions are polynomials in z^n and z^{-n}. It is evident that there will be a delay in getting an output since z^i requires the sample $x(nT + iT)$ and this is only available for the calculation of $y(nT)$ after a delay of iT beyond nT.

One of the difficulties in realizing filters by this method is the slow convergence of the Fourier series when there are discontinuities in the desired transfer function. The convergence can be substantially increased by altering the coefficients in the same manner as in ordinary Fourier analysis, where it is desired to reduce the error when using a truncated series. One useful technique is to multiply the time function $h(nT)$ by a time-limited, even function, $w(t)$. This multiplication in the time domain corresponds to convolution in the frequency domain and results in a smoothing of discontinuities or steep transitions in $H(z)$. With this window function the design equations become

$$H(z) = a_0 w(0) + \frac{1}{2}\sum_{n=1}^{N} a_n w(nT)(z^n + z^{-n}) \qquad \text{m even} \qquad \text{(7-73)}$$

$$H(z) = \frac{1}{2}\sum_{n=D}^{N} b_n w(nT)(z^n - z^{-n}) \qquad \text{m odd} \qquad \text{(7-74)}$$

A practical weighting function is the Hamming-window function, defined as

$$
\begin{aligned}
w(t) &= 0.54 + 0.46 \cos \pi t/T, & |t| &< T \\
&= 0, & |t| &> T
\end{aligned}
\qquad \text{(7-75)}
$$

For this function 99.96 percent of the energy lies in the band $|\omega| \le 2\pi/T$, and the sidelobes are less than 1-percent peak in amplitude. Figure 7-24 shows a multielement filter designed with and without the window function. The effect of the window in removing the Gibbs oscillation is clearly apparent.

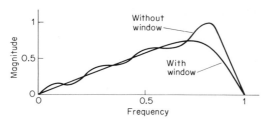

Fig. 7-24 Nonrecursive differentiating filter responses.

7-10   Quantization Effects in Digital Filters

The previous analysis of digital filter performance has implicitly assumed that the variables are continuous functions of their arguments. The arguments, but not the functions, are assumed to be discrete. In an actual implementation of a digital filter, however, the amplitudes must also be discrete; that is, the amplitudes are quantized. This is necessary because of the physical necessity of finite register lengths. Several different effects occur as a result of quantization of the variables; all are undesirable. In the succeeding subsections the following quantization effects are considered individually: register length; deadband effect; parameter quantization; A-D conversion noise; and recursion noise.

Register length.   As digital data proceed through a filter, the number of binary digits required to precisely specify it increases due to the operations performed on the data. For example, when two numbers are multiplied together, the product will require a number of binary digits equal to the sum of the numbers of digits in the two factors. This leads to requirements for very large registers if all of the digits are to be preserved. As an example, consider a fourth-order Butterworth filter. Assuming the poles to have an average distance of 0.1 from the unit circle, the gain through the filter would be 10^4. The output would, therefore, require a register length having 14 digits more than the input. If the input were quantized to one part in a thousand (10 binary digits), then the output would require 24 digits due to the dynamic range required in the system.

One way to reduce requirements for large register sizes is to break up the filter into k-cascaded sections. Each section will then have a gain equal to the kth root of the overall system gain. Following each section an attenuator can be inserted to keep the word lengths from increasing from section to section. Such a configuration is shown in Fig. 7-25 for the fourth-order Butterworth filter mentioned above. The gain in each section of this filter will be 100, and an attenuator of 1/64 will reduce the requirement at the output to $10 + 7 - 6 + 7 - 7 = 11$ bits. A similar result can be obtained for the parallel configuration. If proper allowance is not made for this effect in a digital filter, the output register will overflow, causing loss of the most significant digits in the output signal and completely destroying the filter operation.

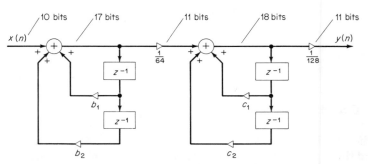

Fig. 7-25 Fourth-order Butterworth filter.

Deadband effect.   Consider a first-order system with a difference equation

$$y(nT) = Ky(nT - T) + x(nT) \qquad \text{(7-76)}$$

Let the input to this system be a single pulse of amplitude 10, and for concreteness let $K = 0.6$. The mathematical solution of (7-76) gives the following result:

$$y(nT) = 10K^n = 10(0.6)^n$$

The values of $y(nT)$ for several samples are given in column 2 of Table 7-3.

Table 7-3   Digital-Filter Response

n	$y(nT)$	$\hat{y}_1(nT)$	$\hat{y}_2(nT)$
0	10	10	10
1	6	6	6
2	3.6	4	3
3	2.16	2	1
4	1.30	1	0
5	0.78	1	0
6	0.47	1	0

Suppose now that in the actual system two different methods are used to obtain quantization: rounding and truncation. In the rounding operation the result of each computation is assigned a value equal to the nearest integer. The output sequence of the system for this type of quantization is shown in column 3 of Table 7-3. It is seen that for $n \geq 4$ the output never changes; that is, the system is stuck at 1. A similar result occurs for truncating, in which case only the nearest lower integer is preserved. The system output for this case is shown in column 4 of Table 7-3. If the output were increasing instead of decreasing, the truncation procedure would lead to a final value different from the correct value, just as did the rounding procedure for a decreasing output. This effect, wherein the output stops changing even though it has not reached the theoretically correct value, is called the *deadband effect*. The size of the deadband depends on the gain of the system. For the first-order system just considered the deadband is approximately $1/[2(1 - K)]$. Whenever the increment between $y(nT)$ and $y(nT + T)$ becomes smaller than this amount no further change takes place. The significance of this result is that the register length should be sufficient so that the deadband is a negligible fraction of the word length.

Parameter quantization.   Quantization of the coefficients in a digital filter leads to slight changes in their value. Thus, the filter is actually altered. This effect only occurs once, and its effect can be reasonably well-predicted. Consider a second-order system specified by the following difference equation:

$$y(nT) = K_1 y(nT - T) - K_2 y(nT - 2T) + x(nT) \qquad \text{(7-77)}$$

Solution of this equation shows that the poles of the transfer function are

$$z = r\underline{/\pm\,\theta} \qquad \text{(7-78)}$$

where

$$r = \sqrt{K_2}, \qquad \cos\theta = \frac{K_1}{2\sqrt{K_2}} \qquad \text{(7-79)}$$

To determine how the poles vary with K_1 and K_2, the total differential is determined.

$$\Delta r = \frac{\partial r}{\partial K_1}\Delta K_1 + \frac{\partial r}{\partial K_2}\Delta K_2 = \frac{1}{2r}\Delta K_2 \qquad \text{(7-80)}$$

$$\Delta\theta = \frac{\partial\theta}{\partial K_1}\Delta K_1 + \frac{\partial\theta}{\partial K_2}\Delta K_2 = \frac{-\Delta K_1}{2r\sin\theta} + \frac{\Delta K_2}{2r^2\tan\theta} \qquad \text{(7-81)}$$

Interpretation of (7-80) and (7-81) can be done by recalling that the unit circle in the z-plane corresponds to the $j\omega$-axis in the s-plane. Accordingly, it is evident that changes in r result in changes in bandwidth, and changes in θ result in changes in resonant frequency. Under certain circumstances the changes in θ can become very large for small changes in the coefficient values. If investigation of a specific design shows these problems to be serious a redesign may be necessary and may require a change in the configuration employed.

A-D conversion noise. When an analog signal is converted to a digital signal, the procedure is to measure the value of the signal at the sampling instant and then assign the amplitude to the nearest quantized level that is to be represented digitally. Except in certain pathological cases this leads to an error between the true signal amplitude and the quantized level that is uniformly distributed over the interval $\pm E_0/2$, where E_0 is the difference between adjacent quantum levels. The variance of this error is readily found from the probability density function to be

$$\sigma_e^2 = \frac{E_0^2}{12} \qquad \text{(7-82)}$$

When truncation is employed, the error will be distributed from $-E_0$ to 0 and will, therefore, have a mean value of $-E_0/2$. The variance will be the same as that given in (7-82). The following analysis will assume a mean of zero for the quantization error.

Analysis of the behavior of a digital filter in the presence of quantization noise can be analyzed by adding to the input $x(nT)$ an additional sequence, $e(nT)$, to account for the noise. A simple system with this modification is shown in Fig. 7-26.

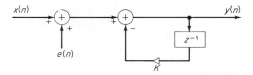

Fig. 7-26 Filter incorporating quantization noise.

If the system unit pulse response is $h(nT)$ the output will be

$$y(nT) = \sum_{m=0}^{n} [x(mT) + e(mT)]h(nT - mT) \qquad \text{7-83})$$

From (7-83) the noise output is seen to be

$$e_0(nT) = \sum_{m=0}^{n} e(nT - mT)h(mT) \qquad (7\text{-}84)$$

Since the samples, $e(nT)$, are assumed to be independent with variance $E_0^2/12$, the variance of the sum in (7-84) is just the weighted sum of the variances. Thus,

$$\sigma_0^2 = \sum_{m=0}^{n} \frac{E_0^2}{12} h^2(mT)$$

$$= \frac{E_0^2}{12} \sum_{m=0}^{n} h^2(mT) \qquad (7\text{-}85)$$

The justification for assuming independent samples in (7-84)) is based on an argument that, if the signal changes many quantum levels between samples, then they will be uncorrelated. This has been shown to be true experimentally for many types of signals. The steady-state value corresponds to $n = \infty$. As an example of (7-85), consider a first-order system with $h(nT) = K^n$.

$$\sigma_0^2 = \sigma_e^2(1 + K^2 + K^4 + \cdots + K^{2n})$$

$$= \sigma_e^2 \left(\frac{1}{1 - K^2} - \sum_{m=2n+1}^{n} K^{2m} \right) \qquad (7\text{-}86)$$

For $n \to \infty$, this becomes

$$\sigma_0^2 = \frac{\sigma_e^2}{1 - K^2} \approx \frac{E_0^2}{24\Delta} \qquad (7\text{-}87)$$

where $\Delta = 1 - K$ (distance from the poles to the unit circle) and $\sigma_e^2 = E_0^2/12$. This gives an idea of the number of bits required to handle a particular signal to keep the signal-to-noise ratio high. Since E_0 represents one bit, it follows that $1/24\Delta$ represents the number of bits of noise. For $\epsilon = 0.01$, there would be about two bits of noise. If a signal-to-noise (power) ratio of 100 was desired, the register length would have to be at least 9 bits, and possibly 11 or 12, to accomodate peak deviations.

Evaluation of (7-85) can also be carried out in the frequency domain by use of the multiplication theorem (see Table 7-1).

$$\sigma_0^2 = \sigma_e^2 \sum_{n=0}^{\infty} h^2(nT) = \sigma_e^2 \sum_{n=0}^{\infty} h(nT)h(nT)z^{-n}\bigg|_{z=1} \tag{7-88}$$

$$= \sigma_e^2 Z[h(nT)h(nT)]|_{z=1}$$

$$= \sigma_e^2 \cdot \frac{1}{j2\pi} \oint \frac{H(\lambda)H\left(\frac{z}{\lambda}\right)}{\lambda}\, d\lambda\bigg|_{z=1} \tag{7-89}$$

$$= \frac{\sigma_e^2}{j2\pi} \oint z^{-1}H(z)H\left(\frac{1}{z}\right) dz$$

As an example of this procedure, consider a second-order system and let it be required to find the variance of the output noise.

$$H(z) = \frac{1}{1 - K_1 z^{-1} + K_2 z^{-2}} = \frac{z^2}{z^2 - K_1 z + K_2} \tag{7-90}$$

The poles are at

$$z = r\epsilon^{\pm j\phi}$$

$$r = \sqrt{K_2}$$

$$\phi = \cos^{-1} \frac{K_1}{2\sqrt{K_2}} \tag{7-91}$$

$$H\left(\frac{1}{z}\right) = \frac{\dfrac{1}{K_2}}{z^2 - \dfrac{K_1}{K_2} z + \dfrac{1}{K_2}}$$

The poles of $H(1/z)$ are at

$$z = \frac{1}{r}\epsilon^{\pm j\phi} \tag{7-92}$$

These poles are outside the unit circle. Determination of σ_0^2 requires evaluating the integral

$$\sigma_0^2 = \frac{\sigma_e^2}{j2\pi K_2} \oint \frac{z^2 z^{-1}\, dz}{(z - r\epsilon^{j\phi})(z - r\epsilon^{-j\phi})\left(z - \dfrac{1}{r}\epsilon^{j\phi}\right)\left(z - \dfrac{1}{r}\epsilon^{-j\phi}\right)} \tag{7-93}$$

This integral can be evaluated by summing the residues at the two poles inside the contour of integration, which is the unit circle. The result is

$$\sigma_0^2 = \frac{\sigma_e^2}{K_2} \frac{1 + r^2}{(1 - r^2)(1 - 2r^2 \cos 2\phi + r^4)} \tag{7-94}$$

For systems having sharp transition regions the poles will be close to the unit circle, and by letting $r = 1 - \Delta$ and ignoring higher-order terms in Δ

the following approximation is obtained:

$$\sigma_0^2 \approx \frac{\sigma_e^2}{4\Delta \sin^2 \phi} = \frac{E_0^2}{48\Delta \sin^2 \phi} \qquad \textbf{(7-95)}$$

Similar analyses can be applied to other configurations to obtain expressions for the output noise due to A–D conversion. It is evident from (7-95) that poles near the unit circle corresponding to low-loss filters lead to increased noise, and that filters having resonances near the point $\phi = 0$ (that is, low-frequency filters) also have increased noise.

Recursion noise. When multiplication is carried out in a digital system errors are introduced. The exact result requires a word length equal to the sum of the individual word lengths, and in order to preserve register lengths it is necessary to round off or truncate. Again it seems reasonable to assume that the errors introduced by this process are independent from sample to sample and from multiplier to multiplier. The noise process can be modeled by putting an additive noise source at the output of each multiplier. Because these noises are generated inside the filter, not all necessarily pass through the same portions of the filter, and consequently, their contributions to the output noise may be quite different. Consider the two filters shown in Fig. 7-27. In Fig. 7-27(a) the noise inputs for each multiplication are seen to be operated on only by the portion of the filter corresponding to the poles. In Fig. 7-27(b) the noise due to the b_1 multiplication is operated on by the entire filter, whereas the noise due to the a_0 and a_1 multiplications is operated on by no part of the filter but merely adds at the output. The contribution to the output noise can be computed by exactly the same methods as used for analyzing A–D conversion noise.

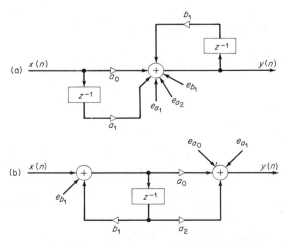

Fig. 7-27 Internal noise sources in digital filters.

All that is required is to be certain that the proper transfer function is used from the point of introduction of the noise to the output. For example, in the case of the filter of Fig. 7-27(a) the digital-transfer function relating the noise inputs to the output is

$$H(z) = \frac{1}{1 - b_1 z^{-1}}$$

Accordingly, the output noise is found from (7-89) to be

$$\sigma_0^2 = (\sigma_{a_1}^2 + \sigma_{a_2}^2 + \sigma_{b_1}^2) \frac{1}{2\pi j} \oint \frac{dz}{(z - b_1)(1 - b_1 z)} \tag{7-96}$$

Assuming that the noise due to each multiplication is $E_0^2/12$, this equation can be evaluated to give

$$\sigma_0^2 = \frac{E_0^2}{4(1 - b_1^2)} \tag{7-97}$$

For the filter of Fig. 7-27(b) the variance of the output noise can be determined in a similar manner. The transfer function applicable to e_{b_1} is

$$H_1(z) = \frac{a_0 + a_1 z^{-1}}{1 - b_1 z^{-1}} \tag{7-98}$$

and the transfer function applicable to e_{a_0} and e_{a_1} is unity. The variance of the output noise is, therefore,

$$\sigma_0^2 = \frac{E_0^2}{12} \left[2 + \frac{1}{2\pi j} \oint H_1(z) H_1\left(\frac{1}{z}\right) z^{-1} dz \right] \tag{7-99}$$

This can be evaluated to give

$$\sigma_0^2 = \frac{E_0^2}{12} \left[2 + \frac{(a_0 + a_1 b_1)(a_1 + a_0 b_1)}{b_1(1 - b_1^2)} \right] \tag{7-100}$$

Similar methods of analysis can be applied to other configurations. Such analyses can become quite involved for complex filters. One configuration that can be analyzed quite easily is the parallel filter. In this filter the noise contributions from each subsection add at the output.

References

1. JURY, E. I., *Theory and Application of the z-Transform Method.* New York: John Wiley & Sons, Inc., 1964.
   This book contains an extensive treatment of the z-transform method.
2. KUO, F. F., and J. F. KAISER, *System Analysis by Digital Computer.* New York: John Wiley & Sons, Inc., 1966.
   Along with much interesting and useful information on the application of digital computers to system analysis and simulation, this book contains an excellent discussion of digital filters.
3. SAUCEDO, R., and E. E. SCHIRING, *Introduction to Continuous and Digital Control Systems.* New York: The Macmillan Co., 1968.

This book contains a good discussion of the z-transform and certain of its modifications and refinements.

4. GOLD, B., and C. M. RADER, *Digital Processing of Signals*. New York: McGraw-Hill Book Company, Inc., 1969.
This book covers a broad range of processing methods for processing discrete signals, including the z-transform and digital filtering. Although at a somewhat higher mathematical level than this text, it contains much useful information not readily available in other texts.
5. RABINER, L. R., J. W. COOLEY, H. D. HELMS, J. F. KAISER, C. M. RADER, R. W. SHAEFER, K. STEIGLITZ, and G. J. WEINSTEIN, "Terminology in Digital Signal Processing," IEEE *Trans. Audio and Electroacoustics*, December 1972, pp. 322–337.

Problems

7-1. Find the z-transform of each of the following sequences:

    a. $\{x(nT)\} = 1, -1, 1, -1$

    b. $\{x(nT)\} = 1, 0, 1, 0, 1$

    c. $\{x(nT)\} = \{\epsilon^{-|nT|}\} - \infty < n < \infty$

    d. $\{x(nT)\} = \{0, 1, 2, 3, \ldots\}$

7-2. Find the Laurent series for the function

$$F(s) = \frac{2s^3 + 4s^2 + 4s + 3}{(s+1)^2}$$

in the vicinity of the poles at $s = -1$. What is the residue at this pole?

7-3. Find the inverse z-transform of each of the following functions by at least two methods:

    a. $\dfrac{z+1}{4z(z - \frac{1}{2})}$

    b. $X(z) = \dfrac{\epsilon^{-\alpha}z^{-1}}{(1 - \epsilon^{-\alpha}z^{-1})^2}$

7-4. Prove the discrete convolution theorem for multiplication of time sequences.

$$Z\{x_1(nT)x_2(nT)\} = \frac{1}{2\pi j}\oint x_1(\xi)x_2\left(\frac{z}{\xi}\right)\xi^{-1}d\xi$$

7-5. Solve the following difference equations and plot the results:

    a. $y(n) = \frac{1}{2}y(n-2) + \epsilon^{-n/10}$ $y(n) = 0$ $n < o$

    b. $y(n+2) + 3y(n+1) - 2y(n) = 2$ $y(1) = 1, \; y(0) = 0$

7-6. Plot the frequency response $|H(\epsilon^{j\omega T})|$ for the system corresponding to the difference equation given below. (Normalize with respect to $f_s = 1/T$.)

$$y(nT) = y(nT - T) + y(nT - 2T) + x(nT)$$

7-7. Show that the following relationship is valid:

$$\frac{b}{(s+a)^2+b^2} \rightarrow \frac{\epsilon^{-aT}(\sin bT)z^{-1}}{1-2\epsilon^{-aT}(\cos bT)z^{-1}+\epsilon^{-2aT}z^{-2}}$$

(*Hint*: Use $\dfrac{1}{s+a} \rightarrow \dfrac{1}{1+\epsilon^{-aT}z^{-1}}$)

7-8. Draw the digital filter sections corresponding to the continuous-transfer functions:

a. $\dfrac{1}{s+a}$ **b.** $\dfrac{s+a}{(s+a)^2+b^2}$ **c.** $\dfrac{s+a}{(s+b)(s+c)}$

7-9. It is desired to design a recursive digital lowpass filter having a third-order Butterworth response with a cutoff frequency of 100 Hz.

a. Find the digital-transfer function for the filter for sampling rates of 1000 Hz and 5000 Hz and sketch the frequency responses of the two filters.

b. Sketch two filter configurations for the filter and select the component values corresponding to a sampling rate of 1000 Hz for each configuration.

7-10. Design a second-order recursive digital bandpass filter having a passband of 100 Hz centered at 1000 Hz and having a sampling rate of 5000 Hz. Compute the component values for the direct realization of the filter and sketch the final filter and the frequency response of the filter.

7-11. Compute a new transfer function for the filter of problem 7-9 using the bilinear transformation to eliminate aliasing for a sampling frequency of 1000 Hz. Sketch the frequency response of the completed filter.

7-12. Design a lowpass nonrecursive digital filter having a cutoff frequency of 100 Hz and a sampling frequency of 1000 Hz. Compute the frequency response for 4 terms and for 6 terms in the approximation. Repeat the design using a Hamming window and compare the response for 4 and 6 terms with that calculated above.

7-13. For the first-order filter shown let the gain K be 0.95 and let the input be a step sequence of amplitude 10. Let the output $y(nT)$ be rounded to the nearest integer. Find the steady-state output of the filter. Compare this to the theoretical value.

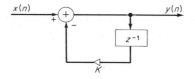

7-14. For the filter of Problem 7-13 assume that the input is a sampled sinusoid of peak value 10 and that the input is quantized to 64 levels. Find the variance of the quantization noise in the output due to the A–D conversion process.

7-15. Derive an expression for the variance of the output noise due to A–D conversion for the digital filter having a transfer function

$$H(z) = \frac{1 - K_3 z^{-1}}{1 - K_1 z^{-1} + K_2 z^{-2}},$$

where $K_1 = 2r \cos \phi$
$\qquad K_2 = r^2$
$\qquad K_3 = r \cos \phi$

7-16. For a digital filter whose transfer function is

$$H(z) = \frac{10}{1 - 0.95 z^{-1}}$$

find the recursion noise at the output if variance associated with each multiplication is $E_0^2 / 12 = 1$.

8

State Space Methods of System Analysis

8-1 Introduction

Each day the number and variety of systems being studied mathematically or simulated on a computer increases. In some instances the overall behavior of the system is of primary concern, whereas in others the details of the internal operation of the system are the primary concern. Examples of this latter case are the necessity of determining the power dissipation in a particular circuit element or the study of the effects of system transients resulting from failure of a specific component. In other types of systems it is often necessary to take into account multiple inputs and multiple outputs. Much of the methodology developed in the previous chapters is directed toward obtaining the overall system response for a single-input, single-output case, and in order to be useful for the situations mentioned above some extensions of the concepts are required. The necessary extensions are quite straightforward and lead to what are generally called *state space* methods of system analysis.

Mathematically, the state space method corresponds to representing a system in terms of a set of differential equations in the normal form, as discussed in Section 2-5. The variables in these equations are called *state variables* and in electrical systems usually correspond to currents and voltages within the system. The techniques are also directly applicable to nonelectrical systems and are used widely in control system analysis where the state variables might be such quantities as displacement, force, velocity, or other similar quantities. These same techniques are also being applied to the study of economic systems, behavioral systems, ecological systems, and many others.

In addition to providing a convenient method for studying the internal behavior of a system and handling multiple inputs and outputs, the state space approach has two other advantages. First, it can be extended readily to handle time-varying systems, and second, the equations always occur in such a form as to be solved readily numerically by standard digital computer routines.

Before discussing the state space method it is helpful to take a more general look at the block-diagram representation of systems and to develop the mathematical models of some simple systems that can be used as the building blocks for more complex systems.

8-2   Block-Diagram Representation of Systems

The concept of representing a system by means of a single block with input and output arrows, as shown in Fig. 8-1, was introduced in Chapter 1 and has been employed numerous times since. This basic concept can be extended to represent complex systems by using a number of blocks with suitable interconnections. This type of representation is particularly useful in modeling systems in terms of their state variables.

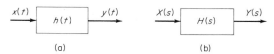

(a)   (b)

Fig. 8-1 System representation by a single block. (a) Time domain. (b) Frequency-domain representation.

The relationship between the output time function and the input time function for the system of Fig. 8-1 is

$$y(t) = x(t) * h(t) \tag{8-1}$$

where $h(t)$ is the inpulse response of the system, and the symbol $*$ denotes convolution. The corresponding representation in the frequency domain is

$$Y(s) = X(s)H(s) \tag{8-2}$$

where $H(s)$ is the system transfer function, and $X(s)$ and $Y(s)$ are the Laplace transforms of $x(t)$ and $y(t)$, respectively.

When two systems are connected in cascade, as shown in Fig. 8-2, the output of the first block serves as the input to the second block. The

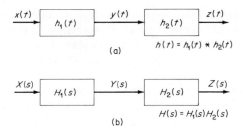

(a)

$$h(t) = h_1(t) * h_2(t)$$

(b)

$$H(s) = H_1(s)H_2(s)$$

Fig. 8-2 Systems connected in cascade. (a) Time-domain representation. (b) Frequency-domain representation.

output of the second block can, therefore, be written as

$$z(t) = y(t) * h_2(t) = [x(t) * h_1(t)] * h_2(t)$$
$$= x(t) * [h_1(t) * h_2(t)]$$

(8-3)

Accordingly, the impulse response of the cascaded system can be written as

$$h(t) = h_1(t) * h_2(t)$$

(8-4)

The transfer function of the cascaded systems is seen to be the product of the individual transfer functions. In deriving (8-3), it was assumed that the connection of the second block does not alter the signal appearing at the output of the first block. This is equivalent to saying that there is no "loading" of the first block by the second block. In many instances this is not a valid assumption, and determination of the overall system response will require that the loading effect be taken into account. This may require solution of the network equations or use of other standard circuit analysis procedures. In general, if the input impedance of the second system is much higher than the output impedance of the first system, the effect of loading can be neglected.

Another commonly encountered interconnection is the parallel connection of two systems. This type of interconnection is shown in Fig. 8-3 and is seen to have an overall response that is the sum of the responses of the individual systems.

The interconnection that is most useful in the state space method of system analysis is one involving the concept of feedback, that is, taking

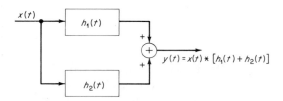

Fig. 8-3 Parallel connection of blocks.

the output, modifying it, and recombining it with the input. This type of interconnection has many practical applications because of the unique capabilities it provides for easily modifying a system's operating characteristics. However, the use of feedback for modifying system characteristics is somewhat incidental to its use in visualizing the normal form equations of a system, and it is this latter property that is of primary concern now.

Consider the system shown in Fig. 8-4. The overall transfer function of the system can be found from

$$Y(s) = [X(s) + Y(s)H_2(s)]H_1(s)$$
$$= X(s)H_1(s) + Y(s)H_1(s)H_2(s)$$

Hence,

$$H(s) = \frac{Y(s)}{X(s)} = \frac{H_1(s)}{1 - H_1(s)H_2(s)} \tag{8-5}$$

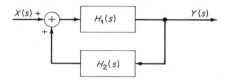

Fig. 8-4 System with feedback.

In order to express this relationship in the time domain, it is convenient to define the concept of the inverse of a system. The inverse of a system (which is not necessarily physically realizable), when cascaded with the original system, has an output that is identical to the input of the original system. The inverse system concept is illustrated in Fig. 8-5. It is clear from the figure that since

$$x(t) = h(t) * h^{-1}(t) * x(t)$$

the convolution of $h(t)$ and $h^{-1}(t)$ is an impulse; that is,

$$h(t) * h^{-1}(t) = \delta(t) \tag{8-6}$$

It should be noted that the symbol $h^{-1}(t)$ does not mean the reciprocal of $h(t)$ but is merely a symbolic notation of the inverse. Because of the multiplicative property of the transfer functions of cascaded systems, it is clear that the transfer function of the inverse system is the reciprocal of

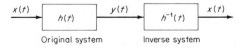

Original system   Inverse system

Fig. 8-5 Cascade connection of a system and its inverse.

the original system transfer function. This result can also be obtained readily by considering the Laplace transform of both sides of (8-6), which leads to

$$H(s) \cdot \mathscr{L}\{h^{-1}(t)\} = 1$$

$$\mathscr{L}\{h^{-1}(t)\} = \frac{1}{H(s)} \tag{8-7}$$

The inverse of a causal system is not necessarily causal; in fact, it may not exist at all in any conventional sense. Frequently, the inverse system need only be approximated over a limited range of frequencies in order to be utilized effectively. The use of the concept of a system inverse in the following sections is primarily for mathematical convenience and does not require that such a system be implementable.

Using the concept of the system inverse as the inverse transform of the reciprocal of a transfer function, it is possible to express the overall system response of a system with feedback in the time domain. Thus, (8-5) can be written in the time domain as

$$h(t) = h_1(t) * [\delta(t) - h_1(t) * h_2(t)]^{-1} \tag{8-8}$$

It is seen readily by comparing the representations of (8-8) and (8-5) why the frequency-domain representation is more widely used in the analysis of feedback systems than is the time-domain representation.

A summary of the most frequently encountered interconnections of system blocks is given in Table 8-1 along with the overall response functions in both the time domain and the frequency domain.

A particular form of feedback system that is useful in drawing block diagrams corresponding to differential equations in the normal form is shown in Fig. 8-6. In this system $H_1(s) = 1/s$ corresponds to an ideal integrator, $H_2(s) = a$ is a constant gain factor, and the block cascaded with the feedback system is a constant gain factor b. From the equivalent transfer functions of Table 8-1, the overall system transfer function is found to be

$$H(s) = \frac{Q(s)}{X(s)} = \frac{b(1/s)}{1 - (1/s)a} = \frac{b}{s - a} \tag{8-9}$$

It is clear from (8-9) that $H(s)$ corresponds to the transfer function of a first-order system and that the system will be stable only if $a < 0$, so that the pole of $H(s)$ is in the left half of the s-plane.

In order to relate the block diagram of Fig. 8-6 to the differential equation representation of the system, it is convenient to consider the time-domain version as shown in Fig. 8-7. The blocks corresponding to the constant gain factors properly should be labeled $a\delta(t)$ and $b\delta(t)$, but the δ functions were omitted to simplify the notation. The differential equation for the system of Fig. 8-7 can be written directly by noting that the input to the integrator (which is the derivative of the output) is the sum of two components; thus,

$$\frac{dq}{dt} = aq(t) + bx(t) \tag{8-10}$$

Table 8-1 Equivalents of Interconnected Systems

System configuration	Equivalent system

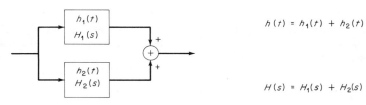

$$h(t) = h_1(t) * h_2(t)$$

$$H(s) = H_1(s) \, H_2(s)$$

Cascade connection

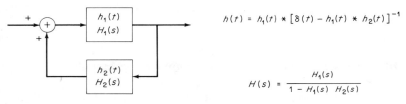

$$h(t) = h_1(t) + h_2(t)$$

$$H(s) = H_1(s) + H_2(s)$$

Parallel connection

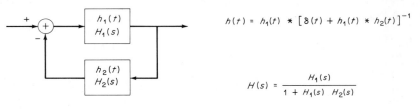

$$h(t) = h_1(t) * \left[\delta(t) - h_1(t) * h_2(t) \right]^{-1}$$

$$H(s) = \frac{H_1(s)}{1 - H_1(s) \, H_2(s)}$$

Positive feedback

$$h(t) = h_1(t) * \left[\delta(t) + h_1(t) * h_2(t) \right]^{-1}$$

$$H(s) = \frac{H_1(s)}{1 + H_1(s) \, H_2(s)}$$

Negative feedback

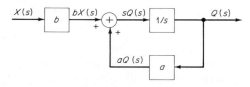

Fig. 8-6 A first-order feedback system.

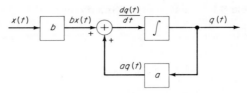

Fig. 8-7 Time-domain representation of a first-order system.

Note that this is a normal-form equation for a first-order system. The application of this representation to more general systems is discussed in the next section.

Exercise 8-2.1
Two systems having impulse responses of $h_1(t) = 10\epsilon^{-2t}u(t)$ and $h_2(t) = 20\epsilon^{-t}u(t)$ are connected in cascade. Find the equivalent impulse response of the combination.

ANS. $h(t) = 200(\epsilon^{-t} - \epsilon^{-2t})u(t)$

Exercise 8-2.2
Find the inverse for the system shown in Fig. 8-7.

ANS. $h^{-1}(t) = \frac{1}{b}\delta'(t) - \frac{a}{b}\delta(t)$

8-3 Block Diagram Representation of Normal-Form Equations

The representation of an nth-order linear system by a set of n simultaneous first-order equations is discussed in Section 2-5. The set of differential equations, called the *normal-form equations*, may be written in the general form

$$\frac{dq_1(t)}{dt} = a_{11}(t)q_1(t) + a_{12}(t)q_2(t) + \cdots + a_{1n}(t)q_n(t) + b_1(t)x(t)$$

$$\frac{dq_2(t)}{dt} = a_{21}(t)q_1(t) + a_{22}(t)q_2(t) + \cdots + a_{2n}(t)q_n(t) + b_2(t)x(t) \qquad \text{(8-11)}$$

$$\frac{dq_n(t)}{dt} = a_{n1}(t)q_1(t) + a_{n2}(t)q_2(t) + \cdots + a_{nn}(t)q_n(t) + b_n(t)x(t)$$

The output equation associated with this system of equations is

$$y(t) = c_1(t)q_1(t) + c_2(t)q_2(t) + \cdots + c_n(t)q_n(t) + d(t)x(t) \qquad \text{(8-12)}$$

The input is $x(t)$, the output is $y(t)$, and the state variables are $q_k(t)$, where $k = 1, 2, \ldots, n$. Note that these equations represent a time-varying system, since the coefficients are shown to be functions of time. In the case of a time-invariant system all of the coefficients are constants.

It is not necessary for the state variables to be physically observable quantities within the system in order for the normal-form representation to be valid. For example, state variables may be sums or differences of currents or voltages and still allow a valid representation of the system. It is convenient from a conceptual standpoint, however, to be able to sketch a block diagram in which the state variables appear explicitly, even if the resulting blocks have no physical counterparts in the actual system. The first-order feedback system of Fig. 8-7 makes it possible to do this.

As an example of the block diagram representation in terms of state variables, consider a simple system with two state variables and the following normal-form equations:

$$\frac{dq_1(t)}{dt} = -10q_1(t) + 5q_2(t) + 3x(t)$$

$$\frac{dq_2(t)}{dt} = 7q_1(t) - q_2(t) - 4x(t)$$

(8-13)

and the output equation is

$$y(t) = 1/2 q_1(t) + 2q_2(t)$$

(8-14)

The block diagram that corresponds to these equations is shown in Fig. 8-8. Although this diagram appears to be fairly complicated, it is relatively easy to compare the block diagram with the equation and to verify the equivalence.

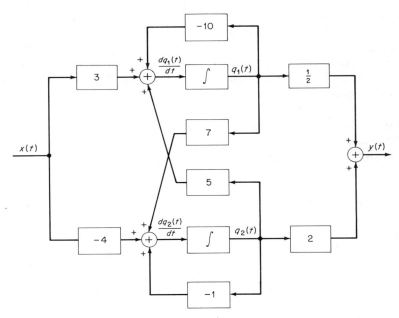

Fig. 8-8 Block diagram for a second-order system.

It is certainly possible, in principle, to draw a similar block diagram for systems of any order. However, such a diagram is likely to be so cluttered with detail that it no longer provides a simple, overall characterization of the system. It is desirable, therefore, to provide some method of simplifying the presentation of both the normal equations and the corresponding block diagram. This is accomplished by matrix notation, as described in the next section. Before we proceed with this discussion, however, it is desirable to consider a more concrete example of system representation and, at the same time, to introduce the concept of multiple-input, multiple-output systems.

All of the discussion of systems so far has assumed that there was only one input signal and one output signal. Actually, however, there are many systems that have several inputs or several outputs or both. It is still possible to represent such systems by a set of normal equations by adding to each equation the appropriate contribution from each input. Likewise, the output is expressed in terms of the state variables by a set of equations instead of by a single equation. Obviously, the complete block diagram for this situation will be even more complicated than in the single-input cases already discussed.

As a simple example, consider the two-input, two-output network shown in Fig. 8-9. The two inputs, $x_1(t)$ and $x_2(t)$, are assumed to be voltage sources here. The two outputs, $y_1(t)$ and $y_2(t)$, are the open-circuit voltages appearing at the outputs of a highpass $R-C$ filter and a lowpass $R-C$ filter, respectively. This circuit might be used to separate the low-frequency and high-frequency components of the input signals into separate outputs regardless of which input contained which component. The resistance and capacitance values are such as to make the cutoff frequency of these two filters about 100 Hz. Thus, a 10-Hz component in either $x_1(t)$ or $x_2(t)$, for example, would appear predominantly in $y_2(t)$, whereas a 1000-Hz component in either input would appear predominantly in $y_1(t)$.

For reasons that will become clearer later on, the state variables, $q_1(t)$ and $q_2(t)$, have been chosen to be the voltages across the two capacitors. By writing the usual equilibrium equations and converting to normal

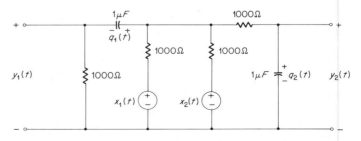

Fig. 8-9 A two-input, two-output network.

form, it is easy to obtain the normal system equations as

$$\frac{dq_1(t)}{dt} = -750q_1(t) + 250q_2(t) + 250x_1(t) + 250x_2(t)$$

$$\frac{dq_2(t)}{dt} = 250q_1(t) - 750q_2(t) + 250x_1(t) + 250x_2(t)$$

(8-15)

Note that both input signals appear in each equation.

Since there are two output signals, there will be two output equations. These are

$$y_1(t) = -0.75q_1(t) + 0.25q_2(t) + 0.25x_1(t) + 0.25x_2(t)$$

and

$$y_2(t) = q_2(t)$$

(8-16)

Note that in this case $y_1(t)$ cannot be expressed solely in terms of the state variables but must also include contributions from the input signals.

Equations (8-15) and (8-16) can now be used to sketch a block diagram for the system of Fig. 8-9 by using first-order feedback systems to create the state variables. The resulting block diagram is shown in Fig. 8-10. The complexity of the diagram for even a simple case like this clearly indicates the need for a more compact notation when there are many inputs, outputs, and state variables.

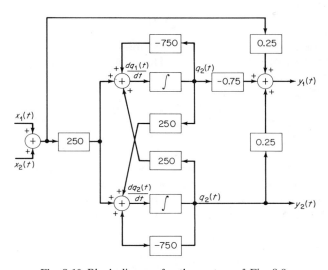

Fig. 8-10 Block diagram for the system of Fig. 8-9.

8-4   Matrix Representations of the Normal Equations

The discussion in this section and throughout the rest of this chapter assumes that the reader is familiar with concepts of vectors, matrices, and elementary matrix manipulations. A brief summary of the necessary matrix manipulations is given in Appendix C.

The first step is to define an appropriate set of vectors and matrices. If there are n state variables, these can be combined in a column vector, or an $(n \times 1)$ matrix, and designated as $\mathbf{q}(t)$.[1] Thus,

$$\mathbf{q}(t) = \begin{bmatrix} q_1(t) \\ q_2(t) \\ \vdots \\ q_n(t) \end{bmatrix} \quad (n \times 1)$$

is known as the *state vector*. If there are m input signals, the *input vector* will be designated as

$$\mathbf{x}(t) = \begin{bmatrix} x_1(t) \\ x_2(t) \\ \vdots \\ x_m(t) \end{bmatrix} \quad (m \times 1)$$

Likewise, if there are p output signals, the *output vector* is

$$\mathbf{y}(t) = \begin{bmatrix} y_1(t) \\ y_2(t) \\ \vdots \\ y_p(t) \end{bmatrix} \quad (p \times 1)$$

The coefficients of the state variables, as given in (8-11), can be represented by an $(n \times n)$ matrix. Thus, for a general time-varying system, let

$$\mathbf{A}(t) = \begin{bmatrix} a_{11}(t) & a_{12}(t) \cdots a_{1n}(t) \\ a_{21}(t) & a_{22}(t) \cdots a_{2n}(t) \\ \vdots \\ a_{n1}(t) \cdots \cdots \cdots a_{nn}(t) \end{bmatrix} \quad (n \times n)$$

The coefficients of the input signal were designated in (8-11) by $b_1(t), \ldots, b_n(t)$, since there was only one input. If there are m inputs, then there will be m such coefficients in each of the n equations. Thus, these coefficients may be represented by an $(n \times m)$ matrix as

$$\mathbf{B}(t) = \begin{bmatrix} b_{11}(t) & b_{12}(t) \cdots b_{1m}(t) \\ b_{21}(t) & b_{22}(t) \cdots b_{2m}(t) \\ \cdot \\ \cdot \\ \cdot \\ b_{n1}(t) & b_{n2}(t) \cdots b_{nm}(t) \end{bmatrix} \quad (n \times m)$$

---

[1]Vectors will be designated by boldface, lower-case letters, such as $\mathbf{a}$ or $\mathbf{x}$. Matrices will be designated by boldface, upper-case letters, such as $\mathbf{A}$ or $\mathbf{X}$.

Again, in the time-invariant case these will all be constants and the matrix becomes just **B**.

If there are p output signals, there will be p output equations, each requiring n coefficients as in (8-11). Thus, define a $(p \times n)$ matrix as

$$\mathbf{C}(t) = \begin{bmatrix} c_{11}(t) & c_{12}(t) \cdots c_{1n}(t) \\ c_{21}(t) & c_{22}(t) \cdots c_{2n}(t) \\ \cdot \\ \cdot \\ \cdot \\ c_{p1}(t) & c_{p2}(t) \cdots c_{pn}(t) \end{bmatrix} \qquad (p \times n)$$

Likewise, each output equation may contain terms from each of the m input signals. Thus, let

$$\mathbf{D}(t) = \begin{bmatrix} d_{11}(t) & d_{12}(t) \cdots d_{1m}(t) \\ d_{21}(t) & d_{22}(t) \cdots d_{2m}(t) \\ \cdot \\ \cdot \\ \cdot \\ d_{p1}(t) & d_{p2}(t) \cdots d_{pm}(t) \end{bmatrix} \qquad (p \times m)$$

Both of these matrices will have constant elements in the time-invariant case and will be designated as **C** and **D**.

Having defined these vectors and matrices, it is now possible to write the complete set of normal equations as

$$\frac{d\mathbf{q}(t)}{dt} = \mathbf{A}(t)\mathbf{q}(t) + \mathbf{B}(t)\mathbf{x}(t) \qquad (8\text{-}17)$$

These equations are frequently referred to as the *state equations*.

Similarly, the set of output equations becomes

$$\mathbf{y}(t) = \mathbf{C}(t)\mathbf{q}(t) + \mathbf{D}(t)\mathbf{x}(t) \qquad (8\text{-}18)$$

The accuracy of these equations can be verified easily by carrying out the indicated multiplications and writing an equation relating corresponding elements of the resulting vectors. This will be illustrated by a specific example.

Consider again the circuit shown in Fig. 8-9 and the corresponding equations (8-15) and (8-16). In this case $n = m = p = 2$, and the system is time-invariant. From the equations it is clear that

$$\mathbf{A} = \begin{bmatrix} -750 & 250 \\ 250 & -750 \end{bmatrix} \qquad \mathbf{B} = \begin{bmatrix} 250 & 250 \\ 250 & 250 \end{bmatrix}$$

$$\mathbf{C} = \begin{bmatrix} -0.75 & 0.25 \\ 0 & 1 \end{bmatrix} \qquad \mathbf{D} = \begin{bmatrix} 0.25 & 0.25 \\ 0 & 0 \end{bmatrix}$$

Thus, (8-17) can be written as

$$
\begin{bmatrix} \dfrac{dq_1(t)}{dt} \\[2mm] \dfrac{dq_2(t)}{dt} \end{bmatrix} = \begin{bmatrix} -750 & 250 \\ 250 & -750 \end{bmatrix}\begin{bmatrix} q_1(t) \\ q_2(t) \end{bmatrix} + \begin{bmatrix} 250 & 250 \\ 250 & 250 \end{bmatrix}\begin{bmatrix} x_1(t) \\ x_2(t) \end{bmatrix} \tag{8-19}
$$

Carrying out the multiplication yields

$$
\begin{bmatrix} \dfrac{dq_1(t)}{dt} \\[2mm] \dfrac{dq_2(t)}{dt} \end{bmatrix} = \begin{bmatrix} -750q_1(t)+250q_2(t) \\ 250q_1(t)-750q_2(t) \end{bmatrix} + \begin{bmatrix} 250x_1(t)+250x_2(t) \\ 250x_1(t)+250x_2(t) \end{bmatrix} \tag{8-20}
$$

Since matrices are equal only when corresponding elements are equal, it follows that

$$
\frac{dq_1(t)}{dt} = -750q_1(t)+250q_2(t)+250x_1(t)+250x_2(t)
$$

results from equating the upper element in each vector. This is exactly the upper equation in (8-15). Equating the lower elements will lead to the lower equation of (8-15).

In a similar way, (8-18) can be written as

$$
\begin{bmatrix} y_1(t) \\ y_2(t) \end{bmatrix} = \begin{bmatrix} -0.75 & 0.25 \\ 0 & 1 \end{bmatrix}\begin{bmatrix} q_1(t) \\ q_2(t) \end{bmatrix} + \begin{bmatrix} 25 & 25 \\ 0 & 0 \end{bmatrix}\begin{bmatrix} x_1(t) \\ x_2(t) \end{bmatrix} \tag{8-21}
$$

The reader may verify that carrying out the multiplication and equating similar elements will lead to equation (8-16).

It should be clear from the foregoing discussion that a considerable notational convenience can be achieved by using matrix notation. Furthermore, increasing the order of the system, the number of inputs, or the number of outputs does not make the state equations more cumbersome. In a later section it is shown that the matrix form of the normal equation can be solved explicitly to obtain a solution in matrix form. Thus, a great

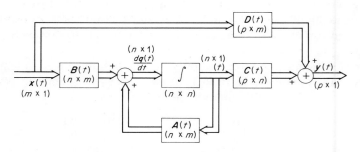

Fig. 8-11 Matrix block diagram.

deal of the analysis of complicated systems can be carried out by procedures that are only slightly more involved than those for first-order systems.

It is also possible to draw a simple block diagram for the state equations. In order to distinguish such a block diagram from the type already
discussed, we shall use double arrows to represent vector quantities and
blocks with thicker edges to represent matrices. Thus, (8-17) and (8-18)
could be represented by the *matrix block diagram* of Fig. 8-11.

Exercise 8-4.1

A system is represented by the third-order differential equation

$$\dddot{y}(t) + 2\ddot{y}(t) - \dot{y}(t) + 2y = x(t)$$

Define the state variables as y, $\dot{y}$, and $\ddot{y}$ and determine the **A** matrix.
Draw the block diagram of the system.

ANS.

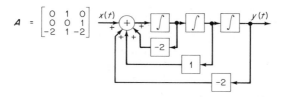

$$A = \begin{bmatrix} 0 & 1 & 0 \\ 0 & 0 & 1 \\ -2 & 1 & -2 \end{bmatrix}$$

8-5   Elementary State Space Concepts

Now that the notational preliminaries are out of the way, it is time to
discuss some of the basic concepts associated with the state space
approach in a little more detail. The most fundamental concept, of course,
is just what is meant by the *state* of a system. For our purposes, it is
sufficient to consider the state of a system at any time to be all of the
necessary information about the past history of the system needed to
determine the future response of the system. This definition glosses over
many subtle points that need to be considered for a full understanding of
the concept of state, but attempting to explain these here would only add
confusion.

The mathematical representation of the state of a system at any time is
the state vector, $\mathbf{q}(t)$. This is not a unique representation, since there are
many different collections of state variables that would convey the same
information about the system. However, if it contains all the essential information, then it does lead to a unique determination of system response
to any excitation $\mathbf{x}(t)$. In most cases it is convenient to pick some time t_0
and assume that all inputs of interest occur for $t \geq t_0$. If this is done, then
$\mathbf{q}(t_0)$ is the initial state of the system and contains all the necessary information about the history of the system prior to the time t_0.

For values of time greater than t_0 the response of the system depends upon both the initial state and the input excitation. For purposes of analysis, it is convenient to separate the total response into the parts dependent upon each. One part, called the *zero-input response*, is the response that would evolve from the initial state if the input excitation were zero. The other part, called the *zero-state response*, is the response that would evolve from the input excitation if the initial state were zero. Each of these concepts requires further clarification.

The input is a vector, $\mathbf{x}(t)$, and in order for it to be zero, all its components must be zero. Thus, zero input is designated as $\mathbf{x}(t) = 0$ and implies that $x_1(t) = 0$, $x_2(t) = 0$, $\ldots$, $x_m(t) = 0$ for all $t \geq t_0$. The concept of zero state is a little more involved, since it does not necessarily imply that $\mathbf{q}(t_0) = 0$. Instead, it implies that the system output $\mathbf{y}(t)$ will be zero when the input is zero. That is, a state $\mathbf{q}_0(t_0)$ is called a *zero state* if, for every t_0, $\mathbf{y}(t) = 0$ when $\mathbf{x}(t) = 0$ for all $t \geq t_0$.

The concept of zero state can be illustrated more concretely by two examples. For the first example, refer to the network shown in Fig. 8-9. A brief examination of this network reveals that $y_1(t)$ and $y_2(t)$ are zero when $x_1(t)$ and $x_2(t)$ are zero if, and only if, both $q_1(t)$ and $q_2(t)$ are zero. Thus, in this case the zero state does, in fact, occur when $\mathbf{q}_0(t_0) = 0$ for any t_0. The second example illustrates that this is not always so, although it is true most of the time. The situation shown in Fig. 8-12 is somewhat contrived, but it is not greatly dissimilar from many situations that arise in connection with active networks.

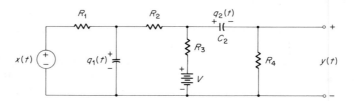

Fig. 8-12 Network with an internal voltage source.

The state variables in the network of Fig. 8-12 are again chosen to be the capacitor voltages. It is clear that $y(t)$ will be zero when $x(t)$ is zero only if

$$q_1(t) = \frac{R_1}{R_1 + R_2 + R_3} V \quad \text{and} \quad q_2(t) = \frac{R_1 + R_2}{R_1 + R_2 + R_3} V$$

Thus, the zero state for this system is

$$\mathbf{q}_0(t_0) = \begin{bmatrix} \dfrac{R_1}{R_1 + R_2 + R_3} V \\[2ex] \dfrac{R_1 + R_2}{R_1 + R_2 + R_3} V \end{bmatrix} \qquad (8\text{-}22)$$

As stated earlier, it is desirable to express the total response of a system as the sum of the zero-input response and the zero-state response. This is possible only if the system is linear. The concept of linearity was introduced in Chapter 2 for single-input, single-output systems and with no consideration of initial conditions. (Although it was not so stated, the definition of linearity employed there is applicable only if the initial state of the system is a zero state.) It is appropriate at this time to extend the concept of linearity in the light of state space methods.

Suppose that the initial state of a system is a zero state $\mathbf{q}_0(t_0)$, the zero-state response to input $\mathbf{x}_1(t)$ is $\mathbf{y}_1(t)$, and the zero-state response to $\mathbf{x}_2(t)$ is $\mathbf{y}_2(t)$. Then the system is *zero-state linear* if, and only if, $[a\mathbf{y}_1(t) + b\mathbf{y}_2(t)]$ is the zero-state response to input $[a\mathbf{x}_1(t) + b\mathbf{x}_2(t)]$ for all a, b, and $t \geq t_0$.

Next, suppose that the zero-input response is $\mathbf{y}_{01}(t)$ when the initial state is $\mathbf{q}_2(t_0)$. Then the system is *zero-input linear* if, and only if, $[a\mathbf{y}_{01}(t) + b\mathbf{y}_{02}(t)]$ is the zero-input response when the initial state is $[a\mathbf{q}_1(t_0) + b\mathbf{q}_2(t_0)]$ for all a, b, and $t \geq t_0$.

In the third place, a system is said to be *decomposable* if, and only if, the response to every input $\mathbf{x}(t)$ for every initial state $\mathbf{q}(t_0)$ can be represented as the sum of the zero-state response for that input and the zero-input response for that initial state.

On the basis of these three concepts, we can now make a more general definition of linearity. A system is *fully linear* if, and only if, it is *decomposable*, *zero-state linear*, and *zero-input linear*. It should be noted that this definition is valid for either time-varying or time-invariant systems.

It is of interest to check on the linearity of the networks shown in Figs. 8-9 and 8-12. It is not difficult to show that the network of Fig. 8-9 satisfies all three criteria and, hence, is fully linear. In the case of Fig. 8-12, however, the system is zero-state linear but is *not* zero-input linear. This is easily verified by noting that the zero-input response to an initial state that is a constant times the zero-state, $\mathbf{q}_0(t_0)$, is not zero.

Exercise 8-5.1
For the system shown determine the zero-state vector. Is this system fully linear?

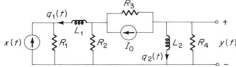

8-6 Time-Domain Solutions of the State Equations: Invariant Systems

The state equations for a linear time-varying system are given in Section 8-5 as Equations (8-17) and (8-18). If the system is time-invariant, these may be written as

$$\frac{d\mathbf{q}(t)}{dt} = \mathbf{A}\mathbf{q}(t) + \mathbf{B}\mathbf{x}(t) \tag{8-23}$$

and

$$\mathbf{y}(t) = \mathbf{C}\mathbf{q}(t) + \mathbf{D}\mathbf{x}(t) \tag{8-24}$$

where $\mathbf{A}$, $\mathbf{B}$, $\mathbf{C}$, and $\mathbf{D}$ are constant matrices that characterize the system. For the purpose of the following discussion the system is assumed to be fully linear as defined in the previous section.

The general problem in system analysis is to find the output $\mathbf{y}(t)$ when the input $\mathbf{x}(t)$ is specified for $t \geq t_0$ and some initial state $\mathbf{q}(t_0)$ is given or can be determined. In order to do this, it is necessary to solve the matrix differential equation (8-23) for the state vector $\mathbf{q}(t)$. Once this has been determined, the evaluation of the output from (8-24) is straightforward.

Since the system is fully linear, it is decomposable, and the total solution can be written as the sum of the zero-input response and zero-state response. This is also true for the state vector as well as the output vector. Thus, the zero-input response for the state vector can be obtained from (8-23) by letting $\mathbf{X}(t) = 0$. The resulting homogeneous equation is

$$\frac{d\mathbf{q}(t)}{dt} = \mathbf{A}\mathbf{q}(t) \tag{8-25}$$

By analogy to the corresponding first-order scalar equation it might be suspected that the solution of this equation would have the form

$$\mathbf{q}_s(t) = \epsilon^{\mathbf{A}(t - t_0)}\mathbf{q}(t_0) \tag{8-26}$$

where the subscript implies that this response is due to the initial state only. It will be shown that this is true by substitution into (8-25). Since $\mathbf{A}$ is a constant matrix, the derivative of (8-26) is[2]

$$\frac{d\mathbf{q}_s(t)}{dt} = \mathbf{A}[\epsilon^{\mathbf{A}(t - t_0)}\mathbf{q}(t_0)] \tag{8-27}$$

Substitution of (8-26) and (8-27) into (8-25) obviously yields an equality.

The quantity $\epsilon^{\mathbf{A}t}$ is an $(n \times n)$ matrix, which can be defined by the series for an exponential. Thus,

$$\epsilon^{\mathbf{A}t} = \mathbf{I} + \mathbf{A}t + \mathbf{A}^2\frac{t^2}{2!} + \mathbf{A}^3\frac{t^3}{3!} + \cdots \qquad (n \times n) \tag{8-28}$$

where $\mathbf{I}$ is the identity matrix. The matrix $\epsilon^{\mathbf{A}t}$ arises so often in the solution of linear equations that it is given a special symbol, $\boldsymbol{\phi}(t)$, and is often called the *fundamental matrix* for the system. Its evaluation and properties are so important to system analysis that the following section is devoted exclusively to discussing them. For the moment we note simply

---

[2]The basic rules of matrix calculus are summarized in Appendix C.

that $\phi(t)$ is a solution to the homogeneous equation, and that the zero-input response for the state vector may also be written as

$$\mathbf{q}_s(t) = \phi(t - t_0)\mathbf{q}(t_0) \tag{8-29}$$

The matrix $\phi(t - t_0)$ is frequently referred to as the *state-transition matrix*, since it specifies the transition from the state at time t_0 to that at some other time $t \geq t_0$.

The zero-state response for the state vector will be determined by the method of variation of parameters. In order to do this, assume that this solution has the form

$$\mathbf{q}_i(t) = \epsilon^{A(t - t_0)}\mathbf{g}(t) \qquad t \geq t_0 \tag{8-30}$$

where the subscript i implies that this is the response due to the input only and where $\mathbf{g}(t)$ is an $(n \times 1)$ vector that is to be determined. The derivative of (8-30) is

$$\frac{d\mathbf{q}_i(t)}{dt} = \mathbf{A}\epsilon^{A(t - t_0)}\mathbf{g}(t) + \epsilon^{A(t - t_0)}\frac{d\mathbf{g}(t)}{dt}$$

and this may be substituted into (8-23) to give, after simplifying,

$$\epsilon^{A(t - t_0)}\frac{d\mathbf{g}(t)}{dt} = \mathbf{Bx}(t)$$

Premultiplying by $\epsilon^{-A(t - t_0)}$ yields[3]

$$\frac{d\mathbf{g}(t)}{dt} = \epsilon^{-A(t - t_0)}\mathbf{Bx}(t) \qquad t \geq t_0$$

which may be integrated to give

$$\mathbf{g}(t) = \int_{t_0}^{t} \epsilon^{-A(\lambda - t_0)}\mathbf{Bx}(\lambda)\,d\lambda \tag{8-31}$$

Substituting (8-31) into (8-30) gives the zero-state response for the state vector as

$$\mathbf{q}_i(t) = \int_{t_0}^{t} \epsilon^{A(t - \lambda)}\mathbf{Bx}(\lambda)\,d\lambda$$

or, using the definition of the state-transition matrix,

$$\mathbf{q}_i(t) = \int_{t_0}^{t} \phi(t - \lambda)\mathbf{Bx}(\lambda)\,d\lambda \tag{8-32}$$

It is now possible to write the total state vector as the sum of the zero-input response, given by (8-29), and the zero-state response, given by (8-32). That is,

$$q(t) = \mathbf{q}_s(t) + \mathbf{q}_i(t)$$
$$= \phi(t - t_0)\mathbf{q}(t_0) + \int_{t_0}^{t} \phi(t - \lambda)\mathbf{Bx}(\lambda)\,d\lambda \tag{8-33}$$

---

[3]This operation is equivalent to premultiplying by the inverse of the matrix $\epsilon^{A(t - t_0)}$ and assumes that the inverse exists.

This result clearly indicates the important role played by the state-transition matrix. It is necessary, therefore, to examine the properties of this matrix and consider ways of evaluating it before we go to a specific example.

8-7 The State-Transition Matrix

The state-transition matrix for time-invariant linear systems was defined to be

$$\boldsymbol{\phi}(t - t_0) = \epsilon^{\mathbf{A}(t - t_0)} \tag{8-34}$$

The most basic method for evaluating the elements of this matrix is using the series expansion for the exponential. Thus,

$$\boldsymbol{\phi}(t - t_0) = \mathbf{I} + \mathbf{A}(t - t_0) + \mathbf{A}^2 \frac{(t - t_0)^2}{2!} + \mathbf{A}^3 \frac{(t - t_0)^3}{3!} + \cdots . \tag{8-35}$$

This method is tedious by hand calculation but can be adapted to computer calculations quite readily. An example that can be carried out easily by hand is shown in the next section.

A second method of evaluating this matrix in the time domain is based on the *Cayley-Hamilton theorem*, which states that a square matrix $\mathbf{A}$ satisfies its own characteristic equation.[4] This makes it possible to express *all* powers of $\mathbf{A}$ in terms of a polynomial in $\mathbf{A}$ of degree no higher than $n - 1$. Thus, the entire series of (8-35) can be written as

$$\boldsymbol{\phi}(t - t_0) = \alpha_0(t - t_0)\mathbf{I} + \alpha_1(t - t_0)\mathbf{A} + \cdots + \alpha_{n-1}(t - t_0)\mathbf{A}^{n-1} \tag{8-36}$$

in which the coefficients $\alpha_i(t - t_0)$ are to be determined by substituting the eigenvalues of $\mathbf{A}$ into the scalar equivalent of (8-36) and obtaining a set of n simultaneous equations. An example will illustrate this procedure.

Suppose that $\mathbf{A}$ is (2×2) and has the form

$$\mathbf{A} = \begin{bmatrix} -3 & 1 \\ 2 & -2 \end{bmatrix}$$

Since the characteristic equation comes from the determinant equation

$$|\mathbf{A} - \lambda \mathbf{I}| = 0$$

it follows that

$$\begin{vmatrix} -3-\lambda & 1 \\ 2 & -2-\lambda \end{vmatrix} = 0$$

and hence,

$$\lambda^2 + 5\lambda + 4 = 0$$

from which the two eigenvalues of $\mathbf{A}$ are $\lambda_1 = -1$ and $\lambda_2 = -4$. The state-

---

[4]See Appendix C for a summary of characteristic equation concepts.

transition matrix can then be expressed as

$$\boldsymbol{\phi}(t - t_0) = \alpha_0(t - t_0)\mathbf{I} + \alpha_1(t - t_0)\mathbf{A} = \epsilon^{A(t-t_0)} \qquad (8\text{-}37)$$

and the scalar form of this is simply

$$\alpha_0(t - t_0) + \alpha_1(t - t_0)\lambda = \epsilon^{\lambda(t-t_0)}$$

Substituting in $\lambda = \lambda_1 = -1$ gives

$$\alpha_0(t - t_0) - \alpha_1(t - t_0) = \epsilon^{-(t-t_0)}$$

while $\lambda = \lambda_2 = -4$ yields

$$\alpha_0(t - t_0) - 4\alpha_1(t - t_0) = \epsilon^{-4(t-t_0)}$$

from which

$$\alpha_0(t - t_0) = \frac{4}{3}\epsilon^{-(t-t_0)} - \frac{1}{3}\epsilon^{-4(t-t_0)}$$

$$\alpha_1(t - t_0) = \frac{1}{3}\epsilon^{-(t-t_0)} - \frac{1}{3}\epsilon^{-4(t-t_0)}$$

Inserting these values into (8-37) yields the results

$$\boldsymbol{\phi}(t - t_0) = \left[\frac{4}{3}\epsilon^{-(t-t_0)} - \frac{1}{3}\epsilon^{-4(t-t_0)}\right]\mathbf{I} + \left[\frac{1}{3}\epsilon^{-(t-t_0)} - \frac{1}{3}\epsilon^{-4(t-t_0)}\right]\mathbf{A}$$

$$= \begin{bmatrix} \frac{1}{3}\epsilon^{-(t-t_0)} + \frac{2}{3}\epsilon^{-4(t-t_0)} & \frac{1}{3}\epsilon^{-(t-t_0)} - \frac{1}{3}\epsilon^{-4(t-t_0)} \\ \frac{2}{3}\epsilon^{-(t-t_0)} - \frac{2}{3}\epsilon^{-4(t-t_0)} & \frac{2}{3}\epsilon^{-(t-t_0)} + \frac{1}{3}\epsilon^{-4(t-t_0)} \end{bmatrix} \qquad (8\text{-}38)$$

The foregoing procedures for evaluating the state-transition matrix are general in the sense that they can be carried out in almost every case, although it obviously becomes much more difficult if n is large. Another procedure, which may be very simple and straightforward in some cases (even for large n), but which may offer no simplification at all in other cases, is based on the physical interpretation of the state-transition matrix in terms of the impulse response of the state variables. In order to investigate this possibility, consider the zero-state response for the state vector given in (8-32) as

$$\mathbf{q}_i(t) = \int_{t_0}^{t} \boldsymbol{\phi}(t - \lambda)\mathbf{B}\mathbf{x}(\lambda)d\lambda$$

It is now desired to pick $\mathbf{B}$ and $\mathbf{x}(\lambda)$ so that they correspond to applying a unit impulse directly into the integrator associated with any one state variable, say the jth one. This can be done in several ways, but the easiest is to let the input vector be

$$\mathbf{x}(\lambda) = \begin{bmatrix} \delta(\lambda - t_0) \\ 0 \\ \vdots \\ 0 \end{bmatrix} \quad (m \times 1)$$

and let the **B** matrix be all zeros except b_{j1}; that is,

$$\mathbf{B} = \begin{bmatrix} 0 \cdots 0 \\ 0 \\ 0 \\ 1 \cdots 0 \\ 0 \\ \vdots \\ 0 \quad 0 \end{bmatrix} \leftarrow j\text{th row} \quad (n \times m)$$

Then

$$\mathbf{Bx}(\lambda) = \begin{bmatrix} 0 \cdots 0 \\ \vdots \\ 1 \\ \vdots \\ 0 \cdots 0 \end{bmatrix} \begin{bmatrix} \delta(\lambda - t_0) \\ 0 \\ \vdots \\ \vdots \\ 0 \end{bmatrix} = \begin{bmatrix} 0 \\ \vdots \\ \delta(\lambda - t_0) \\ \vdots \\ 0 \end{bmatrix} \leftarrow j\text{th element} \quad (n \times 1)$$

Multiplying by the state-transition matrix gives

$$\boldsymbol{\phi}(t - \lambda)\mathbf{Bx}(\lambda) = \begin{bmatrix} \phi_{11}(t - \lambda) \cdots \phi_{1m}(t - \lambda) \\ \vdots \\ \vdots \\ \phi_{n1}(t - \lambda) \cdots \phi_{nm}(t - \lambda) \end{bmatrix} \begin{bmatrix} 0 \\ \vdots \\ \delta(\lambda - t_0) \\ \vdots \\ 0 \end{bmatrix}$$

$$= \begin{bmatrix} \phi_{1j}(t - \lambda)\delta(\lambda - t_0) \\ \phi_{2j}(t - \lambda)\delta(\lambda - t_0) \\ \vdots \\ \phi_{nj}(t - \lambda)\delta(\lambda - t_0) \end{bmatrix} \tag{8-39}$$

Introducing (8-39) into (8-32) and integrating over the δ functions leads immediately to

$$\mathbf{q}_i(t) = \begin{bmatrix} \phi_{1j}(t - t_0) \\ \phi_{2j}(t - t_0) \\ \vdots \\ \phi_{nj}(t - t_0) \end{bmatrix} \tag{8-40}$$

The interpretation of the elements of the state-transition matrix is now clear. The element $\phi_{jk}(t - t_0)$ is the zero-state response for the kth state variable when a unit impulse occurs in the *derivative* of the jth state variable, that is, at the input to the jth integrator in the block diagram.

As an illustration of the procedure, consider the system represented by the block diagram of Fig. 8-13 and, in particular, the portion inside the dotted lines. If a unit impulse at $t = t_0$ is introduced at point A, the

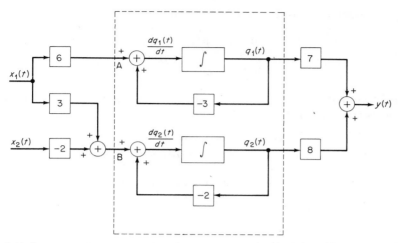

Fig. 8-13 System to illustrate the impulse-response interpretation of the state-transition matrix.

resulting $q_1(t)$ and $q_2(t)$ are clearly just

$$q_1(t) = \epsilon^{-3(t-t_0)} \qquad t \geq t_0$$

$$q_2(t) = 0 \qquad t \geq t_0$$

since there is no cross-coupling between state variables in this example. Likewise, if a unit impulse at $t = t_0$ is introduced at point B, the state response is

$$q_1(t) = 0 \qquad t \geq t_0$$

$$q_2(t) = \epsilon^{-2(t-t_0)} \qquad t \geq t_0$$

Hence, the state-transition matrix can be written immediately as

$$\boldsymbol{\phi}(t - t_0) = \begin{bmatrix} \epsilon^{-3(t-t_0)} & 0 \\ 0 & \epsilon^{-2(t-t_0)} \end{bmatrix} \tag{8-41}$$

For the case of time-invariant systems, the state-transition matrix can also be obtained quite straightforwardly by the Laplace transform. This procedure is discussed in Section 8-9.

A few of the properties of the state-transition matrix are frequently useful in analysis or for the purpose of checking results. These are summarized as follows:

1. $\boldsymbol{\phi}(0) = \mathbf{I}$.
2. $\boldsymbol{\phi}(t_2 - t_0) = \boldsymbol{\phi}(t_2 - t_1)\boldsymbol{\phi}(t_1 - t_0)$.
3. $\boldsymbol{\phi}(t_0 - t) = \boldsymbol{\phi}^{-1}(t - t_0)$.

The first property may be obtained either by setting $t = t_0$ in the series expansion (8-35) for $\boldsymbol{\phi}(t - t_0)$ or from the uniqueness of the solution of the homogeneous state equation, as given in (8-23). That is, since

$$\mathbf{q}_s(t) = \boldsymbol{\phi}(t - t_0)\mathbf{q}(t_0)$$

and

$$\mathbf{q}_s(t_0) = \mathbf{q}(t_0)$$

it follows that $\boldsymbol{\phi}(0) = \mathbf{I}$. The second property can be obtained from the same equation, since

$$\begin{aligned}\mathbf{q}_s(t_2) &= \boldsymbol{\phi}(t_2 - t_0)\mathbf{q}(t_0)\\ &= \boldsymbol{\phi}(t_2 - t_1)\mathbf{q}_s(t_1)\end{aligned}$$

But

$$\mathbf{q}(t_1) = \boldsymbol{\phi}(t_1 - t_0)\mathbf{q}(t_0)$$

so

$$\mathbf{q}_s(t_2) = \boldsymbol{\phi}(t_2 - t_1)\boldsymbol{\phi}(t_1 - t_0)\mathbf{q}(t_0)$$

from which the second property is apparent.

The third property, concerning the inverse of the state-transition matrix, is obtained by premultiplying (8-33) by $\boldsymbol{\phi}^{-1}(t - t_0)$. Thus,

$$\boldsymbol{\phi}^{-1}(t - t_0)\mathbf{q}_s(t) = \boldsymbol{\phi}^{-1}(t - t_0)\boldsymbol{\phi}(t - t_0)\mathbf{q}(t_0) = \mathbf{q}(t_0)$$

But

$$\mathbf{q}(t_0) = \boldsymbol{\phi}(t_0 - t)\mathbf{q}_s(t)$$

from which it follows that

$$\boldsymbol{\phi}^{-1}(t - t_0) = \boldsymbol{\phi}(t_0 - t) \tag{8-42}$$

An important consideration with regard to the inverse of the state-transition matrix is whether or not it exists. It can be shown, by methods that are too lengthy to include here, that $\boldsymbol{\phi}(t - t_0)$ is always nonsingular, and hence, its inverse will always exist. In addition, (8-30) is true for *all* values of t, not just $t \geq t_0$. The inverse can be interpreted as being the state-transition matrix for another system known as the *adjoint system*. The use of the adjoint system is an important aspect of system analysis but is beyond the scope of the present discussion.

Exercise 8-7.1

Use the eigenvalue method to determine the state-transition matrix of a system whose **A** matrix is

$$\mathbf{A} = \begin{bmatrix} -1 & 1 \\ 0 & -2 \end{bmatrix}$$

ANS

$$\boldsymbol{\phi}(t - t_0) = \begin{bmatrix} \epsilon^{-(t - t_0)} & \epsilon^{-(t - t_0)} - \epsilon^{-2(t - t_0)} \\ 0 & \epsilon^{-2(t - t_0)} \end{bmatrix}$$

8-8   The Impulse-Response Matrix

In the previous discussion of single-input, single-output linear systems, the impulse response was a convenient way of characterizing systems and could be used for determining the system output for any input by means of convolution. As a matter of convenience, the impulse response was defined to be the response to an impulse of unit area applied at time $t = 0$.

A similar concept can be applied to multiple-input, multiple-output systems by defining the *impulse-response matrix* for such a system. If there are m inputs and p outputs, the impulse-response matrix will be a $(p \times m)$ matrix whose (i, j) element is the zero-state response at the ith output due to a unit impulse at $t = 0$, applied at the jth input, with all other inputs zero. This matrix, which will be designated as $\mathscr{H}\{t\}$, can then be used in a multidimensional form of the convolution integral to obtain the system output for any input.

For purposes of obtaining the impulse-response matrix, it will be assumed that the system output equation is of the form

$$\mathbf{y}(t) = \mathbf{Cq}(t) \tag{8-43}$$

thus assuming that there are no contributions of the form $\mathbf{Dx}(t)$ in $\mathbf{y}(t)$.[5] Since only the zero-state response is being considered, the state vector can be obtained from (8-32) with $t_0 = 0$. Then the output vector becomes

$$\mathbf{y}(t) = \int_0^t \mathbf{C}\boldsymbol{\phi}(t - \lambda)\mathbf{Bx}(\lambda)\,d\lambda \tag{8-44}$$

When a unit impulse is applied to the jth input only, then

$$\mathbf{Bx}(\lambda) = \begin{bmatrix} b_{11} \cdots b_{1m} \\ \vdots \\ \\ \vdots \\ b_{n1} \cdots b_{nm} \end{bmatrix} \begin{bmatrix} 0 \\ \vdots \\ \delta(\lambda) \\ \vdots \\ 0 \end{bmatrix} = \begin{bmatrix} b_{1j} \\ \vdots \\ \\ \vdots \\ b_{nj} \end{bmatrix} \delta(\lambda) \tag{8-45}$$

which is simply the jth column of $\mathbf{B}$ times the impulse. The corresponding output vector, which is the jth column of $\mathscr{H}\{t\}$, is, after integrating over the δ function, just

$$\mathbf{y}_j(t) = \begin{bmatrix} h_{1j}(t) \\ \vdots \\ h_{pj}(t) \end{bmatrix} = \mathbf{C}\boldsymbol{\phi}(t) \begin{bmatrix} b_{1j} \\ \vdots \\ b_{nj} \end{bmatrix} \tag{8-46}$$

---

[5]The case in which $\mathbf{D}$ is not zero can be handled also by using additional state variables. It is being eliminated here simply for ease in discussion.

A similar result can be obtained for every $1 \le j \le m$, and this will give every column of $\mathcal{H}\{t\}$. Then the complete impulse-response matrix is given by

$$\mathcal{H}\{t\} = \mathbf{C}\boldsymbol{\phi}(t)\mathbf{B} \qquad t \ge 0 \qquad (8\text{-}47)$$

When the input vector $\mathbf{x}(t)$ is not an impulse but a set of general time functions, the output vector can be obtained by using (8-47) in (8-44) to give

$$\mathbf{y}(t) = \int_0^t \mathcal{H}\{t - \lambda\}\mathbf{x}(\lambda)d\lambda \qquad (8\text{-}48)$$

The similarity of this result to the previous convolution integral is apparent.

It may be useful to consider a simple example to illustrate these concepts. The system represented by the block diagram of Fig. 8-14 will be used for this purpose. The various matrices that characterize this system are

$$\mathbf{A} = \begin{bmatrix} -\alpha & 1 \\ 0 & -\alpha \end{bmatrix} \qquad \mathbf{B} = \begin{bmatrix} 1 & 0 \\ 0 & 1 \end{bmatrix}$$

$$\mathbf{C} = \begin{bmatrix} 1 & 1 \end{bmatrix}$$

The first step is to determine the state-transition matrix. For illustrative purposes, this will be done directly from the series expansion. The important thing needed here is a general expression for any power of $\mathbf{A}$; this will be determined by induction. Consider first

$$\mathbf{A}^2 = \begin{bmatrix} -\alpha & 1 \\ 0 & -\alpha \end{bmatrix}\begin{bmatrix} -\alpha & 1 \\ 0 & -\alpha \end{bmatrix} = \begin{bmatrix} \alpha^2 & -2\alpha \\ 0 & \alpha^2 \end{bmatrix}$$

and also

$$\mathbf{A}^3 = \begin{bmatrix} -\alpha & 1 \\ 0 & -\alpha \end{bmatrix}\begin{bmatrix} \alpha^2 & -2\alpha \\ 0 & \alpha^2 \end{bmatrix} = \begin{bmatrix} -\alpha^3 & 3\alpha^2 \\ 0 & -\alpha^3 \end{bmatrix}$$

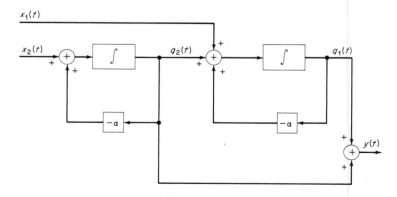

Fig. 8-14 Example of system analysis.

From these it can be deduced that:

$$\mathbf{A}^k = \begin{bmatrix} (-\alpha)^k & k(-\alpha)^{k-1} \\ 0 & (-\alpha)^k \end{bmatrix}$$

Using this result in the series expansion (8-35)

$$\boldsymbol{\phi}(t - t_0) = \mathbf{I} + \mathbf{A}(t - t_0) + \mathbf{A}^2 \frac{(t - t_0)^2}{2!} + \cdots$$

leads to

$$\boldsymbol{\phi}(t - t_0) = \begin{bmatrix} \displaystyle\sum_{k=0} (-\alpha)^k \frac{(t - t_0)^k}{k!} & \displaystyle\sum_{k=0} k(-\alpha)^{k-1} \frac{(t - t_0)^{k-1}}{k!} \\ 0 & \displaystyle\sum_{k=0}^{\infty} (-\alpha)^k \frac{(t - t_0)^k}{k!} \end{bmatrix} \tag{8-49}$$

Each of the summations in (8-49) can be recognized as the series for an exponential. Thus, it can be written as

$$\boldsymbol{\phi}(t - t_0) = \begin{bmatrix} \epsilon^{-\alpha(t-t_0)} & (t - t_0)\epsilon^{-\alpha(t-t_0)} \\ 0 & \epsilon^{-\alpha(t-t_0)} \end{bmatrix} \tag{8-50}$$

In this example it was a simple matter to recognize the closed-form summation of the series. In other cases a closed form may not exist or may be too complicated to recognize easily. Hence, the use of the series has only limited application in attempts to find state-transition matrices analytically.

Once the state-transition matrix is found, the impulse-response matrix can be found from (8-47). Thus, in this case,

$$\mathcal{H}\{t\} = \mathbf{C}\boldsymbol{\phi}(t)\mathbf{B} = [1 \quad 1]\begin{bmatrix} \epsilon^{-\alpha t} & te^{-\alpha t} \\ 0 & \epsilon^{-\alpha t} \end{bmatrix}\begin{bmatrix} 1 & 0 \\ 0 & 1 \end{bmatrix}$$

$$= [\epsilon^{-\alpha t} \quad (t + 1)\epsilon^{-\alpha t}] \qquad t \geq 0 \tag{8-51}$$

In order to illustrate a particular case of system analysis, let the input signals be

$$x_1(t) = u(t)$$

$$x_2(t) = \epsilon^{-t}u(t)$$

The integrand of (8-48) then becomes

$$\mathcal{H}\{t - \lambda\}\mathbf{x}(\lambda) = [\epsilon^{-\alpha(t-\lambda)}(t - \lambda + 1) \quad \epsilon^{-\alpha(t-\lambda)}]\begin{bmatrix} u(\lambda) \\ \epsilon^{-\lambda}u(\lambda) \end{bmatrix}$$

$$= \epsilon^{-\alpha(t-\lambda)}u(\lambda) + (t - \lambda + 1)\epsilon^{-\alpha t}\epsilon^{(\alpha-1)\lambda}u(\lambda) \tag{8-52}$$

Since there is only one output signal in in this example, (8-48) represents a scalar. In general, it would be a $(p \times 1)$ vector, and each component would have to be integrated to obtain the corresponding output component. In the present case, however,

$$y(t) = \int_0^t [\epsilon^{-\alpha(t-\lambda)} u(\lambda) + (t-\lambda+1)\epsilon^{-\alpha t}\epsilon^{(\alpha-1)\lambda} u(\lambda)]d\lambda$$

$$= \epsilon^{-\alpha t}\frac{\epsilon^{\alpha\lambda}}{\alpha}\Big|_0^t + (t+1)\epsilon^{-\alpha t}\frac{\epsilon^{(\alpha-1)}}{\alpha-1}\Big|_0^t - \epsilon^{-\alpha t}\frac{\epsilon^{(\alpha-1)\lambda}[(\alpha-1)\lambda-1]}{(\alpha-1)^2}\Big|_0^t \qquad \text{(8-53)}$$

$$= \frac{u(t)}{\alpha} + \frac{\alpha}{(\alpha-1)^2}\epsilon^{-t}u(t) - \left[\frac{t+2}{\alpha-1} + \frac{1}{\alpha(\alpha-1)^2}\right]\epsilon^{-\alpha t}u(t)$$

Exercise 8-8.1

For the system shown find the impulse-response matrix and determine the output when the input is

$$x(t) = \begin{bmatrix} 0 \\ u(t) \end{bmatrix}$$

(**A** is the same as in Exercise 8-7.1).

ANS.

$$\mathcal{H}\{\mathbf{t}\} = \begin{bmatrix} \epsilon^{-t} & 2\epsilon^{-t}-\epsilon^{-2t} \\ 0 & 0 \end{bmatrix}$$

$$y(t) = \left[\frac{3}{2} - 2\epsilon^{-t} + \frac{1}{2}\epsilon^{-2t}\right]u(t)$$

8-9 Frequency-Domain Solution of the State Equations

The use of Fourier and Laplace transforms in linear, time-invariant system analysis has been discussed in previous chapters. It was shown there that the use of such transforms converts differential equations into algebraic equations and, thus, often simplifies the mechanics of carrying out a solution. It should be expected, therefore, that a similar simplification would result from using transform methods in connection with the state equations.

In order to examine this possibility, recall that the state equations for time-invariant systems [(8-23) and (8-24)] had the form

$$\frac{d\mathbf{q}(t)}{dt} = \mathbf{A}\mathbf{q}(t) + \mathbf{B}\mathbf{x}(t)$$

$$\mathbf{y}(t) = \mathbf{C}\mathbf{q}(t) + \mathbf{D}\mathbf{x}(t)$$

Let the one-sided Laplace transform[6] of $\mathbf{q}(t)$ be designated as

---

[6]The use of the one-sided Laplace transform is equivalent to setting $t_0 = 0$. It is possible to avoid this restriction by using the two-sided Laplace transform, but it is seldom necessary to do so.

$$\mathbf{Q}(s) = \begin{bmatrix} Q_1(s) \\ Q_2(s) \\ \vdots \\ Q_n(s) \end{bmatrix} = \begin{bmatrix} \mathcal{L}\{q_1(t)\} \\ \mathcal{L}\{q_2(t)\} \\ \vdots \\ \mathcal{L}\{q_n(t)\} \end{bmatrix}$$

and let the Laplace transform of $\mathbf{x}(t)$ and $\mathbf{y}(t)$ be defined in a similar manner to be $\mathbf{X}(s)$ and $\mathbf{Y}(s)$. Then the Laplace transform of (8-23) can be written as

$$s\mathbf{Q}(s) - \mathbf{q}(0) = \mathbf{A}\mathbf{Q}(s) + \mathbf{B}\mathbf{X}(s) \tag{8-54}$$

and the Laplace transform of (8-24) as

$$\mathbf{Y}(s) = \mathbf{C}\mathbf{Q}(s) + \mathbf{D}\mathbf{X}(s) \tag{8-55}$$

Equation (8-54) can be written as

$$(s\mathbf{I} - \mathbf{A})\mathbf{Q}(s) = \mathbf{q}(0) + \mathbf{B}\mathbf{X}(s)$$

where $\mathbf{I}$ is an $(n \times n)$ identity matrix. This can be solved for $\mathbf{Q}(s)$ by premultiplying by $(s\mathbf{I} - \mathbf{A})^{-1}$. Thus,

$$\mathbf{Q}(s) = (s\mathbf{I} - \mathbf{A})^{-1}\mathbf{q}(0) + (s\mathbf{I} - \mathbf{A})^{-1}\mathbf{B}\mathbf{X}(s) \tag{8-56}$$

The Laplace transform of the output can then be obtained by substituting into (8-55) to yield

$$\mathbf{Y}(s) = \mathbf{C}(s\mathbf{I} - \mathbf{A})^{-1}\mathbf{q}(0) + [\mathbf{C}(s\mathbf{I} - \mathbf{A})^{-1}\mathbf{B} + \mathbf{D}]\mathbf{X}(s) \tag{8-57}$$

This result can be used to determine the system response to any causal input when the initial state is known.

The transfer function for a linear system was previously defined as the ratio of the output and input transforms when the initial condition was zero. Thus, if $\mathbf{q}(0) = \mathbf{O}$, one can define a *transfer function matrix* by

$$\mathbf{Y}(s) = \mathbf{H}(s)\mathbf{X}(s)$$

Upon comparison with (8-57) it is clear that

$$\mathbf{H}(s) = \mathbf{C}(s\mathbf{I} - \mathbf{A})^{-1}\mathbf{B} + \mathbf{D} \tag{8-58}$$

The transfer function matrix is a $(p x m)$ matrix whose (i, j) element is the transfer function relating the ith output and the jth input. Hence, this element must be the Laplace transform of the impulse response between these two terminals; that is,

$$H_{ij}(s) = \mathcal{L}\{h_{ij}(t)\}$$

or, more completely,

$$\mathbf{H}(s) = \mathcal{L}\{\mathcal{H}(t)\}$$
$$= \int_0^\infty \mathcal{H}(t)\epsilon^{-st} dt \tag{8-59}$$

Thus, the transfer-function matrix is just the Laplace transform of the impulse-response matrix.

It is also possible to obtain an explicit expression for the state-transition matrix. The total state response was shown in (8-33) to be given (for $t_0 = 0$) by

$$\mathbf{q}(t) = \boldsymbol{\phi}(t)\mathbf{q}(0) + \int_0^t \boldsymbol{\phi}(t - \lambda)\mathbf{B}\mathbf{X}(\lambda)d\lambda \qquad \text{(8-60)}$$

Since the integral in (8-60) is a convolution integral, its Laplace transform is simply a product of transforms. Thus, the Laplace transform of (8-60) is

$$\mathbf{Q}(s) = \boldsymbol{\Phi}(s)\mathbf{q}(0) + \boldsymbol{\Phi}(s)\mathbf{B}\mathbf{X}(s) \qquad \text{(8-61)}$$

where $\boldsymbol{\Phi}(s)$ is the Laplace transform of the state-transition matrix, $\boldsymbol{\phi}(t)$, and is also an $(n \times n)$ matrix. Upon comparison of (8-61) with (8-56), it is clear that

$$\boldsymbol{\Phi}(s) = (s\mathbf{I} - \mathbf{A})^{-1} \qquad \text{(8-62)}$$

This result leads to another procedure for evaluating the state-transition matrix, namely, taking the inverse Laplace transform of (8-62). Thus,

$$\boldsymbol{\phi}(t) = \mathscr{L}^{-1}\{(s\mathbf{I} - \mathbf{A})^{-1}\} \qquad \text{(8-63)}$$

Although this procedure is straightforward, finding the inverse may become cumbersome if n is large.

In order to illustrate the foregoing procedure for obtaining the state-transition matrix, consider the system shown in Fig. 8-14, for which the $\mathbf{A}$ matrix was

$$\mathbf{A} = \begin{bmatrix} -\alpha & 1 \\ 0 & -\alpha \end{bmatrix}$$

Hence,

$$s\mathbf{I} - \mathbf{A} = \begin{bmatrix} s + \alpha & -1 \\ 0 & s + \alpha \end{bmatrix}$$

and its inverse is

$$\boldsymbol{\Phi}(s) = (s\mathbf{I} - \mathbf{A})^{-1} = \begin{bmatrix} s + \alpha & -1 \\ 0 & s + \alpha \end{bmatrix}^{-1}$$

$$= \begin{bmatrix} \dfrac{1}{s + \alpha} & \dfrac{1}{(s + \alpha)^2} \\ 0 & \dfrac{1}{s + \alpha} \end{bmatrix}$$

The inverse Laplace transform of $\boldsymbol{\Phi}(s)$ yields the state-transition matrix as

$$\boldsymbol{\phi}(t) = \mathscr{L}^{-1} \begin{bmatrix} \dfrac{1}{s + \alpha} & \dfrac{1}{(s + \alpha)^2} \\ 0 & \dfrac{1}{s + \alpha} \end{bmatrix} = \begin{bmatrix} \epsilon^{-\alpha t} & t\epsilon^{-\alpha t} \\ 0 & \epsilon^{-\alpha t} \end{bmatrix} \qquad \text{(8-64)}$$

which is identical to the previous result, (8-50), when $t_0 = 0$.

The transfer-function matrix for a system in the zero state can also be related to the Laplace transform of the state-transition matrix. It was shown in (8-47) that

$$\mathcal{H}\{t\} = \mathbf{C}\phi(t)\mathbf{B}$$

when $\mathbf{D} = 0$. The Laplace transform of this immediately yields

$$\mathbf{H}(s) = \mathbf{C}\Phi(s)\mathbf{B} \tag{8-65}$$

As an illustration of this result return to the example considered above. For the system of Fig. 8-14 the $\mathbf{B}$ and $\mathbf{C}$ matrices were

$$\mathbf{B} = \begin{bmatrix} 1 & 0 \\ 0 & 1 \end{bmatrix} \qquad \mathbf{C} = [1 \quad 1]$$

Hence,

$$\mathbf{H}(s) = \mathbf{C}\Phi(s)\mathbf{B} = [1 \quad 1] \begin{bmatrix} \dfrac{1}{s+\alpha} & \dfrac{1}{(s+\alpha)^2} \\ 0 & \dfrac{1}{s+\alpha} \end{bmatrix} \begin{bmatrix} 1 & 0 \\ 0 & 1 \end{bmatrix} \tag{8-66}$$

$$= \left[\dfrac{1}{s+\alpha} \quad \dfrac{1}{(s+\alpha)^2} + \dfrac{1}{s+\alpha} \right]$$

This also leads directly to the Laplace transform of the output since

$$\mathbf{Y}(s) = \mathbf{H}(s)\mathbf{X}(s)$$

In this example the input vector is

$$\mathbf{x}(t) = \begin{bmatrix} u(t) \\ \epsilon^{-t}u(t) \end{bmatrix}$$

so that its Laplace transform is

$$\mathbf{X}(s) = \begin{bmatrix} \dfrac{1}{s} \\ \dfrac{1}{s+1} \end{bmatrix} \tag{8-67}$$

Hence,

$$\mathbf{Y}(s) = \left[\dfrac{1}{s+\alpha} \quad \dfrac{1}{(s+\alpha)^2} + \dfrac{1}{s+\alpha} \right] \begin{bmatrix} \dfrac{1}{s} \\ \dfrac{1}{s+1} \end{bmatrix}$$

$$= \dfrac{1}{s(s+\alpha)} + \dfrac{1}{(s+1)(s+\alpha)^2} + \dfrac{1}{(s+1)(s+\alpha)} \tag{8-68}$$

$$= \dfrac{2s^2 + 2(\alpha+1)s + \alpha}{s(s+1)(s+\alpha)^2}$$

The output time function may be obtained by first making a partial fraction expansion of (8-68). Thus,

$$\mathbf{Y}(s) = \frac{1}{\alpha s} + \frac{\alpha}{(\alpha - 1)^2(s + 1)} + \frac{1}{(1 - \alpha)(s + \alpha)^2}$$

$$- \left(\frac{2}{\alpha - 1} + \frac{1}{\alpha(\alpha - 1)^2} \right) \frac{1}{s + \alpha}$$

from which

$$y(t) = \frac{u(t)}{\alpha} + \frac{1}{(\alpha - 1)^2} \epsilon^{-t} u(t) - \left(\frac{t + 2}{\alpha - 1} + \frac{1}{\alpha(\alpha - 1)^2} \right) \epsilon^{-\alpha t} u(t)$$

which agrees with the previous result (8-53).

Exercise 8-9.1

Starting with the state equations for the system of Exercise 8-8.1 and using transform methods, determine the state-transition matrix, the impulse-response matrix, and the system response for an input

$$\mathbf{x}(t) = \begin{bmatrix} 0 \\ u(t) \end{bmatrix}$$

8-10 Selection of State Variables for Network Analysis

Since this text is intended primarily as an introduction to general system analysis, it is not appropriate to consider network analysis in great detail. Nevertheless, it is of interest to have some idea of how to select an appropriate set of state variables in typical networks. This section briefly outlines a procedure which, while not infallible, will work in a majority of the cases.

Although it is not necessary that state variables represent physical quantities in the network, it is a great conceptual advantage to have them do so. Furthermore, from the block-diagram representation of the normal equations, such as Fig. 8-10, it is clear that each state variable should also be the *integral* of another physical quantity, since it always appears as the output of an integrator. This implies that it might be reasonable to associate state variables with the energy-storage elements of the network, that is, with the inductances and capacitances.

In line with this discussion, therefore, it is appropriate to let the state variables consist of the capacitor voltages and the inductor currents. Note that a capacitor voltage is the integral of the corresponding capacitor current, and that an inductor current is the integral of the corresponding inductor voltage. In certain cases this simple procedure leads to too many state variables, and some have to be eliminated. This occurs, for example, if there is a capacitance loop, so that the sum of the capacitor voltages is constrained to be zero. It is also true if there is an inductance cut-set, so that the sum of the inductor currents is constrained to be zero. In such cases, it is necessary to eliminate one capacitance voltage in each capacitance loop and one inductance current in each inductance cut-set.

Choosing state variables in this way also makes it relatively easy to determine the initial state of the system. In particular, the initial state vector will have as its elements the initial capacitor voltages and the initial inductor currents; these are the quantities that can most easily be determined by inspection of the circuit.

Although several different networks have been considered in previous examples, it may be desirable to consider still another network in order to emphasize some of the points mentioned above. Hence, consider the R-L-C network shown in Figure 8-15 and designate the state variables as the capacitor voltages and the inductor currents. The first step in the analysis is to determine the **A** matrix for the network. There are several standard procedures for doing this almost by inspection. However, since it is not appropriate in this brief discussion to develop the theoretical basis for such a procedure, we shall resort to a less formal approach.

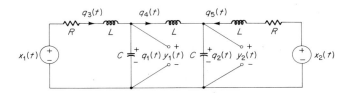

Fig. 8-15 An R-L-C network with two inputs and two outputs.

In the first place, it should be recalled that the first row of the **A** matrix comes from the relationship between $q_1(t)$ and the other state variables; that is

$$\frac{dq_1(t)}{dt} = a_{11}q_1(t) + a_{12}q_2(t) + \cdots + a_{15}q_5(t) + \cdots \qquad \text{(8-69)}$$

In the present example, $C[dq_1(t)/dt]$ is the current in the left-hand capacitor, and this current must be related to the inductor currents $q_3(t)$ and $q_4(t)$ by

$$C\frac{dq_1(t)}{dt} = q_3(t) - q_4(t)$$

from which

$$\frac{dq_1(t)}{dt} = \frac{1}{C}q_3(t) - \frac{1}{C}q_4(t) \qquad \text{(8-70)}$$

It is clear from this that $a_{11} = a_{12} = a_{15} = 0$, $a_{13} = 1/C$, and $a_{14} = -1/C$.

The second row of **A** relates to the current in the right-hand capacitor, the third row relates to the voltage across the left-hand inductor, and the fourth and fifth rows relate to the other two inductor voltages. Thus, by using only elementary reasoning of this sort, it is possible to write down the **A** matrix immediately. It is

$$\mathbf{A} = \begin{bmatrix} 0 & 0 & 1/C & -1/C & 0 \\ 0 & 0 & 0 & 1/C & 1/C \\ -1/L & 0 & -R/L & 0 & 0 \\ 1/L & -1/L & 0 & 0 & 0 \\ 0 & -1/L & 0 & 0 & -R/L \end{bmatrix} \qquad \text{(8-71)}$$

The **B** matrix, which is (5×2) in this case, can also be determined by inspection by noting that $x_1(t)$ enters only into the voltage relationship involving $q_3(t)$, and $x_2(t)$ enters only into the voltage relationship involving $q_5(t)$. Hence,

$$\mathbf{B} = \begin{bmatrix} 0 & 0 \\ 0 & 0 \\ 1 & 0 \\ 0 & 0 \\ 0 & 1 \end{bmatrix} \qquad \text{(8-72)}$$

Likewise, the **C** matrix, which is (2×5) in this example, is obtained readily by noting that

$$y_1(t) = q_1(t)$$

and

$$y_2(t) = q_2(t)$$

Hence,

$$\mathbf{C} = \begin{bmatrix} 1 & 0 & 0 & 0 & 0 \\ 0 & 1 & 0 & 0 & 0 \end{bmatrix} \qquad \text{(8-73)}$$

These relationships also make it clear that $\mathbf{D} = \mathbf{0}$ in this case. Hence, the state equations can be written as

$$\frac{d\mathbf{q}(t)}{dt} = \mathbf{A}\mathbf{q}(t) + \mathbf{B}\mathbf{x}(t) \qquad \text{(8-74)}$$

$$\mathbf{y}(t) = \mathbf{C}\mathbf{q}(t) \qquad \text{(8-75)}$$

where

$$\mathbf{q}(t) = \begin{bmatrix} q_1(t) \\ q_2(t) \\ q_3(t) \\ q_4(t) \\ q_5(t) \end{bmatrix} \qquad \mathbf{x}(t) = \begin{bmatrix} x_1(t) \\ x_2(t) \end{bmatrix} \qquad \mathbf{y}(t) = \begin{bmatrix} y_1(t) \\ y_2(t) \end{bmatrix}$$

and the other matrices are as defined above.

For the circuit as shown in Figure 8-15, the zero-state vector is

$$\mathbf{q}_0(t_0) = \mathbf{0}$$

The initial-state vector, on the other hand, need not be zero (although it may be), but the individual elements of this vector cannot be adjusted arbitrarily and independently unless additional inputs are permitted.

The foregoing discussion of state-variable selection avoids many subtle problems which may arise with some network configurations or in the case of active networks or networks with negative elements. These problems are discussed in the literature, however, and there are straight-forward procedures for dealing with them. Some appropriate references are given at the end of this chapter for the interested reader.

8-11 State Space Methods for Discrete Systems

The state space methods considered in the preceding sections can be applied directly to discrete-time systems. Difference equations replace differential equations, and analysis in the frequency domain is carried out by means of the z-transform instead of the Laplace transform. The normal-form difference equations for a discrete-time system can be written in matrix form as

$$\mathbf{q}[(n+1)T] = \mathbf{A}\mathbf{q}(nT) + \mathbf{B}\mathbf{x}(nT) \tag{8-76}$$

The corresponding output equation is

$$\mathbf{y}(nT) = \mathbf{C}\mathbf{q}(nT) + \mathbf{D}\mathbf{x}(nT) \tag{8-77}$$

In (8-76) and (8-77) the system is assumed to be time-invariant so that $\mathbf{A}$, $\mathbf{B}$, $\mathbf{C}$, and $\mathbf{D}$ are constant matrices. A solution can be obtained by recursively solving for the succeeding values of the state variables in terms of the preceding values. Starting with *initial state* of the system $\mathbf{q}(0)$, the solution is obtained as follows:

$$\mathbf{q}(T) = \mathbf{A}\mathbf{q}(0) + \mathbf{B}\mathbf{x}(0)$$

$$\begin{aligned}\mathbf{q}(2T) &= \mathbf{A}\mathbf{q}(T) + \mathbf{B}\mathbf{x}(T) \\ &= \mathbf{A}[\mathbf{A}\mathbf{q}(0) + \mathbf{B}\mathbf{x}(0)] + \mathbf{B}\mathbf{x}(T) \\ &= \mathbf{A}^2\mathbf{q}(0) + \mathbf{A}\mathbf{B}\mathbf{x}(0) + \mathbf{B}\mathbf{x}(T)\end{aligned}$$

$$\begin{aligned}\mathbf{q}(3T) &= \mathbf{A}\mathbf{q}(2T) + \mathbf{B}\mathbf{x}(2T) \\ &= \mathbf{A}^3\mathbf{q}(0) + \mathbf{A}^2\mathbf{B}\mathbf{x}(0) + \mathbf{A}\mathbf{B}\mathbf{x}(T) + \mathbf{B}\mathbf{x}(2T)\end{aligned}$$

Continuing this procedure leads to the following general expression:

$$\mathbf{q}(nT) = \mathbf{A}^n\mathbf{q}(0) + \sum_{m=0}^{n-1} \mathbf{A}^{n-m-1}\mathbf{B}\mathbf{x}(mT) \tag{8-78}$$

This is called the *discrete state-transition equation* of the system and is analogous to (8-33), which is the state-transition equation for a continuous system. The system output can be written as

$$\mathbf{y}(nT) = \mathbf{C}\mathbf{A}^n\mathbf{q}(0) + \mathbf{C}\sum_{m=0}^{n-1} \mathbf{A}^{n-m-1}\mathbf{B}\mathbf{x}(mT) + \mathbf{D}\mathbf{x}(nT) \tag{8-79}$$

The matrix $\mathbf{A}^n$ is called the *discrete state-transition matrix* and is written as

$$\boldsymbol{\phi}(nT) = \mathbf{A}^n \tag{8-80}$$

Considerable simplification in the analytical solution of the state equations for discrete systems can frequently be obtained through use of transform methods. Taking the z-transform of both sides of (8-76) leads to

$$\mathscr{Z}\{\mathbf{q}[(n+1)T]\} = \mathbf{A}\mathscr{Z}\{\mathbf{q}(nT)\} + \mathbf{B}\mathscr{Z}\{\mathbf{x}(nT)\}$$

$$z\mathbf{Q}(z) - z\mathbf{q}(0) = \mathbf{A}\mathbf{Q}(z) + \mathbf{B}\mathbf{X}(z)$$

Solving for $\mathbf{Q}(z)$,

$$[z\mathbf{I} - \mathbf{A}]\mathbf{Q}(z) = z\mathbf{q}(0) + \mathbf{B}\mathbf{X}(z)$$

$$\mathbf{Q}(z) = z[z\mathbf{I} - \mathbf{A}]^{-1}\mathbf{q}(0) + [z\mathbf{I} - \mathbf{A}]^{-1}\mathbf{B}\mathbf{X}(z) \tag{8-81}$$

Taking the inverse z-transform of (8-81) gives

$$\mathbf{q}(nT) = \mathscr{Z}^{-1}\{z[z\mathbf{I} - \mathbf{A}]^{-1}\}\mathbf{q}(0) + \mathscr{Z}^{-1}\{[z\mathbf{I} - \mathbf{A}]^{-1}\mathbf{B}\mathbf{X}(z)\} \tag{8-82}$$

Comparing (8-82) with (8-78), the following relationships can be seen:

$$\mathbf{A}^n = \mathscr{Z}^{-1}\{z[z\mathbf{I} - \mathbf{A}]^{-1}\} \tag{8-83}$$

$$\sum_{m=0}^{n-1} \mathbf{A}^{n-m-1}\mathbf{B}\mathbf{x}(mT) = \mathscr{Z}^{-1}\{[z\mathbf{I} - \mathbf{A}]^{-1}\mathbf{B}\mathbf{X}(z)\} \tag{8-84}$$

These identities can be verified by taking the z-transform of each side of the equation and establishing the equivalence of the various terms that result. Through use of (8-83) and (8-84) the matrices involved in the time-domain solution of the state equations can be evaluated for the general case.

As an example of the application of state variable techniques in the analysis of discrete systems consider a system characterized by the following equation:

$$y[(n+2)T] - 3y[(n+1)T] + 2y(nT) = x(nT) \tag{8-85}$$

Define the state variables as

$$q_1(nT) = y(nT) \tag{8-86}$$

$$q_2(nT) = y[(n+1)T] \tag{8-87}$$

The state equations are then

$$q_1[(n+1)T] = q_2(nT) \tag{8-88}$$

$$q_2[(n+1)T] = -2q_1(nT) + 3q_2(nT) + x(nT) \tag{8-89}$$

and the output equation is

$$y(nT) = q_1(nT) \tag{8-90}$$

The defining matrices are then

$$A = \begin{bmatrix} 0 & 1 \\ -2 & 3 \end{bmatrix} \tag{8-91}$$

$$B = \begin{bmatrix} 0 & 0 \\ 0 & 1 \end{bmatrix} \tag{8-92}$$

$$C = \begin{bmatrix} 1 & 0 \\ 0 & 0 \end{bmatrix} \tag{8-93}$$

$$D = \begin{bmatrix} 0 & 0 \\ 0 & 0 \end{bmatrix} \tag{8-94}$$

The state transition matrix is

$$\boldsymbol{\phi}(nT) = \mathcal{L}^{-1}\{z[z\mathbf{I} - \mathbf{A}]^{-1}\}$$

$$= \mathcal{L}^{-1}\left\{ z\left[\begin{bmatrix} z & 0 \\ 0 & z \end{bmatrix} - \begin{bmatrix} 0 & 1 \\ -2 & 3 \end{bmatrix} \right]^{-1} \right\}$$

$$= \mathcal{L}^{-1}\left\{ z\begin{bmatrix} z & -1 \\ 2 & z-3 \end{bmatrix}^{-1} \right\}$$

$$= \mathcal{L}^{-1}\left\{ z\begin{bmatrix} \dfrac{z-3}{z^2 - 3z + 2} & \dfrac{1}{z^2 - 3z + 2} \\ \dfrac{-2}{z^2 - 3z + 2} & \dfrac{z}{z^2 - 3z + 2} \end{bmatrix} \right\}$$

$$= \begin{bmatrix} 2 - 2^n & -1 + 2^n \\ 2 - 2^{n+1} & -1 + 2^{n+1} \end{bmatrix} \tag{8-95}$$

The general expression for $y(nT)$ can then be written from (8-79) and (8-95) as

$$y(nT) = \begin{bmatrix} 1 & 0 \\ 0 & 0 \end{bmatrix} \begin{bmatrix} 2 - 2^n & -1 + 2^n \\ 2 - 2^{n+1} & -1 + 2^{n+1} \end{bmatrix} \begin{bmatrix} q_1(0) \\ q_2(0) \end{bmatrix}$$

$$+ \begin{bmatrix} 1 & 0 \\ 0 & 0 \end{bmatrix} \sum_{m=0}^{n-1} \begin{bmatrix} 2 - 2^{n-m-1} & -1 + 2^{n-m-1} \\ 2 - 2^{n-m} & -1 + 2^{n-m} \end{bmatrix} \begin{bmatrix} 0 & 0 \\ 0 & 1 \end{bmatrix} \begin{bmatrix} 0 \\ x(mT) \end{bmatrix}$$

$$= (2 - 2^n)q_1(0) + (-1 + 2^n)q_2(0) \tag{8-96}$$

$$+ \sum_{m=0}^{n-1} (-1 + 2^{n-m-1})x(mT)$$

In order to make the example more specific let $q_1(0) = 1$, $q_2(0) = 0$, and $x(nT) = \delta_{n,0}$. Substituting these values into (8-96) gives

$$y(nT) = 2 - 2^n = 1 \qquad n = 0 \tag{8-97}$$

$$= 2 - 2^n - 1 + 2^{n-1} = 1 - 2^{n-1} \qquad n > 0 \tag{8-98}$$

This result can be checked readily with the iterative solution of the original equation starting with the specified initial conditions.

Exercise 8-11.1

Find the general expression for $q_n(nT)$ from (8-95) and (8-78) and check the result by substituting for $q_1(nT)$ in (8-88).

References

1. DE RUSSO, P. M., R. J. ROY, and D. C. CLOSE, *State Variables for Engineers*. New York: McGraw-Hill Book Company, Inc., 1965.

This book is an intermediate level text that discusses many of the details of system analysis using state variable methods.

2. KUO, B. C., *Linear Networks and Systems*. New York: McGraw-Hill Book Company, Inc., 1967.

Written at about the same level as this text, this book provides a good elementary discussion of state variable techniques in circuit analysis problems. Both continuous and discrete systems are considered, and the use of signal-flow graphs is given extensive coverage.

3. LEON, B. J., and P. A. WINTZ, *Basic Linear Networks for Electrical and Electronics Engineers*. New York: Holt, Rinehart and Winston, Inc., 1970.

This book presents comprehensive procedures for writing the equations of active and passive electrical circuits in normal-form in addition to many other topics in modern circuit analysis.

4. SCHWARTZ, R. J., and B. FRIEDLAND, *Linear Systems*. New York: McGraw-Hill Book Company, Inc., 1965.

Although generally written at a more advanced level than this text, this book contains a good discussion of state space analysis techniques that should be understandable to junior and senior students in electrical engineering.

Problems

8-1. a. Prove that the equivalent-impulse response of two linear systems connected in cascade is the convolution of the individual-impulse responses.

b. Prove that the equivalent-transfer function of two linear systems connected in cascade is the product of the individual-transfer functions.

8-2. a. Show that the equivalent system of two-cascaded, *invariant* linear systems is independent of the sequence in which they are connected.

b. Show by means of a simple example that the statement in (a) does *not* hold for cascaded, time-varying linear systems.

8-3. Two systems having impulse responses of

$$h_1(t) = 10\epsilon^{-2t}u(t) \quad \text{and} \quad h_2(t) = 20\epsilon^{-t}u(t)$$

are connected in cascade. Find the equivalent-impulse response of the combination.

8-4. a. Prove that the equivalent-impulse response of two linear systems connected in parallel is the sum of the individual-impulse responses.

b. Repeat part a for transfer functions.

8-5. Find the equivalent-transfer function for each of the three block diagrams shown.

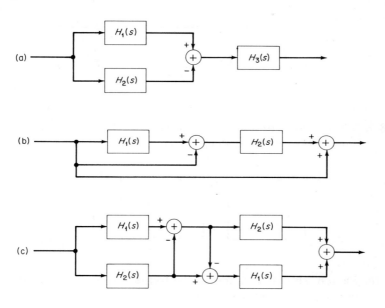

(a)

(b)

(c)

8-6. a. A linear system has an impulse response of

$$h(t) = 0.1\epsilon^{-2t}u(t)$$

Find the impulse response of the inverse system.

b. Is the inverse system causal?

8-7. Prove the results given for the equivalent of the feedback system shown in Fig. 8-3(d).

8-8. Find the equivalent-transfer function for each of the three block diagrams shown.

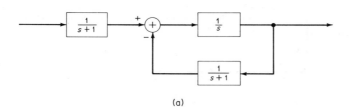

(a)

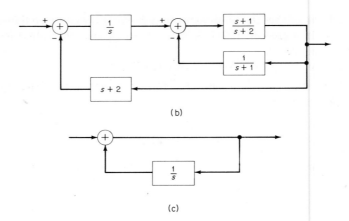

(b)

(c)

8-9. Write a single differential equation that relates the output $y(t)$ to the input $x(t)$ for the system shown.

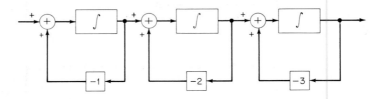

8-10. Sketch a block diagram for the following set of normal differential equations and output equation:

$$\frac{dq_1(t)}{dt} = -5q_1(t) + q_2(t)$$

$$\frac{dq_2(t)}{dt} = -3q_2(t) + x(t)$$

$$y(t) = q_1(t)$$

8-11. Write a set of normal differential equations and the output equation corresponding to the block diagram shown.

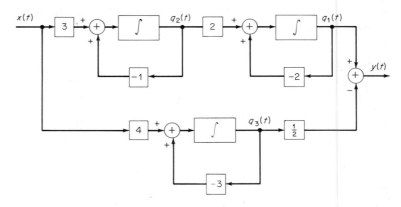

8-12. a. Write the normal equations and output equation for the two-input, one-output network shown, using the state variables indicated.

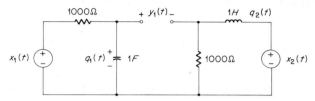

b. Sketch a block diagram for this network based on first-order feedback systems.

8-13. a. Write the matrix form of the normal equations for the network of Problem 8-12 and specify the elements of all vectors and matrices.

b. Sketch the matrix block diagram for this network.

8-14. Write the matrix equations for the block diagram shown. Specify the elements of all vectors and matrices.

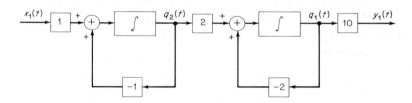

8-15.

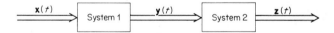

System 1 is defined by the matrix equations

$$\frac{d\mathbf{q}_1(t)}{dt} = \mathbf{A}_1(t)\mathbf{q}_1(t) + \mathbf{B}_1(t)\mathbf{x}(t)$$

$$\mathbf{y}(t) = \mathbf{C}_1(t)\mathbf{q}_1(t)$$

and system 2 by the equations

$$\frac{d\mathbf{q}_2(t)}{dt} = \mathbf{A}_2(t)\mathbf{q}_2(t) + \mathbf{B}_2(t)\mathbf{y}(t)$$

$$\mathbf{z}(t) = \mathbf{C}_2\mathbf{q}_2(t)$$

If the cascaded combination of these two systems is defined by

$$\frac{d\mathbf{q}(t)}{dt} = \mathbf{A}(t)\mathbf{q}(t) + \mathbf{B}(t)\mathbf{x}(t)$$

$$\mathbf{z}(t) = \mathbf{C}(t)\mathbf{q}(t)$$

find $\mathbf{q}(t)$, $\mathbf{A}(t)$, $\mathbf{B}(t)$, and $\mathbf{C}(t)$ in terms of the similar quantities for the separate systems.

8-16. Repeat Problem 8-15 if the two systems are connected in parallel.

8-17.

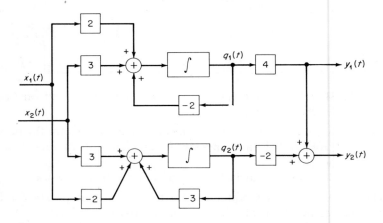

Write the matrix state equations for the system shown and specify the elements of all vectors and matrices.

8-18. A system is characterized by state equations with the following matrices:

$$\mathbf{A} = \begin{bmatrix} -1 & 1 \\ 0 & -1 \end{bmatrix} \quad \mathbf{B} = \begin{bmatrix} 1 & 0 & 1 \\ 0 & 1 & 0 \end{bmatrix}$$

$$\mathbf{C} = [-1 \quad 1] \qquad \mathbf{D} = [1 \quad 0 \quad 0]$$

Sketch the complete block diagram for this system.

8-19.

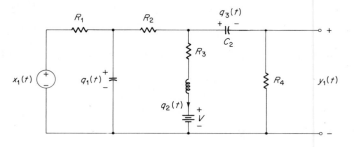

a. Find a zero-state vector for the system shown.
b. Is this system fully linear?

8-20. Use the series expansion for the exponential to show that

a. $\dfrac{d\epsilon^{At}}{dt} = \mathbf{A}\epsilon^{At}$

b. $\epsilon^{At}\epsilon^{-At} = \epsilon^{-At}\epsilon^{At} = \mathbf{I}$

8.21. A time-invariant system has a fundamental matrix of the form

$$\phi(t) = \begin{bmatrix} \epsilon^{-\alpha t} & t\epsilon^{-\alpha t} \\ 0 & \epsilon^{-\alpha t} \end{bmatrix}$$

a. Find the **A** matrix for the system.
b. If the initial state for the system is

$$\mathbf{q}(t_0) = \begin{bmatrix} 10 \\ -5 \end{bmatrix}$$

write the zero-input response for the state vector for all $t \geq t_0$.

8-22. For the system of Problem 8-21, the **B** matrix is

$$\mathbf{B} = \begin{bmatrix} 2 \\ -2 \end{bmatrix}$$

If the input is a δ-function at $t = t_0 + 1$, write the total state vector for all $t > t_0$.

8-23. Use the series expansion for the exponential to determine the state-transition matrix for a system whose **A** matrix is $(n \times n)$ and has the form

$$\mathbf{A} = -\mathbf{I} \quad (n \times n)$$

8.24. A particular class of nth-order system has a block diagram consisting of n integrators with feedback connected in parallel with *no* cross connections from the output of one integrator to the input of another.

a. Show that the **A** matrix for such a system has the form

$$\mathbf{A} = \begin{bmatrix} -\alpha_1 & 0 & \cdots\cdots & & & 0 \\ 0 & -\alpha_2 & 0 & \cdots\cdots & & 0 \\ 0 & 0 & -\alpha_3 & 0 & \cdots & 0 \\ \cdot & & & & & \cdot \\ \cdot & & & & & \\ \cdot & & & & & 0 \\ 0 & & \cdots\cdots & & 0 & -\alpha_n \end{bmatrix}$$

b. Find the state-transition matrix for this class of system.

8-25. Use the eigenvalue method to evaluate the state-transition matrix for a system whose **A** matrix is

$$\mathbf{A} = \begin{bmatrix} -1 & 1 & 0 \\ 0 & -2 & 1 \\ 0 & 0 & -3 \end{bmatrix}$$

8-26. Using the impulse-response interpretation, find the state-transition matrix for the system of Problem 8-12.

8-27. Using the impulse-response method, find the state-transition matrix for the system shown.

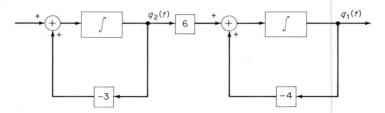

8-28. For each of the following matrices. specify whether or not it is a valid state-transition matrix for a linear, time-invariant system. If it is not, state the reason.

a. $\begin{bmatrix} 2\epsilon^{-2(t-t_0)} & 0 \\ 0 & \epsilon^{-3(t-t_0)} \end{bmatrix}$

b. $\begin{bmatrix} \epsilon^{-(t-t_0)} & \epsilon^{-2(t-t_0)} \\ \epsilon^{-3(t-t_0)} & \epsilon^{-4(t-t_0)} \end{bmatrix}$

c. $\begin{bmatrix} \epsilon^{-(t-t_0)} & (t-t_0)\,\epsilon^{-2(t+t_0)} \\ 0 & \epsilon^{-3(t-t_0)} \end{bmatrix}$

d. $\begin{bmatrix} 1 & 0 \\ 0 & 1 \end{bmatrix}$

8-29. **a.** Find the impulse-response matrix for the system of Problem 8-17.

b. Find the output vector for this system if the input vector is

$$\mathbf{x}(t) = \begin{bmatrix} \epsilon^{-100t}u(t) \\ 5\delta(t) \end{bmatrix}$$

and the initial state is a zero state.

8-30. All of the impulse responses indicated in the block diagram shown are physically realizable.

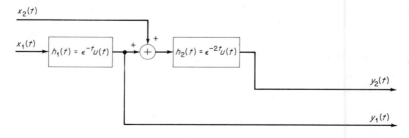

a. Find the impulse-response matrix for this system.

b. Find the zero-state response of this system if the input vector is

$$\mathbf{x}(t) = \begin{bmatrix} u(t) \\ u(t-1/2) \end{bmatrix}$$

8-31. Find the state-transition matrix for the system of Problem 8-24 by using Laplace transform methods.

8-32. Find the state-transition matrix for the system of Problem 8-25 by using Laplace transform methods.

8-33. Find the transfer-function matrix for the system of Problem 8-27.

8-34. a. Find the transfer-function matrix for the system of Problem 8-30.

  b. Find the zero-state response for the specified input by using frequency-domain methods.

8.35. Write the state equations for the network shown and define all matrices and vectors.

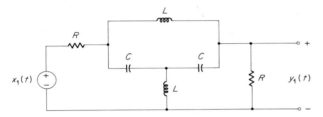

8-36. Find the impulse response of the network shown using state space methods.

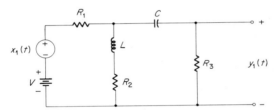

8-37. a. Find the state-transition matrix for the time-varying system shown.

  b. Find the impulse response for this system.

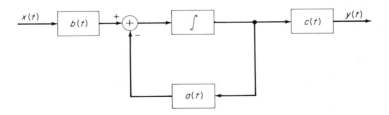

APPENDIX

A
Mathematical Tables

Table A-1   Trigonometric Identities

$\sin (A \pm B) = \sin A \cos B \pm \cos A \sin B$
$\cos (A \pm B) = \cos A \cos B \mp \sin A \sin B$

$\cos A \cos B = 1/2[\cos (A + B) + \cos (A - B)]$
$\sin A \sin B = 1/2[\cos (A - B) - \cos (A + B)]$
$\sin A \cos B = 1/2[\sin (A + B) + \sin (A - B)]$

$\sin A + \sin B = 2 \sin 1/2(A + B) \cos 1/2(A - B)$
$\sin A - \sin B = 2 \sin 1/2(A - B) \cos 1/2(A + B)$
$\cos A + \cos B = 2 \cos 1/2(A + B) \cos 1/2(A - B)$
$\cos A - \cos B = - 2 \sin 1/2(A + B) \sin 1/2(A - B)$

$\sin 2A = 2 \sin A \cos A$
$\cos 2A = 2 \cos^2 A - 1 = 1 - 2 \sin^2 A = \cos^2 A - \sin^2 A$

$\sin 1/2A = \sqrt{1/2(1 - \cos A)}$ $\sin^2 A = 1/2(1 - \cos 2A)$

$\cos 1/2A = \sqrt{1/2(1 + \cos A)}$ $\cos^2 A = 1/2(1 + \cos 2A)$

$\sin x = \dfrac{\epsilon^{ix} - \epsilon^{-ix}}{2j}$ $\cos x = \dfrac{\epsilon^{ix} + \epsilon^{-ix}}{2}$

$\epsilon^{ix} = \cos x + j \sin x$

$A \cos (\omega t + \phi_1) + B \cos (\omega t + \phi_2) = C \cos (\omega t + \phi_3)$

where

$$C = \sqrt{A^2 + B^2 - 2AB \cos (\phi_2 - \phi_1)}$$

$$\phi_3 = \tan^{-1} \left\{ \frac{A \sin \phi_1 + B \sin \phi_2}{A \cos \phi_1 + B \cos \phi_2} \right\}$$

$\sin (\omega t + \phi) = \cos (\omega t + \phi - 90°)$

360

Table A-2 Indefinite Integrals

$$\int \sin ax dx = -1/a \cos ax \qquad \int \cos ax dx = 1/a \sin ax$$

$$\int \sin^2 ax dx = x/2 - \frac{\sin 2ax}{4a}$$

$$\int x \sin ax dx = 1/a^2(\sin ax - ax \cos ax)$$

$$\int x^2 \sin ax dx = 1/a^3(2ax \sin ax + 2 \cos ax - a^2x^2 \cos ax)$$

$$\int \cos^2 ax dx = x/2 + \frac{\sin 2ax}{4a}$$

$$\int x \cos ax dx = 1/a^2(\cos ax + ax \sin ax)$$

$$\int x^2 \cos ax dx = 1/a^3(2ax \cos ax - 2 \sin ax + a^2x^2 \sin ax)$$

$$\int \sin ax \sin bx dx = \frac{\sin (a-b)x}{2(a-b)} - \frac{\sin (a+b)}{2(a+b)} \qquad a^2 \neq b^2$$

$$\int \sin ax \cos bx dx = - \left[\frac{\cos (a-b)x}{2(a-b)} + \frac{\cos (a+b)x}{2(a+b)} \right] \qquad a^2 \neq b^2$$

$$\int \cos ax \cos bx dx = \frac{\sin (a-b)x}{2(a-b)} + \frac{\sin (a+b)x}{2(a+b)} \qquad a^2 \neq b^2$$

$$\int \epsilon^{ax} dx = 1/a\epsilon^{ax}$$

$$\int x\epsilon^{ax} dx = \frac{\epsilon^{ax}}{a^2}(ax - 1)$$

$$\int x^2\epsilon^{ax} dx = \frac{\epsilon^{ax}}{a^3}(a^2x^2 - 2ax + 2)$$

$$\int \epsilon^{ax} \sin bx dx = \frac{\epsilon^{ax}}{a^2 + b^2}(a \sin bx - b \cos bx)$$

$$\int \epsilon^{ax} \cos bx dx = \frac{\epsilon^{ax}}{a^2 + b^2}(a \cos bx + b \sin bx)$$

Table A-3 Definite Integrals

$$\int_0^\infty x^n \epsilon^{-ax} dx = \frac{n!}{a^{n+1}} = \frac{\Gamma(n+1)}{a^{n+1}}$$

$$\int_0^\infty \epsilon^{-r^2x^2} dx = \frac{\sqrt{\pi}}{2r}$$

$$\int_0^\infty x\epsilon^{-r^2x^2} dx = \frac{1}{2r^2}$$

$$\int_0^\infty x^2\epsilon^{-r^2x^2} dx = \frac{\sqrt{\pi}}{4r^3}$$

$$\int_0^\infty x^n\epsilon^{-r^2x^2} dx = \frac{\Gamma[(n+1)/2]}{2r^{n+1}}$$

$$\int_0^\infty \frac{\sin ax}{x} dx = \frac{\pi}{2}, 0, -\frac{\pi}{2} \quad \text{for} \quad a > 0, a = 0; a < 0$$

$$\int_0^\infty \frac{\sin^2 x}{x} dx = \frac{\pi}{2}$$

$$\int_0^\infty \frac{\sin^2 ax}{x^2} dx = |a|\frac{\pi}{2}$$

$$\int_0^\pi \sin^2 mx dx = \int_0^\pi \sin^2 x dx = \int_0^\pi \cos^2 mx dx = \int_0^\pi \cos^2 x dx = \frac{\pi}{2}, m \text{ an integer}$$

$$\int_0^\pi \sin mx \sin nx dx = \int_0^\pi \cos mx \cos nx dx = 0 \qquad m \neq n \quad m, n \text{ integers}$$

$$\int_0^\pi \sin mx \cos nx dx = \begin{cases} \frac{2m}{m^2 - n^2} & \text{if } m + n \text{ odd} \\ 0 & \text{if } m + n \text{ even} \end{cases}$$

Table A-4 $sinc\ (x) = \dfrac{\sin(\pi x)}{\pi x}$

x	sinc (x)	x	sinc (x)	x	sinc (x)
0.0	1.00000	1.0	0.00000	2.0	0.00000
0.1	0.98363	1.1	− 0.08942	2.1	0.04684
0.2	0.93549	1.1	− 0.15592	2.2	0.08504
0.3	0.85839	1.3	− 0.19809	2.3	0.11196
0.4	0.75683	1.4	− 0.21624	2.4	0.12614
0.5	0.63662	1.5	− 0.21221	2.5	0.12732
0.6	0.50455	1.6	− 0.18921	2.6	0.11644
0.7	0.36788	1.7	− 0.15148	2.7	0.09538
0.8	0.23387	1.8	− 0.10394	2.8	0.06682
0.9	0.10929	1.9	− 0.05177	2.9	0.03392

For values of sinc (x) outside the range of x given in the table the following procedure may be used:

1. Subtract the largest integer, n, less than x from x.
2. Look up in the table below the value of $(1/\pi)\sin[\pi(x - n)]$.
3. Multiply the tabular value by $(-1)^n$ and divide by x to give sinc (x).

$(x - n)$		$\dfrac{1}{\pi}\sin \pi x$
0.0	1.0	0.000000
0.1	0.9	0.098363
0.2	0.8	0.187098
0.3	0.7	0.257518
0.4	0.6	0.302731
0.5	0.5	0.318310

Table A-5   Fourier Series Representation of Common Waveforms

Waveform	Fourier series
 General periodic wave $f(t)$	$\displaystyle\sum_{n=-\infty}^{\infty} a_n \epsilon^{j(2\pi n/T)t}$ $\displaystyle a_n = \frac{1}{T}\int_{t_1}^{t_1+T} f(t)\,\epsilon^{-j(2\pi n/T)t}$ $\displaystyle \frac{a_0}{2} + \sum_{n=1}^{\infty} a_n \cos\frac{2\pi n}{T}t + b_n \sin\frac{2\pi n}{T}t$ $\displaystyle a_n = \frac{2}{T}\int_{t_1}^{t_1+T} f(t)\cos\frac{2\pi n}{T}t\,dt$ $\displaystyle b_n = \frac{2}{T}\int_{t_1}^{t_1+T} f(t)\sin\frac{2\pi n}{T}t\,dt$
 Odd square wave	$\displaystyle \frac{4}{\pi}\sum_{n=1}^{\infty}\frac{1}{2n-1}\sin\frac{2\pi(2n-1)}{T}t$
 Even square wave	$\displaystyle \frac{4}{\pi}\sum_{n=1}^{\infty}\frac{(-1)^{n+1}}{2n-1}\cos\frac{2\pi(2n-1)t}{T}$
 Rectangular pulse train	$\displaystyle \frac{2t_a}{T}+\frac{4t_a}{T}\sum_{n=1}^{\infty}\frac{\sin\left(\frac{2\pi n t_a}{T}\right)}{\frac{2\pi n t_a}{T}}\cos\frac{2\pi n t}{T}$
 Triangular wave	$\displaystyle \frac{8}{\pi^2}\sum_{n=1}^{\infty}\frac{1}{(2n-1)^2}\cos\frac{2\pi(2n-1)}{T}t$
 Sawtooth wave	$\displaystyle \frac{2}{\pi}\sum_{n=1}^{\infty}\frac{(-1)^{n+1}}{n}\sin\frac{2\pi n}{T}t$
 Half-wave rectified cosine wave	$\displaystyle \frac{2}{\pi}\left(\frac{1}{2}+\frac{\pi}{4}\cos\frac{2\pi t}{T}-\sum_{n=1}^{\infty}\frac{(-1)^n}{4n^2-1}\cos\frac{4\pi n t}{T}\right)$

Table A-6   Fourier Transforms of Mathematical Operations

Operation	$f(t)$	$F(\omega)$	$F(f)$
Superposition	$a_1 f_1(t) + a_2 f_2(t)$	$a_1 F_1(\omega) + a_2 F_2(\omega)$	$a_1 F_1(f) + a_2 F_2(f)$
Reversal	$f(-t)$	$F(-\omega)$	$F(-f)$
Symmetry	$F(t)$	$2\pi f(-\omega)$	$f(-f)$
Scaling	$f(at)$	$\dfrac{1}{|a|} F\left(\dfrac{\omega}{a}\right)$	$\dfrac{1}{|a|} F\left(\dfrac{f}{a}\right)$
Delay	$f(t-t_0)$	$\epsilon^{-j t_0 \omega} F(\omega)$	$\epsilon^{-j 2\pi f t_0} F(f)$
Complex conjugate	$f^*(t)$	$F^*(-\omega)$	$F^*(-f)$
Modulation	$\epsilon^{j\omega_0 t} f(t)$	$F[(\omega - \omega_0)]$	$F\left(f - \dfrac{\omega_0}{2\pi}\right)$
Time differentiation	$\dfrac{d^n}{dt^n} f(t)$	$(j\omega)^n F(\omega)$	$(j2\pi f)^n F(f)$
Frequency differentiation	$t^n f(t)$	$(j)^n \dfrac{d^n}{d\omega^n} F(\omega)$	$\left(\dfrac{j}{2\pi}\right)^n \dfrac{d^n}{df^n} F(f)$
Integration	$\displaystyle\int_0^t f_e(t)dt + \int_{-\infty}^t f_0(t)dt$	$\dfrac{1}{j\omega} F(\omega)$	$\dfrac{1}{j2\pi f} F(f)$
Integration	$\displaystyle\int_{-\infty}^t f(t)dt$	$\dfrac{1}{j\omega} F(\omega) + \pi F(0)\delta(\omega)$	$\dfrac{1}{j2\pi f} F(f) + \dfrac{1}{2} F(0)\delta(f)$
Convolution	$f_1 * f_2 = \displaystyle\int_{-\infty}^\infty f_1(\lambda) f_2(t-\lambda)d\lambda$	$F_1(\omega)F_2(\omega)$	$F_1(f)F_2(f)$
Multiplication	$f_1(f)f_2(t)$	$\dfrac{1}{2\pi}\displaystyle\int_{-\infty}^\infty F_1(\xi)F_2(\omega - \xi)d\xi$	$\displaystyle\int_{-\infty}^\infty F_1(\xi)F_2(f - \xi)d\xi$
Correlation	$\displaystyle\int f_1(\lambda)f_2^*(\lambda + t)d\lambda$	$F_1(\omega)F_2^*(\omega)$	$F_1(f)F_2^*(f)$

Definitions: $F(\omega) = \displaystyle\int_{-\infty}^\infty f(t)\epsilon^{-j\omega t}dt$

$$F(f) = \int_{-\infty}^\infty f(t)\epsilon^{-j2\pi ft}dt = F(\omega)\big|_{\omega=2\pi f}$$

$$f(t) = \frac{1}{2\pi}\int_{-\infty}^\infty F(\omega)\epsilon^{j\omega t}\,d\omega = \int_{-\infty}^\infty F(f)\epsilon^{j2\pi ft}\,df$$

Table A-7 Fourier Transforms of Energy Signals

$f(t)$		$F(\omega)$	$|F(f)|$
Rectangular pulse	$u(t+\frac{T}{2})-u(t-\frac{T}{2})$	$T\,\dfrac{\sin(\omega T/2)}{\omega T/2}$	$T\,\dfrac{\sin(\pi Tf)}{\pi Tf}$
Exponential	$\epsilon^{-at}u(t)$	$\dfrac{1}{j\omega+a}$	$\dfrac{1}{j2\pi f+a}$
Triangular	$1-2\,\dfrac{|t|}{T},\ |t|<\dfrac{T}{2}$ O elsewhere	$\dfrac{T}{2}\left[\dfrac{\sin(\omega T/4)}{\omega T/4}\right]^2$	$\dfrac{T}{2}\left[\dfrac{\sin(\pi Tf/2)}{\pi Tf/2}\right]^2$
Gaussian	$\epsilon^{-a^2t^2}$	$\dfrac{\sqrt{\pi}}{a}\epsilon^{-(\omega^2/4a^2)}$	$\dfrac{\sqrt{\pi}}{a}\epsilon^{-(\pi^2f^2/a^2)}$
Double exponential	$\epsilon^{-a|t|}$	$\dfrac{2a}{a^2+\omega^2}$	$\dfrac{2a}{a^2+4\pi^2f^2}$
Damped sine	$\epsilon^{-at}\sin(\omega_0 t)u(t)$	$\dfrac{\omega_0}{(a+j\omega)^2+\omega_0^2}$	$\dfrac{\omega_0}{(a+j2\pi f)^2+\omega_0^2}$
Damped cosine	$\epsilon^{-at}\cos(\omega_0 t)u(t)$	$\dfrac{a+j\omega}{(a+j\omega)^2+\omega_0^2}$	$\dfrac{a+j2\pi f}{(a+j2\pi f)^2+\omega_0^2}$
	$\dfrac{1}{\beta-a}[\epsilon^{-at}-\epsilon^{-\beta t}]u(t)$	$\dfrac{1}{(j\omega+a)(j\omega+\beta)}$	$\dfrac{1}{(j2\pi f+a)(j2\pi f+\beta)}$
Cosine pulse	$\cos\omega_0 t[u(t+\frac{T}{2})-u(t-\frac{T}{2})]$	$\dfrac{T}{2}\left[\dfrac{\sin(\omega-\omega_0)T/2}{(\omega-\omega_0)T/2}\right.$ $\left.+\dfrac{\sin(\omega+\omega_0)T/2}{(\omega+\omega_0)T/2}\right]$	$\dfrac{T}{2}\left[\dfrac{\sin\pi T(f-f_0)}{\pi T(f-f_0)}\right.$ $\left.+\dfrac{\sin\pi T(f+f_0)}{\pi T(f+f_0)}\right]$

Table A-8 Fourier Transforms of Power Signals

$f(t)$		$F(\omega)$	$F(f)$
Unit impulse	$\delta(t)$	1	1
Unit step	$u(t)$	$\pi\delta(\omega) + \dfrac{1}{j\omega}$	$\dfrac{1}{2}\delta(f) + \dfrac{1}{j2\pi f}$
Signum function	$\operatorname{sgn} t = \dfrac{t}{\lvert t\rvert}$	$\dfrac{2}{j\omega}$	$\dfrac{1}{j\pi f}$
Constant	K	$2\pi K\delta(\omega)$	$K\,\delta(f)$
Cosine wave	$\cos \omega_0 t$	$\pi\left[\delta(\omega-\omega_0) + \delta(\omega+\omega_0)\right]$	$\dfrac{1}{2}\left[\delta(f-f_0) + \delta(f+f_0)\right]$
Sine wave	$\sin \omega_0 t$	$-j\pi\left[\delta(\omega-\omega_0) - \delta(\omega+\omega_0)\right]$	$\dfrac{-j}{2}\left[\delta(f-f_0) - \delta(f+f_0)\right]$
Periodic wave	$\displaystyle\sum_{n=-\infty}^{\infty} f_T(t-nt)$ $\displaystyle\sum_{n=-\infty}^{\infty} a_n \epsilon^{j2\pi nt/\tau}$	$\dfrac{2\pi}{T}\displaystyle\sum_{n=-\infty}^{\infty} F_T\!\left(\dfrac{2\pi n}{T}\right)$ $\delta\!\left(\omega - \dfrac{2\pi n}{T}\right)$	$\displaystyle\sum_{n=-\infty}^{\infty} a_n \delta\!\left(f-\dfrac{n}{T}\right)$
Impulse train	$\displaystyle\sum \delta(t-nt)$	$\dfrac{2\pi}{T}\displaystyle\sum \delta\!\left(\omega - \dfrac{2\pi n}{T}\right)$	$\dfrac{1}{T}\displaystyle\sum \delta\!\left(f-\dfrac{n}{T}\right)$
Complex sinusoid	$\epsilon^{j\omega_0 t}$	$2\pi\delta(\omega-\omega_0)$	$\delta(f-f_0)$
Unit ramp	$t\,u(t)$	$j\pi\delta'(\omega) - \dfrac{1}{\omega^2}$	$\dfrac{j}{4\pi}\delta'(f) - \dfrac{1}{4\pi^2 f^2}$
	t^n	$2\pi(j)^n \delta^{(n)}(\omega)$	$\left(\dfrac{j}{2\pi}\right)^n \delta^{(n)}(f)$

Table A-9 Laplace Transforms of Mathematical Operations

	Time Domain		s-Domain	
Property	Property	Time function	Laplace transform	Property
Linearity		$a_1 f_1(t) + a_2 f_2(t)$	$a_1 F_1(s) + a_2 F_2(s)$	Linearity
Time differentiation		$f'(t)$	$sF(s) - f(0)$	s-Multiplication
Time integration		$\int_0^t f(\xi)d\xi$	$\dfrac{1}{s}F(s)$	Division by s
t multiplication		$tf(t)$	$-\dfrac{dF(s)}{ds}$	Differentiation
Division by t		$\dfrac{1}{t}f(t)$	$\int_s^\infty F(\xi)d\xi$	Integration
Delay (t-shift)		$f(t - t_0)u(t - t_0)$	$\epsilon^{-st_0}F(s)$	Multiplication by exponential
Multiplication by exponential		$\epsilon^{-at}f(t)$	$F(s + a)$	s-shift
Scale change		$f(at)\ a > 0$	$\dfrac{1}{a}F\left(\dfrac{s}{a}\right)$	Scale change
Time convolution		$f_1 * f_2 = \int_0^t f_1(\lambda)f_2(t - \lambda)d\lambda$	$F_1(s)F_2(s)$	Multiplication of transforms
Initial value		$f(0^+)$	$\lim_{s\to\infty} sF(s)$	Limit as $s \to \infty$
Final value		$f(\infty)$	$\lim_{s\to 0} sF(s)$ [$F(s)$ left half-plane poles only]	Limit as $s \to 0$
Second derivative		$f''(t)$	$s^2 F(s) - sf(0) - f'(0)$	s-Multiplication
Multiplication of time functions		$f_1(t)f_2(t)$	$\dfrac{1}{2\pi j}\displaystyle\int_{c-j\infty}^{c+j\infty} F_1(s - \lambda)F_2(\lambda)d\lambda$	Complex convolution

Table A-10 Laplace Transforms
of Elementary Functions

Time function	Transform
$\delta(t)$	1
$u(t)$	$\dfrac{1}{s}$
$tu(t)$	$\dfrac{1}{s^2}$
$t^n u(t)$	$\dfrac{n!}{s^{n+1}}$
$\epsilon^{-\alpha t} u(t)$	$\dfrac{1}{s+\alpha}$
$t\epsilon^{-\alpha t} u(t)$	$\dfrac{1}{(s+\alpha)^2}$
$\sin(\beta t)u(t)$	$\dfrac{\beta}{s^2+\beta^2}$
$\cos(\beta t)u(t)$	$\dfrac{s}{s^2+\beta^2}$
$\epsilon^{-\alpha t}\sin(\beta t)u(t)$	$\dfrac{\beta}{(s+\alpha)^2+\beta^2}$
$\epsilon^{-\alpha t}\cos(\beta t)u(t)$	$\dfrac{s+\alpha}{(s+\alpha)^2+\beta^2}$
$\sinh atu(t)$	$\dfrac{a}{s^2-a^2}$
$\cosh atu(t)$	$\dfrac{s}{s^2-a^2}$

Table A-11   z-Transform Operations

Operation	Time function	z-Transform
Definition	$f(n)$	$\displaystyle\sum_{n=0}^{\infty} f(n)z^{-n}$
Inversion	$\dfrac{1}{2\pi j}\oint F(z)z^{n-1}dz$	$F(z)$
Linearity	$af_1(n)+bf_2(n)$	$aF_1(z)+bF_2(z)$
Delay (right shift)	$f(n-k)u(n-k)$	$z^{-k}F(z)$
Advance (left shift)	$f(n+k)u(n)$	$z^k F(z) - z^k \displaystyle\sum_{i=0}^{k-1} f(i)z^{-i}$
Multiplication by a^n	$a^n f(n)$	$F(za^{-1})$
Multiplication of functions	$f_1(n)f_2(n)$	$\dfrac{1}{2\pi j}\oint \dfrac{F_1(\lambda)F_2(z\lambda^{-1})}{\lambda}d\lambda$
Convolution of functions	$\displaystyle\sum_{k=0}^{n} f_1(n-k)f_2(k)$	$F_1(z)F_2(z)$
Initial value	$\displaystyle\lim_{n\to 0} f(n)$	$\displaystyle\lim_{z\to\infty} F(z)$
Final value	$\displaystyle\lim_{n\to\infty} f(n)$	$\displaystyle\lim_{z\to 1} (1-z^{-1})F(z)$

Table A-12 z-Transforms

$f(nT)$	$F(z) = \mathscr{L}\{f(nT)\}$
$1, 0, 0, \cdots$	1
$1, 1, 1, \cdots$	$\dfrac{1}{1 - z^{-1}}$
nT	$\dfrac{Tz^{-1}}{(1 - z^{-1})^2}$
$(nT)^2$	$\dfrac{T^2 z^{-1}(1 + z^{-1})}{(1 - z^{-1})^3}$
ϵ^{-anT}	$\dfrac{1}{1 - \epsilon^{-aT} z^{-1}}$
$(K)^n$	$\dfrac{1}{1 - Kz^{-1}}$
nTK^{nT}	$\dfrac{TK^T z^{-1}}{(1 - K^T z^{-1})^2}$
$\sin \omega nT$	$\dfrac{z^{-1} \sin \omega T}{1 - 2z^{-1} \cos \omega T + z^{-2}}$
$\cos \omega nT$	$\dfrac{(1 - z^{-1} \cos \omega T)}{1 - 2z^{-1} \cos \omega T + z^{-2}}$

B

Contour Integration

Integrals of the following types are frequently encountered in the analysis of linear systems.

$$\frac{1}{2\pi j} \oint F(z)z^{n-1}dz \tag{B-1}$$

$$\frac{1}{2\pi j} \int_{c-j\infty}^{c+j\infty} F(s)\epsilon^{st}ds \tag{B-2}$$

The integral (B-1) is the inversion integral for the z-transform, while (B-2) is the inversion integral for the Laplace transform. In both cases the integration is with respect to a complex variable, and only in very special cases can these integrals be evaluated by elementary methods. However, because of the generally well-behaved nature of their integrands, these integrals can frequently be evaluated very simply by the method of residues. This method of evaluation is based on the following theorem from complex variable theory: If a function $G(s)$ is analytic on and interior to a closed contour, C, except at a number of poles, then the integral of $G(s)$ taken counterclockwise around the contour is equal to $2\pi j$ times the sum of the residues at the poles within the contour. In equation form this becomes

$$\oint_C G(s)ds = 2\pi j \, \Sigma \text{ residues at poles enclosed} \tag{B-3}$$

What is meant by the left-hand side of (B-3) is that the value of $G(s)$ at each point on the contour, C, is to be multiplied by the differential path length and summed over the complete contour. As indicated by the arrow, the contour is to be traversed in a counterclockwise direction. Reversing the direction introduces a minus sign on the right-hand side of (B-3).

In order to utilize (B-3) for the evaluation of integrals such as (B-1) and (B-2) two further steps are required: First we must learn how to find the residue at a pole, and second we must reconcile the closed contour in (B-3) with the apparently open path of integration in (B-2).

Consider the problem of poles and residues first. A single-valued function $G(s)$ is *analytic* at a point, $s = s_0$, if its derivative exists at every point in the neighborhood of (and including) s_0. A function is *analytic in a region* of the s-plane if it is analytic at every point in that region. If a function is analytic at every point in the neighborhood of s_0, but not at s_0 itself, then s_0 is called a *singular point*. For example, the function $G(s) = 1/(s - 2)$ has a derivative $G'(s) = -1/(s - 2)^2$. It is readily seen by inspection that this function is analytic everywhere except at $s = 2$, where it has a singularity. An *isolated singular point* is a point interior to a region throughout which the function is analytic except at that point. It is evident that the above function has an isolated singularity at $s = 2$. The most frequently encountered singularity is the *pole*. If a function $G(s)$ becomes infinite at $s = s_0$ in such a manner that, by multiplying $G(s)$ by a factor of the form $(s - s_0)^n$, the singularity is removed, then $G(s)$ is said to have a pole of order n at $s = s_0$. For example, the function $1/\sin s$ has a pole at $s = 0$ and can be written as

$$G(s) = \frac{1}{\sin s} = \frac{1}{s - \dfrac{s^3}{3!} + \dfrac{s^5}{5!} - \cdots}$$

Multiplying by s [that is, the factor $(s - 0)$], we obtain

$$\phi(s) = \frac{s}{s - \dfrac{s^3}{3!} + \dfrac{s^5}{5!} + \cdots} = \frac{1}{1 - \dfrac{s^2}{3!} + \dfrac{s^4}{5!} + \cdots}$$

which is seen to be well-behaved near $s = 0$. It may, therefore, be concluded that $1/\sin s$ has a simple (that is, first-order) pole at $s = 0$.

It is an important property of analytic functions that they can be represented by convergent power series throughout their region of analyticity. By a simple extension of this property it is possible to represent functions in the vicinity of a singularity. Consider a function $G(s)$ having an nth order pole at $s = s_0$. Define a new function $\phi(s)$ such that

$$\phi(s) = (s - s_0)^n G(s) \tag{B-4}$$

Now $\phi(s)$ will be analytic in the region of s_0, since the singularity of $G(s)$ has been removed. Therefore, $\phi(s)$ can be expanded in a Taylor series as follows:

$$\phi(s) = A_{-n} + A_{-n+1}(s - s_0) + A_{-n+2}(s - s_0)^2 + \cdots + A_{-1}(s - s_0)^{n-1}$$

$$+ \sum_{k=0}^{\infty} B_k (s - s_0)^{n+k} \qquad \text{(B-5)}$$

Substituting (B-5) into (B-4) and solving for $G(s)$ gives

$$G(s) = \frac{A_{-n}}{(s - s_0)^n} + \frac{A_{n-1}}{(s - s_0)^{n-1}} + \cdots + \frac{A_{-1}}{s - s_0} + \sum_{k=0}^{\infty} B_k (s - s_0)^k \qquad \text{(B-6)}$$

This expansion is valid in the vicinity of the pole at $s = s_0$. The series converges in a region around s_0 that extends out to the nearest singularity. Equation (B-6) is called the *Laurent expansion*, or *Laurent series*, for $G(s)$ about the singularity at $s = s_0$. There are two distinct parts to the series: The first, called the *principal part*, consists of the terms containing $(s - s_0)$ raised to negative powers; the second part has no special name and consists of terms containing $(s - s_0)$ raised to zero or positive powers. It should be noted that the second part is analytic throughout the s-plane (except at infinity) and assumes the value B_0 at $s = s_0$. If there were no singularity in $G(s)$, only the second part of the expansion would be present and would just be the power series expansion. The coefficient of $(s - s_0)^{-1}$, which is A_{-1} in (B-6), is called the *residue* of $G(s)$ in the pole at $s = s_0$.

Formally, the Laurent series expansion of $G(s)$ can be obtained from the usual Taylor series expansion of the function $\phi(s)$ and the subsequent division by $(s - s_0)^n$. For most cases of engineering interest, simpler methods can be employed. Because the uniqueness of the properties of analytic functions, it follows that any series of the proper form [that is, the form given in (B-6)] must, in fact, be the Laurent series. When $G(s)$ is a ratio of two polynomials in s, a simple procedure for finding the Laurent series is as follows: Form $\phi(s) = (s - s_0)^n G(s)$; let $s - s_0 = v$, or $s = v + s_0$; expand $\phi(v)$ around the origin by dividing the denominator into the numerator; and replace v by $s - s_0$. As an example, consider the following:

$$G(s) = \frac{2}{s^2(s^2 - 1)}$$

Let it be required to find the Laurent series for $G(s)$ in the vicinity of $s = -1$.

$$\phi(s) = \frac{2}{s^2(s - 1)}$$

Let $s = v - 1$.

$$\phi(v) = \frac{2}{(v^2 - 2v + 1)(v - 2)}$$

$$= \frac{2}{v^3 - 4v^2 - 3v - 2}$$

$$-2 - 3v - 4v^2 + v^3 \overline{\Big)} \; \frac{-1 + \dfrac{3v}{2} - \dfrac{1}{4}v^2}{2 + 3v + 4v^2 - v^3}$$

$$\frac{-3v - 4v^2 + v^3}{}$$

$$-3v - \frac{9v^2}{2} - 6v^3 + \frac{3v^4}{2}$$

$$\frac{1}{2}v^2 + 7v^3 - \frac{3}{2}v^4$$

$$\frac{1}{2}v^2 + \frac{3}{4}v^3 + v^4 - \frac{1}{4}v^5$$

$$\phi(v) = -1 + \frac{3}{2}v - \frac{1}{4}v^2 - \cdots$$

$$\phi(s) = -1 + \frac{3}{2}(s+1) - \frac{1}{4}(s+1)^2 - \cdots$$

$$G(s) = -\frac{1}{s+1} + \frac{3}{2} - \frac{1}{4}(s+1) - \cdots$$

The residue is seen to be -1.

A useful formula for finding the residue at an nth-order pole, $s = s_0$, is as follows:

$$k_{s_0} = \frac{\phi^{(n-1)}(s_0)}{(n-1)!} \tag{B-7}$$

where $\phi(s) = (s - s_0)^n G(s)$. This formula is valid for $n = 1$ and is not restricted to rational functions.

When $G(s)$ is not a ratio of polynomials, it is permissible to replace transcendental terms by series valid in the vicinity of the pole. For example,

$$G(s) = \frac{\sin s}{s^2} = \frac{1}{s^2}\left[s - \frac{s^3}{3!} + \frac{s^5}{5!} - \cdots \right]$$

$$= \frac{1}{s} - \frac{s}{3!} + \frac{s^3}{5!} - \cdots$$

In this instance the residue in the pole at the origin is 1.

There is a direct connection between the Laurent series and the partial fraction expansion of a function $G(s)$. In particular, if $H_i(s)$ is the principal part of the Laurent series at the pole $s = s_i$, then the partial fraction expansion of $G(s)$ can be written as

$$G(s) = H_1(s) + H_2(s) \cdots H_k(s) + q(s) \tag{B-8}$$

where the first k terms are the principal parts of the Laurent series about the poles, and $q(s)$ is a polynomial $a_0 + a_1 s + a_2 s^2 + \cdots + a_m s^m$ representing the behavior of $G(s)$ for large s. The value of m is the difference between the degrees of the numerator and denominator polynomials. In general $q(s)$ can be determined by dividing the denominator polynomial into the numerator polynomial until the remainder is of lower order than the denominator. The remainder can then be expanded in its principal parts.

With the question of how to determine residues out of the way, the only remaining question is how to relate the closed contour of (B-3) with the open (straight-line) contour of (B-2). This is handled quite easily by restricting consideration to intergrands that approach zero rapidly enough for large values of the variable, so that there will be no contribution to the integral from distant portions of the contour. Thus, although the specified path of integration in the s-plane may be from $s = c - j\infty$ to $c + j\infty$, the integral that will be evaluated will have a path of integration as shown in Fig. B-1. The path of integration will be taken as the $\lim_{R_0 \to \infty}$ of the contour along C_0 from $c - jR_0$ to $c + jR_0$ and then along either contour C_1 or C_2, which are circular arcs closing to the left or right. The vertical portion of the contour must lie in the region of convergence of the Laplace integral.

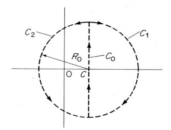

Fig. B-1 Path of integration in the s-plane.

For a one-sided Laplace transform c must lie to the right of σ_0, the abscissa of absolute convergence. For the two-sided Laplace transform the vertical path must lie in the strip of convergence. As will be discussed shortly, the contour is always closed to the left for positive time. Therefore, the inversion integral for the one-sided Laplace transform (or positive-time portion of the two-sided Laplace transform) can be written as

$$\frac{1}{2\pi j} \int_{c-j\infty}^{c+j\infty} F(s)\epsilon^{st}\,ds = \frac{1}{2\pi j} \lim_{R_0 \to \infty} \int_{c-jR_0}^{c+jR_0} F(s)\epsilon^{st}\,ds \qquad \text{(B-9)}$$

If in the limit as $R_0 \to \infty$ the contribution along the contour C_2 is zero, then (B-9) can be written as an integral around the closed contour $C_0 + C_2$, giving

$$\frac{1}{2\pi j}\int_{c-j\infty}^{c+j\infty}F(s)\epsilon^{st} = \lim_{R_0\to\infty}\frac{1}{2\pi j}\int_{C_0+C_2}F(s)\epsilon^{st} \tag{B-10}$$

$$= \Sigma \text{ residues of } F(s)\epsilon^{st} \text{ to the} \\ \text{left of the vertical path of} \tag{B-11} \\ \text{integration}$$

As an example, consider the Laplace transform $F(s) = \dfrac{1}{s(s+1)}$. Formal application of the inversion integral gives

$$f(t) = \frac{1}{2\pi j}\int_{c-j\infty}^{c+j\infty}\frac{1}{s(s+1)}\epsilon^{st}ds \tag{B-12}$$

where c is to the right of all poles. The only poles of the integrand are at $s = 0, -1$, and the residues are readily found [for example, by equation (B-7)], giving

$$f(t) = \sum \text{residues } \frac{1}{s(s+1)}\epsilon^{st} \quad \text{at} \quad s = 0, 1 \tag{B-13}$$

$$= (1 - \epsilon^{-t})u(t)$$

In any specific case the validity of assuming that the contribution over the closure arc is zero can be investigated. However, there are some general conditions that can be stated that encompass many of the situations encountered in system analysis problems. Consider the contours again as shown in Fig. B-2. Here a new coordinate system is defined as having its vertical axis coincident with the vertical portion of the integration contour. This coordinate system has the complex variable ξ related to s by the equation

$$\xi = s - c \tag{B-14}$$

The inversion integral now becomes

$$f(t) = \frac{\epsilon^{ct}}{2\pi j}\lim_{R\to\infty}\int_C F(\xi + c)\epsilon^{\xi t}d\xi \tag{B-15}$$

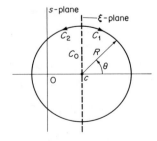

Fig. B-2 New coordinate system.

A theorem from complex variable theory (Jordan's lemma) may now be applied to (B-15) to give the following results:

1. If $\lim\limits_{|\xi| \to \infty} F(\xi + c) = 0$ uniformly for $(\pi/2) \le \theta \le (3\pi/2)$, and if $t > 0$,

then the integral (B-15) over C_2 approaches zero as R approaches infinity.

2. If $\lim\limits_{|\xi| \to \infty} F(\xi + c) = 0$ uniformly for $-(\pi/2) \le \theta \le (\pi/2)$, and if $t < 0$,

then the integral (B-15) over C_1 approaches zero as R approaches infinity.

Note that $\lim\limits_{|\xi| \to \infty} F(\xi + c_1) = \lim\limits_{|s| \to \infty} F(s)$, and in many practical cases the following more general condition is met:

3. $\lim\limits_{|s| \to \infty} F(s) = 0$ uniformly for all θ which meets the requirements of

both 1 and 2.

The procedure for inverting the Laplace transform can now be stated. If conditions 1 or 2 are met, the value of $f(t)$ for positive time is found by closing the contour to the left and evaluating the integral as the sum of the residues at the poles enclosed. If conditions 2 or 3 are met, the value of $f(t)$ for negative time is found by closing the contour to the right and evaluating the integral as the sum of the residues at the poles enclosed. These results may be stated more compactly as

$$\mathcal{L}^{-1}\{F(s)\} = f(t)u(t) = \begin{cases} 0 & t < 0 \\ \Sigma \text{ residues to left of } C_0 & t > 0 \end{cases} \quad \text{(B-16)}$$

$$\mathcal{L}^{-1}\{F_{\text{II}}(s)\} = f(t) = \begin{cases} -\Sigma \text{ residues to right of } C_0 & t < 0 \\ \Sigma \text{ residues to left of } C_0 & t > 0 \end{cases} \quad \text{(B-17)}$$

It is clear that (B-16) is a special case of (B-17) since the one-sided transform $F(s)$ never has poles to the right of C_0 and, therefore, is identically zero for $t < 0$.

The above procedures are directly applicable and reasonably straightforward when $F(s)$ is single-valued. However, when $F(s)$ is multivalued, as for example $F(s) = s^{-1/2}$, the integration is more complicated. Such integrals can be evaluated by deforming the path of integration to avoid the lines along which $\lim\limits_{|s| \to \infty} F(s)$ dose not converge uniformly to zero. The evaluation of integrals of this type is discussed in the references and will not be considered further.

The following two examples illustrate the procedures discussed above.

Example B-1.   Given $F(z) = 1/z(z + 1)$, find the inverse z-transform,

$$f(nT) = \frac{1}{2\pi j} \oint F(z)z^{n-1} dz$$

$$= \frac{1}{2\pi j} \oint \frac{z^{n-1}}{z(z + 1)} dz$$

The pole configuration depends on n. There are three cases to consider.

1. For $n = 0$ the integral is

$$\frac{1}{2\pi j} \oint \frac{1}{z^2(z+1)} dz = \Sigma \text{ residues at } 0, -1$$

$$K_0 = \frac{d}{dz}(z+1)^{-1}\Big|_{z=0} = \frac{-1}{(z+1)^2}\Big|_{z=0} = -1$$

$$K_{-1} = \frac{1}{z^2}\Big|_{z=-1} = 1$$

$$\therefore f(0) = -1 + 1 = 0$$

2. For $n = 1$ the integral is

$$\frac{1}{2\pi j} \oint \frac{1}{z(z+1)} dz = \Sigma \text{ residues at } 0, -1$$

$$K_0 = \frac{1}{z+1}\Big|_{z=0} = 1$$

$$K_{-1} = \frac{1}{z}\Big|_{z=-1} = -1$$

$$\therefore f(T) = 0$$

3. For $n \geq 2$ the integral becomes

$$\frac{1}{2\pi j} \oint \frac{z^{n-2}}{(z+1)} dz = \text{residue at } -1$$

$$K_{-1} = z^{n-2}\big|_{z=-1} = (-1)^{n-2} = (-1)^n$$

$$\therefore f(nT) = (-1)^n \quad \text{for} \quad n \geq 2$$
$$= 0 \quad\quad \text{for} \quad n < 2$$

Example B-2. Find the inverse transform of $F_{\mathrm{II}}(s) = 1/s^2(s-2)$, $0 < \sigma_0 < 2$.

$$f(t) = \frac{1}{2\pi j} \int_{c-j\infty}^{c+j\infty} \frac{\epsilon^{st}}{s^2(s-2)} ds$$

$$= \begin{cases} \text{residue at } s = 0 & \text{for } t > 0 \\ (-1) \text{ residue at } s = 2 & \text{for } t < 0 \end{cases}$$

From B-7

$$k_0 = \frac{d}{ds}\left[\frac{\epsilon^{st}}{s-2}\right]_{s=0} = \frac{t\epsilon^{st}(s-2) - \epsilon^{st}}{(s-2)^2}\Big|_{s=0}$$

$$= \frac{-2t - 1}{4}$$

$$k_2 = \frac{\epsilon^{st}}{s^2}\Big|_{s=+2} = \frac{1}{4}\epsilon^{+2t}$$

Therefore,

$$f(t) = = \left(-\frac{1}{2}t - \frac{1}{4}\right)u(t) - \frac{1}{4}\epsilon^{2t}u(-t)$$

References

1. PAPOULIS, A., *The Fourier Integral and its Applications.* New York: McGraw-Hill Book Company, Inc., 1962. In Chapter 9 and Appendix II, a readily understandable treatment of evaluation of transforms by contour integration is given.
2. CHURCHILL, R., *Operational Mathematics* 2nd ed., New York: McGraw-Hill Book Company, Inc., 1958. A thorough treatment of the mathematics relating particularly to the Laplace transform is given, including an introduction to complex variable theory and contour integration.
3. LE PAGE, W. *Complex Variables and the Laplace Transform for Engineers.* New York: McGraw-Hill Book Company Inc., 1961. This book presents a detailed and very readable treatment of the mathematics of transform analysis. It provides an excellent discussion of contour integration as applied to the inversion integral of the Laplace transform.

APPENDIX

C
Matrices

Definitions

A rectangular array of elements having m rows and n columns is called an $(m \times n)$ matrix. A matrix is denoted symbolically by a boldface capital letter in the following manner:

$$\mathbf{A} \equiv \begin{bmatrix} a_{11} & a_{12} & . & . & . & . & . & a_{1n} \\ a_{21} & a_{22} & . & . & . & . & . & a_{2n} \\ . & . & . & . & . & . & . & . \\ a_{m1} & a_{m2} & . & . & . & . & . & a_{mn} \end{bmatrix} \equiv [a_{jk}]$$

An $(n \times 1)$ matrix is called a *column matrix*, a *column vector*, or oftentimes just a *vector*. Such a vector will be designated by boldface lowercase letters as follows:

$$\mathbf{a} = \begin{bmatrix} a_1 \\ a_2 \\ \vdots \\ a_n \end{bmatrix}$$

A similar notation is also used for a $(1 \times n)$ matrix, which is called a *row matrix*, or *row vector*. However, in this text, row vectors will always be designated as the transpose of a column vector (see below). Accordingly, boldface lowercase letters will always refer to column vectors.

An ($n \times n$) matrix is called a *square matrix*, of order n. A square matrix is *diagonal* if, and only if, $j \neq k$ implies $a_{jk} = 0$; that is, all elements not on the main diagonal (from upper left to lower right) are zero.

A *scalar* is an ordinary number. It is also sometimes identified with a matrix of order (1×1)—that is, a single element. Scalars will be denoted by ordinary (not boldface) type.

The *identity matrix* is a diagonal matrix that has all its diagonal elements equal to one. It will be denoted by $\mathbf{I}$.

The *null, or zero, matrix* (either square or rectangular) has all its elements equal to zero. It is denoted by $\mathbf{0}$.

The *transpose*, $\mathbf{A}^T$, of a matrix $\mathbf{A}$ is formed by interchanging the rows and columns of $\mathbf{A}$. For example, if

$$\mathbf{A} = \begin{bmatrix} a_{11} & a_{12} & a_{13} \\ a_{21} & a_{22} & a_{23} \end{bmatrix} \quad \text{then } \mathbf{A}^T = \begin{bmatrix} a_{11} & a_{21} \\ a_{12} & a_{22} \\ a_{13} & a_{23} \end{bmatrix}$$

From this definition it is seen that the transpose of a column vector is a row vector, and vice versa.

The *determinant* of a matrix (necessarily square) is the determinant of the elements of the matrix and is accordingly a number or scalar. The determinant of $\mathbf{A}$ is denoted $|\mathbf{A}|$. For example, if

$$\mathbf{A} = \begin{bmatrix} 1 & 1 & 2 \\ 2 & 1 & 1 \\ 1 & 2 & 2 \end{bmatrix} \quad \text{then } |\mathbf{A}| = \begin{vmatrix} 1 & 1 & 2 \\ 2 & 1 & 1 \\ 1 & 2 & 2 \end{vmatrix} = 3$$

A matrix (necessarily square) for which the determinant of its elements is zero is called a *singular matrix*; that is, if $|\mathbf{A}| = 0$, then $\mathbf{A}$ is a singular matrix. If $|\mathbf{A}| \neq 0$, then $\mathbf{A}$ is *nonsingular*.

The *cofactor* of the element a_{jk} of the square matrix $\mathbf{A}$ is the scalar element formed by considering the matrix as a determinant, striking out the jth row and kth column and multiplying by $(-1)^{j+k}$. For example, if

$$\mathbf{A} = \begin{bmatrix} a_{11} & a_{12} & a_{13} \\ a_{21} & a_{22} & a_{23} \\ a_{31} & a_{32} & a_{33} \end{bmatrix}$$

then

$$\text{cofactor of } a_{31} = (-1)^4 \begin{vmatrix} a_{12} & a_{13} \\ a_{22} & a_{23} \end{vmatrix} = a_{12}a_{23} - a_{13}a_{22}$$

The *adjoint matrix*, adj $\mathbf{A}$, of a (necessarily) square matrix $\mathbf{A}$ is formed by replacing each element of $\mathbf{A}$ by its cofactor and transposing. Thus, if

$$\mathbf{A} = \begin{bmatrix} 1 & 2 & 1 \\ 0 & 1 & 1 \\ 0 & 1 & 0 \end{bmatrix}$$

$$\text{adj } \mathbf{A} = \begin{bmatrix} (-1)^2 \begin{vmatrix} 1 & 1 \\ 1 & 0 \end{vmatrix} & (-1)^3 \begin{vmatrix} 0 & 1 \\ 0 & 0 \end{vmatrix} & (-1)^4 \begin{vmatrix} 0 & 1 \\ 0 & 1 \end{vmatrix} \\ (-1)^3 \begin{vmatrix} 2 & 1 \\ 1 & 0 \end{vmatrix} & (-1)^4 \begin{vmatrix} 1 & 1 \\ 0 & 0 \end{vmatrix} & (-1)^5 \begin{vmatrix} 1 & 2 \\ 0 & 1 \end{vmatrix} \\ (-1)^4 \begin{vmatrix} 2 & 1 \\ 1 & 1 \end{vmatrix} & (-1)^5 \begin{vmatrix} 1 & 1 \\ 0 & 1 \end{vmatrix} & (-1)^6 \begin{vmatrix} 1 & 2 \\ 0 & 1 \end{vmatrix} \end{bmatrix}^T$$

$$= \begin{bmatrix} -1 & 0 & 0 \\ 1 & 0 & -1 \\ 1 & -1 & 1 \end{bmatrix}^T = \begin{bmatrix} -1 & 1 & 1 \\ 0 & 0 & -1 \\ 0 & -1 & 1 \end{bmatrix}$$

Matrix Algebra

Equality of matrices.   Two matrices $\mathbf{A}$ and $\mathbf{B}$ are equal if, and only if, they are of the same order and all their corresponding elements are equal; that is, $\mathbf{A} = \mathbf{B}$ provided $\mathbf{A}$ and $\mathbf{B}$ are of the same order and $a_{jk} = b_{jk}$ for all j and k.

Addition and subtraction of matrices.   In order for two matrices to be added or subtracted they must be of the same order. If $\mathbf{A}$ and $\mathbf{B}$ are of the same order, then their sum, $\mathbf{C}$, is defined as the matrix whose elements are the sums of the corresponding elements in $\mathbf{A}$ and $\mathbf{B}$. Thus,

$$\mathbf{C} = \mathbf{A} + \mathbf{B} \qquad \text{where } c_{jk} = a_{jk} + b_{jk}$$

The subtraction of two matrices is defined similarly; thus,

$$\mathbf{D} = \mathbf{A} - \mathbf{B} \qquad \text{where } d_{jk} = a_{jk} - b_{jk}$$

Scalar multiplication.   The multiplication of a matrix $\mathbf{A}$ by a scalar (ordinary number) k is called *scalar multiplication* and results in a new matrix, each element of which is multiplied by the scalar. Thus,

$$\mathbf{B} = k\mathbf{A} = k \begin{bmatrix} a_{11} & a_{12} & \cdots & a_{1n} \\ a_{21} & a_{22} & \cdots & a_{2n} \\ \cdots\cdots\cdots\cdots\cdots\cdots \\ a_{m1} & a_{m2} & \cdots & a_{mn} \end{bmatrix} = \begin{bmatrix} ka_{11} & ka_{12} & \cdots & ka_{1n} \\ ka_{21} & ka_{22} & \cdots & ka_{2n} \\ \cdots\cdots\cdots\cdots\cdots\cdots \\ ka_{m1} & ka_{m2} & \cdots & ka_{mn} \end{bmatrix}$$

Scalar multiplication is commutative, and therefore, $k\mathbf{A} = \mathbf{A}k$.

Matrix multiplication.   Two matrices can be multiplied together only when the number of columns of the first matrix is equal to the number of rows of the second. Such matrices are said to be *conformable* in that order. If two matrices are conformable, their product is defined as

$$\mathbf{AB} = \mathbf{C}$$

where

$$c_{jk} = \sum_{i=1}^{q} a_{ji} b_{ik}$$

and the orders of the matrices **A**, **B**, and **C** are $(m \times q)$, $(q \times n)$, and $(m \times n)$, respectively. The operation of matrix multiplication can be visualized as follows: To obtain the jkth element in the product matrix, take the jth row of the first matrix and the kth column of the second matrix and multiply them term by term and add the result. This is illustrated below for the product of a (4×3) matrix and a (3×2) matrix.

$$\mathbf{A} = \begin{bmatrix} 1 & 2 & 2 \\ 1 & -3 & 1 \\ 2 & 0 & 1 \\ -1 & 1 & -2 \end{bmatrix} \qquad \mathbf{B} = \begin{bmatrix} 2 & 1 \\ 1 & 1 \\ 2 & 2 \end{bmatrix} \qquad \mathbf{AB} = \mathbf{C}$$

The product matrix **C** will be a (4×2) matrix. To find the element c_{32} of the product matrix we must multiply the elements in the third row of **A** by the elements in the second column of **B** and add

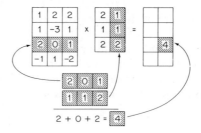

The final result is as follows:

$$\begin{bmatrix} 1 & 2 & 2 \\ 1 & -3 & 1 \\ 2 & 0 & 1 \\ -1 & 1 & -2 \end{bmatrix} \times \begin{bmatrix} 2 & 1 \\ 1 & 1 \\ 2 & 2 \end{bmatrix} = \begin{bmatrix} 8 & 7 \\ 1 & 0 \\ 6 & 4 \\ -5 & -4 \end{bmatrix}$$

In general it is not possible to reverse the order of multiplication in a product. Only for an $(m \times n)$ and an $(n \times m)$ matrix would both products be defined, since otherwise the matrices would not be conformable in both orders. Over and above that, however, reversing the order of a product when possible leads, in general, to a different result. Thus,

$$\mathbf{AB} \neq \mathbf{BA}$$

If two matrices do have the property that their product is independent of their order of multiplication, they are said to *commute*. In particular the identity matrix **I** *commutes* with any square matrix of the same order; that is,

$$\mathbf{IA} = \mathbf{AI}$$

The product **AB** can be referred to in two ways: We may say **B** is premultiplied by **A**, or that **A** is postmultiplied by **B**. In the notation we are using, a vector **a** is a column matrix; therefore, when a vector premultiplies a matrix **B** the product will always be in the form $\mathbf{a}^T\mathbf{B}$ in order that the matrices be conformable. When the order of multiplication is reversed the transpose is not required, and the product will be **Ba**. Premultiplication and postmultiplication of a matrix by a vector leads to the scalar form $\mathbf{a}^T\mathbf{Ba}$.

The inverse matrix. The matrix operation corresponding to division is defined in terms of the inverse matrix. For a given matrix **A**, the inverse matrix $\mathbf{A}^{-1}$, when it exists, is defined by the following equation:

$$\mathbf{AA}^{-1} = \mathbf{I} = \mathbf{A}^{-1}\mathbf{A}$$

Thus, any matrix multiplied by its inverse leads to the identity matrix. Only a nonsingular matrix (that is, one whose determinant does not vanish) has an inverse. The inverse of a nonsingular (necessarily square) matrix is

$$\mathbf{A}^{-1} = \frac{\text{adj }\mathbf{A}}{|\mathbf{A}|}$$

As an example, consider

$$\mathbf{A} = \begin{bmatrix} a_{11} & a_{12} \\ a_{21} & a_{22} \end{bmatrix}$$

$$\text{adj }\mathbf{A} = \begin{bmatrix} a_{22} & -a_{21} \\ -a_{12} & a_{11} \end{bmatrix}^T = \begin{bmatrix} a_{22} & -a_{12} \\ -a_{21} & a_{11} \end{bmatrix}$$

$$\det \mathbf{A} = a_{11}a_{22} - a_{12}a_{21}$$

Therefore,

$$\mathbf{A}^{-1} = \frac{1}{a_{11}a_{22} - a_{12}a_{21}} \begin{bmatrix} a_{22} & -a_{12} \\ -a_{21} & a_{11} \end{bmatrix}$$

Characteristic equation of a matrix. A frequently occurring problem in matrix analysis is to find values of a scalar parameter λ for which there exist vectors $\mathbf{x} \neq \mathbf{0}$ that satisfy the matrix equation

$$\mathbf{Ax} = \lambda\mathbf{x}$$

where **A** is a given nth-order square matrix. Such a problem is called a *characteristic value*, or *eigenvalue*, problem. This equation can be rewritten by means of the identity matrix as

$$\mathbf{Ax} = \lambda\mathbf{Ix}$$
$$(\mathbf{A} - \lambda\mathbf{I})\mathbf{x} = \mathbf{0}$$

There will be a solution if, and only if,

$$|\mathbf{A} - \lambda\mathbf{I}| = 0$$

that is, if and only if

$$
\begin{vmatrix}
a_{11} - \lambda & a_{12} & \cdots & a_{1n} \\
a_{21} & a_{22} - \lambda & & \\
\vdots & \vdots & & \\
a_{n1} & a_{n2} & \cdots & a_{nn} - \lambda
\end{vmatrix} = 0
$$

Either of these last two equations is called the *characteristic equation* of the matrix **A** and the expression

$$ f(\lambda) = |\mathbf{A} - \lambda \mathbf{I}| = \lambda^n + \alpha_{n-1}\lambda^{n-1} + \cdots + \alpha_1\lambda + \alpha_0 = 0 $$

is called the *characteristic polynomial* for **A**.

The roots of the characteristic equation, denoted by λ_i (where $i = 1, 2, \ldots, n$) are called the *characteristic values*, or *eigenvalues*, of the matrix **A**. The vectors **x** that satisfy the equation $\mathbf{Ax} = \lambda\mathbf{x}$ are called *characteristic vectors*, or *eigenvectors*, of the matrix **A**.

If there exists a nonsingular matrix **T** such that $\mathbf{B} = \mathbf{TAT}^{-1}$, the square matrices **A** and **B** are said to be *similar*. All similar matrices have the same eigenvalues and eigenvectors.

If **A** is a symmetric matrix, then all of its eigenvalues are real, and the eigenvectors corresponding to different eigenvalues are orthogonal; that is, if $\mathbf{x}_j$ is the eigenvector corresponding to λ_j (where $\lambda_j \neq \lambda_k$), then

$$ \mathbf{x}_j{}^T \mathbf{x}_k = 0 $$

Cayley-Hamilton theorem. The Cayley–Hamilton theorem states that: "Every square matrix satisfies its own characteristic equation in a matrix sense." This means that if an nth-order square matrix **A** has as its characteristic equation

$$ f(\lambda) = \lambda^n + \alpha_{n-1}\lambda^{n-1} + \cdots + \alpha_1\lambda + \alpha_0 = 0 $$

then

$$ f(\lambda) = \mathbf{A}^n + \alpha_{n-1}\mathbf{A}^{n-1} + \cdots + \alpha_1\mathbf{A} + \alpha_0\mathbf{I} = 0 $$

Matrix calculus. If the elements of a matrix **A** are differentiable functions $a_{jk}(t)$ of a scalar parameter t, then the derivative of **A** with respect to t is defined as

$$
\frac{d}{dt}\mathbf{A} =
\begin{bmatrix}
\dfrac{da_{11}(t)}{dt} & \dfrac{da_{12}(t)}{dt} & \cdots & \dfrac{da_{1n}(t)}{dt} \\[2mm]
\dfrac{da_{21}(t)}{dt} & \dfrac{da_{22}(t)}{dt} & \cdots & \dfrac{da_{2n}(t)}{dt} \\[2mm]
\vdots & \vdots & & \vdots \\[2mm]
\dfrac{da_{m1}(t)}{dt} & \dfrac{da_{m2}(t)}{dt} & \cdots & \dfrac{da_{mn}(t)}{dt}
\end{bmatrix}
$$

Partial differentiation and integration of matrices are defined in an analogous manner.

Functions of matrices.   The definition of a function can be extended to matrix arguments. This concept is particularly useful when the function can be expanded in a convergent power series. Such a representation takes the following form:

$$g(\mathbf{A}) = \sum_{m=0}^{\infty} \mathbf{B}_m \mathbf{A}^m \qquad \mathbf{A}^0 = \mathbf{I}$$

For example, the exponential function of the square matrix $\mathbf{A}$

$$\epsilon^A = \mathbf{I} + \mathbf{A} + \tfrac{1}{2}\mathbf{A}^2 + \cdots$$

This series converges for all square matrices. From the Cayley–Hamilton theorem it is seen that if $\mathbf{A}$ is an nth-order matrix we can express $\mathbf{A}^n$ as a polynomial in $\mathbf{A}$ where the highest power of $\mathbf{A}$ is $n - 1$. Using this relationship it is possible to systematically reduce a polynomial of any order in $\mathbf{A}$ to an equivalent expression containing no terms of power higher than $n - 1$. The resulting expression will be of the form

$$g(\mathbf{A}) = \frac{\Delta_{n-1}}{\Delta}\mathbf{A}^{n-1} + \frac{\Delta_{n-2}}{\Delta}\mathbf{A}^{n-2} + \cdots + \frac{\Delta_1}{\Delta}\mathbf{A} + \frac{\Delta_0}{\Delta}\mathbf{I}$$

where Δ is the determinant

$$\Delta = \begin{vmatrix} 1 & \cdots & & 1 \\ \lambda_1 & \lambda_2 & \cdots & \lambda_n \\ \lambda_1^{2} & \lambda_2^{2} & \cdots & \lambda_n^{2} \\ \vdots & \vdots & & \\ \lambda_1^{n-1} & \lambda_2^{n-1} & \cdots & \lambda_n^{n-1} \end{vmatrix}$$

and $\lambda_1, \lambda_2, \ldots, \lambda_n$ are the eigenvalues of $\mathbf{A}$. The determinant Δ_{r-1} is derived from Δ by replacing the elements of the rth row by $g(\lambda), g(\lambda), g(\lambda_2), \ldots, g(\lambda_n)$.

If the n eigenvalues of $\mathbf{A}$ are distinct, then the expression for the matrix function can be written in the following alternative form:

$$g(\mathbf{A}) = \sum_{j=1}^{n} f(\lambda_j) \frac{\prod_{j \neq k} (\mathbf{A} - \lambda_k \mathbf{I})}{\prod_{j \neq k} (\lambda_j - \lambda_k)}$$

This is known as *Sylvester's theorem.*

Miscellaneous matrix relationships.   It is assumed in the following that the various matrix operations are permissible; that is, matrices being multipled are conformable, matrices being inverted are nonsingular, and so on.

$$(\mathbf{ABC}) = (\mathbf{AB})\mathbf{C} = \mathbf{A}(\mathbf{BC})$$
$$\mathbf{A}^n = \mathbf{AA} \ldots \mathbf{A} \text{ to } n \text{ factors}$$
$$\text{If } \mathbf{D} = \mathbf{ABC}, \text{ then } \mathbf{D}^{-1} = \mathbf{C}^{-1}\mathbf{B}^{-1}\mathbf{A}^{-1}$$
$$\mathbf{A}^{-n} = (\mathbf{A}^{-1})^n$$
$$\text{If } \mathbf{C} = \mathbf{AB}, \text{ then } \mathbf{C}^T = (\mathbf{AB})^T = \mathbf{B}^T\mathbf{A}^T$$
$$\text{If } \mathbf{D} = \mathbf{ABC}, \text{ then } \mathbf{D}^T = (\mathbf{ABC})^T = \mathbf{C}^T\mathbf{B}^T\mathbf{A}^T$$

Premultiplication (postmultiplication) of $\mathbf{A}$ by the matrix obtained by replacing the 1 in the jth row (column) of the identity matrix by α multiplies all elements in the jth row (column) of $\mathbf{A}$ by α.

Premultiplication (postmultiplication) of $\mathbf{A}$ by the matrix obtained by replacing the zero in the nondiagonal element δ_{jk} of the identity matrix by 1 adds the kth row (column) of $\mathbf{A}$ to jth row (column) of $\mathbf{A}$.

Premultiplication (postmultiplication) of $\mathbf{A}$ by the permutation matrix obtained by a permutation of the rows (columns) of the identity matrix results in an identical permutation of the rows (columns) of $\mathbf{A}$.

Index

Index